VENUS

The geological story

Peter Cattermole
Department of Earth Science, University of Sheffield

The Johns Hopkins University Press
Baltimore

Printed in Great Britain

First published in 1994 by
UCL Press Limited
The name of University College London (UCL) is a registered
trade mark used by UCL Press with the consent of the owner.

First published in the United States of America in 1994 by
The Johns Hopkins University Press
2715 North Charles Street
Baltimore, Maryland 21218-4319

Library of Congress Cataloging-in-Publication Data

Cattermole, Peter John.
Venus, the geological story / Peter Cattermole.
p. cm.
Includes bibliographical references and index.
ISBN 0-8018-4787-7 (alk. paper)
1. Venus (Planet)—Geology. I. Title.
QB621.C38 1994
559.9'22—dc20 93-11636
CIP

Contents

Preface

This book has one simple objective: to introduce the reader to the geology of Venus as it emerges from research currently being undertaken in the wake of the recent Magellan Mission. Because there is so much new information, and because its analysis is still at a formative stage, the story that follows can at best be considered the author's modest attempt to abstract, from a wealth of data, the bare bones of the geology of Venus, almost as it is being written. There is little doubt that, by the time the book appears in print, even more data will have been processed, newer maps will have been published, and planetary scientists will have either modified old ideas or come up with new ones. Such is the exciting stage at which Venus research finds itself. All that I can hope, therefore, is that this first offering will be seen as an introduction to a fast-developing field of enquiry, and that it will find a readership among those who wish to delve into the fascinating story that is unfolding.

PETER CATTERMOLE 1 March 1993

Acknowledgements

I have received continual support from R. Stephen Saunders of JPL, Project Scientist for the Magellan Mission. Despite his obvious immense work burden, he has always responded to my enquiries. I would like to tender my thanks here. Similarly, I wish to acknowledge the enthusiastic support of David Okerson at NASA headquarters. His was an important part in this enterprise, as it enabled me to obtain most of the excellent Magellan images that grace this book.

Many other people have supplied me with copies of their research publications, given permission for the use of their illustrative material, or supplied me with fair copies of maps and diagrams, or imagery. I hope that I have acknowledged all in the appropriate way within the text; should there be any omissions, I apologize to those individuals for my error. I would like to make specific mention of Alexander Basilevsky, Ron Greeley, John Guest, Jim Head, Kathy Hoyt and Gerry Schaber: thank you for your interest and support over the past several years.

List of tables

To those who died aboard Challenger

---

1 Introduction

Now came still evening on, and twilight gray
Had in her sober livery all things clad;
Silence accompanied, for beast and bird,
They to their grassy couch, these to their nests,
Were slunk, all but the wakeful nightingale;
She all night long her amorous descant sung;
Silence was pleas'd: now glow'd the firmament
With living sapphires: Hesperus that led
The starry host, rode brightest, till the moon,
Rising in clouded majesty, at length
Apparent queen unveil'd her peerless light,
And o'er the dark her silver mantle threw.

John Milton (*Paradise lost*, book iv)

1.1 The veiled planet

There can be few more beautiful celestial sights than that of the planet Venus. The stunning brilliance of our neighbour world was recorded by ancient astronomers: the Babylonians knew her as Istar, "the bright torch of heaven" – the embodiment of all things womanly, and Mother of the Gods. The Greeks and Romans gave her the name of the Goddess of Beauty. A female association is almost universal. The Chinese called her Tai-pe – "the Beautiful White One". Homer refers to the planet as "Hesperos, which is the most beautiful star in the sky". This name is preserved for posterity in peerless literature, with John Milton's use of it in *Paradise lost*.

Oddly enough, the "clouded majesty" afforded to our own satellite in Milton's poem is actually more apposite to Venus, for it, of all the inner planets, has proved shy in yielding the secrets of its permanently cloud-shrouded surface. The brilliance of Venus seen by the naked eye is not matched by the telescopic appearance; frankly, it is a marked disappointment. All that even the most powerful telescope will reveal is a whitish disc with nebulous darker shadings (Fig. 1.1). The first record of these can probably be ascribed to a Neapolitan lawyer, one Francesco Fontana, a keen amateur observer who, in 1645, reported a dark area

Figure 1.1 Telescopic view of Venus
A drawing by the author, made on 5 May 1956 with a 16 cm reflector.

almost in the centre of the disc. Since that time, generations of both professional and amateur astronomers have reported ephemeral markings, but in vain have sought a glimpse of the solid surface. All that we have ever seen from the Earth is the reflective cloud tops of a dense opaque atmosphere.

1.2 Astronomical observations

Until well into the present century, nearly all our knowledge of the planet was gleaned from observations at the eye end of telescopes. In the 17th century, Galileo Galilei discovered that Venus showed phases, a fact that was to contribute significantly to the demise of the Ptolemaic view of the universe. Later in the same century, Francesco Fontana (1645) and Giovanni Cassini (1667) recorded both bright and dusky patches on the disc. Somewhat later, Francesco Bianchini, an Italian lawyer and keen amateur astronomer, observed Venus with his 6.4 cm refractor, noting many disc markings; and, in 1726, he published a map of Venus's "oceans" and "continents", followed by a book in the next year. He also claimed a rotation period of 24 d 8 h. Subsequently, during the late 18th century, two of the truly great planetary observers, Johann Schröter and William Herschel, both saw dusky shadings. Furthermore, Schröter also believed that some bright spots he had recorded were high mountains. While Herschel placed no credence in his German contemporary's idea of mountains, he shared the belief that the visible surface was nothing more than a layer of cloud. Confirmation of this came in 1761, when M. V. Lomonosov observed the transit of Venus across the Sun. He noted that the outline of the planet had an indistinct blurred appearance as it crossed in front of the solar disc, and quite rightly attributed this to the presence of an atmosphere that, incidentally, he believed to be at least as deep as our own.

Schröter also observed the "ashen light", a faint luminosity observed on the night side of the planet, inside the crescent. This effect has been seen by many hundreds of observers since and there can be little doubt as to its authenticity. Explaining exactly what it is, however, is more difficult. In the light of what we now know about Venus, there is every likelihood that the effect is a response to electrical effects in Venus's atmosphere, that is, to aurorae. Early support for this idea came from spectrograms of the night side of the planet obtained in 1955 by N. Kozyrev at the Crimean Astrophysical Observatory. Kozyrev reported spectral bands due to ionized nitrogen at wavelengths of 3914 and 4278 Å (angstroms) – exactly what is observed during terrestrial airglow. Three years later, Newkirk, at the High Altitude Observatory in Colorado, obtained further spectrograms of the night side of Venus, which went some way to corroborating the earlier results.

Spacecraft images returned from Venus do show streaky clouds that are darker than the generally bright disc; however, whether or not these are what Earth-based observers purported to record as "dusky shadings" – and indeed, inspire some observers actually to draw maps of Venus – remains something of an enigma. The much-respected French astronomer, Adouan Dollfus, considers that drifting and relatively dark, lower-lying clouds may at times be perceived through the uppermost cloud deck, giving rise to the dusky shadings recorded. Even more recently it has been suggested that some of the dusky markings may be due to volcanic smoke or dust.

Whether or not the telescopic markings prove to be illusory or real does not change the fact that the polar regions frequently appear bright through a telescope. When Venus shows a crescent phase, the cusps often are particularly bright, but not necessarily at every apparition. All kinds of experiments, many using filters, have been conducted to establish the nature of these polar "caps". To date, no consensus view exists, but an excellent discussion of the topic can be found in *The planet Venus* by Hunt & Moore (1982), a book that also includes a bibliography of early Venus observations. Only in 1962, when Mariner 2 reached Venus, was it established that the caps were truly polar, for only then was the axis of rotation shown to be inclined to the perpendicular to the orbital plane by just a few degrees.

Venus orbits the Sun in a near-circular orbit, at a mean distance of 107 400 000 km. At its closest it approaches to within 39 000 000 km of Earth, making it the closest natural object apart from the Moon and meteoroids. The Venusian year is 224.7 Earth days in length, and Venus rotates once on its axis every 243 Earth days; in other words the day on Venus is longer than the year! This bizarre fact was not discovered until the early 1960s, when powerful radar telescopes sited at Goldstone (Jet Propulsion Laboratory [JPL] facility), Haystack (Massachusetts Institute of Technology [MIT] facility) and Arecibo (Cornell facility) were able to identify persistent features on the surface. Venus is unique among the planets in this respect.

To compound this oddness, it also rotates in a retrograde direction and is one of only two worlds in the Solar System to do so (the other is Uranus).

1.3 Venus and Earth compared

Venus is similar to the Earth in terms of size. It is almost spherical and has a mean diameter of 12 104 km, compared with 12 742 km for the Earth. The mass is 4881×10^{24} kg, which is just over 80% that of Earth and it is only slightly less dense at 5.134×10^{3} kg m^{-3} (Earth is 5.517×10^{3} kg m^{-3}). However, it has evolved in such a way that conditions within its atmosphere and on its surface are very different indeed. When Mariner 2 bypassed the planet at a distance of only 35 000 km in 1962, it found the surface temperature to be very high indeed at 468°C; the currently accepted figure is 482°C. This is above the **triple point** for water and also the melting point of lead. Add to this the fact that the atmospheric pressure is 90 times that of the Earth, and you have a very inhospitable environment indeed. Mariner 10, sent by the Americans in 1974, later found that the atmosphere rotates in the direction opposite to the solid body, taking about 5 Earth days to complete one revolution.

By the early 20th century, some scientists had suggested that water vapour might be present in the Venusian atmosphere, but this had not been confirmed. In 1921, V. M. Slipher, working at the famous Lowell Observatory, tried to determine whether earlier reports of atmospheric oxygen on Venus were correct. His attempts failed, and it was not until 1932 that the first successful spectroscopic work was undertaken by Adams and Dunham who, while failing to find oxygen, recorded strong infrared (IR) absorptions that subsequently were correlated with carbon dioxide. It is this gas that dominates the air of Venus.

The atmosphere itself is highly reflective and it returns 75% of the incident solar radiation (as compared to 16% for Mars and around 7% for the Moon). This is in large part attributable to the behaviour of carbon dioxide (CO_2) which comprises approximately 97% of the total atmospheric volume. This is very different from the Earth's air, in which CO_2 accounts only for a mere 0.031%. This marked difference in atmospheric composition leads to significantly different surface environments on the two sister worlds.

The proportion of incoming radiation that succeeds in penetrating the dense Venusian cloud mantle is, understandably, less than that which hits the Earth's surface. Venus, however, owing to its relative proximity to the Sun, receives about twice as much solar energy as Earth; consequently, the actual amount that gets through is still very high. This energy is used to heat the solid surface, after which much of it is re-radiated at longer (infrared) wavelengths; however, whereas on Earth much of the re-radiated solar energy escapes into space, things are very different on Venus. Thus, although the Venusian carbon-dioxide-rich atmosphere is relatively transparent to the incoming solar radiation, it cannot escape so readily owing to re-emission at longer wavelengths. The upshot is that a great deal of this thermal energy becomes trapped and gives rise to the notorious **greenhouse effect**. (If what we read is true, then we appear to be endeavouring to emulate the appalling conditions on our nearest planetary neighbour!)

We have already noted that, in terms of size, density and mass, Venus is almost a twin of the Earth; for this reason it might be expected to have a similar internal structure. Now Earth, with a dense iron-rich **core**, a hot mobile **mantle** and a thin rigid **crust**, has remained dynamic throughout geological time, a period exceeding 4700 million years (Ma). Furthermore, its **lithosphere** is segmented into slowly moving **plates** whose motions are driven by mantle activity, a process known as **plate tectonics**. Geologists are anxious to establish whether or not Venus has developed in a similar way and, if it has, whether plate motions and/or volcanicity continue at the present time.

The resultant very high **ambient** temperatures prevalent in its crust dictate that this cannot behave as rigidly as does the Earth's. This is an important difference, for the rigidity of the terrestrial lithosphere is the factor that deters plates from deforming internally, and allows them to transmit elastic stresses over long periods of time. It is a key factor in the predictability of plate tectonics. Then again, the thermal gradient within Venus is very different from that of the Earth and its lower crust is likely to be of relatively low strength. The most important implication of these differences is that, if lithospheric plates do occur on Venus, they would be poorer stress guides than their terrestrial counterparts and, as a consequence, deformation would be less concentrated at plate boundaries and more widely dispersed throughout plates. The lower strength of the Venusian crust means that it is more akin to the Earth's continents than to rigid oceanic crust, at least as far as its deformational history is concerned. This would be reflected in a rather different distribution of landforms.

Earth and Venus are disparate in another way: the Earth has oceans full of water, Venus does not. Water, too, is locked into the minerals within many of Earth's primary rocks and in intergranular pores within sedimentary rocks. Indeed the latter owe their origins to the activity of the volatile-laden atmosphere and to running water. The fold mountains that have risen over the sites of ancient convergent plate margins on Earth are built largely from these stratified and easily deformed rocks. On Venus, where the rocks presumably are dry, no sedimentary rocks are being laid down by water today. If plate movements are taking place there at the present time, fluvially deposited sedimentary rocks cannot form a part of the planet's plate tectonic equation.

Water and other volatiles also play a vital part in the internal workings of the Earth. For instance, it is the presence of water in the lower crust and upper mantle that allows the formation of **partial melts** at relatively shallow depths. Such melts rise to form new oceanic crust. Dry rocks simply could not melt at the temperatures prevailing at these modest depths. Then again, the water trapped in sediments scraped off the ocean floor and carried down with slabs of oceanic crust during **subduction** fuels a whole variety of mineralogical reactions that are vital to Earth's dynamism and it influences the processes that occur in the outermost layers. Venus must be different in these respects.

Earth's considerable size has determined that it has remained highly active throughout geological time, whereas smaller worlds, such as the Moon and Mercury, lost their original heat relatively quickly and have been effectively "dead" for at least three billion years. On Venus, while the very high crustal temperatures militate against efficient terrestrial-type plate recycling at present, Venus might not have been so hot and dry in the distant past, in which case the situation could have been very different. However, even if Venus has always been hot and dry, its large size suggests that it should have remained dynamic throughout much of geological time. Therefore, it may have developed a segmented crust and lithosphere in some measure like the Earth's.

Planetary exploration has shown that there are several different ways by which thermal energy is transferred from the inside of a planet to its surface. For instance, Mercury, the Moon and Mars apparently generated their primary crusts by accretional heating; then, over the rest of geological history, they slowly lost their heat by conduction through a globally continuous crustal shell. At some time during their evolution, there was a stage of secondary crustal development due to partial melting of their mantles, with the production on Mercury of the inter-crater plains, on the Moon of the mare basalts, and on Mars of the extensive volcanic plains and shield volcanoes of the northern hemisphere. In this way a smaller amount of heat was lost advectively. On this basis, volcanism can be predicted to have been important on Venus too.

On Earth, roughly 62% of the thermal energy derived from the mantle layer is used to drive a different type of secondary crust formation, which involves crustal generation, spreading and recycling. The production of new oceanic crust and lithosphere occurs along the crests of submarine ridges such as the Mid-Atlantic Ridge and the East Pacific Rise. Such sources of new oceanic crust are located along divergent plate boundaries where mantle motion pulls the crust apart. The newly formed basaltic crust then cools and becomes denser, subsiding as it is dragged laterally by mantle convection (Fig. 1.2). Eventually such cool oceanic crust is recycled by subduction, whereupon it plunges down beneath more buoyant continental crust at convergent plate boundaries such as those that girdle the Pacific.

Because of the segmented nature of the Earth's crust and lithosphere, volcanic, **seismic** and **tectonic** activity are strongly concentrated along plate boundaries. If Earth-style plate tectonics had operated on Venus, a similar global distribution of linear belts would be expected. Again, on Earth, fewer major volcanic centres may develop within individual plates, where they grow above major mantle **hotspots**; Hawaii is one such example. Occasionally a divergent plate boundary and hotspot may coincide, as currently is happening beneath Iceland. In this way a kind of mini-continent or "plume plateau" may develop over a divergent plate boundary. In the light of what we know about the planet, these kinds of features might be detected on Venus. If so, something can be inferred about the way in which the planet is transferring its internal energy towards the surface.

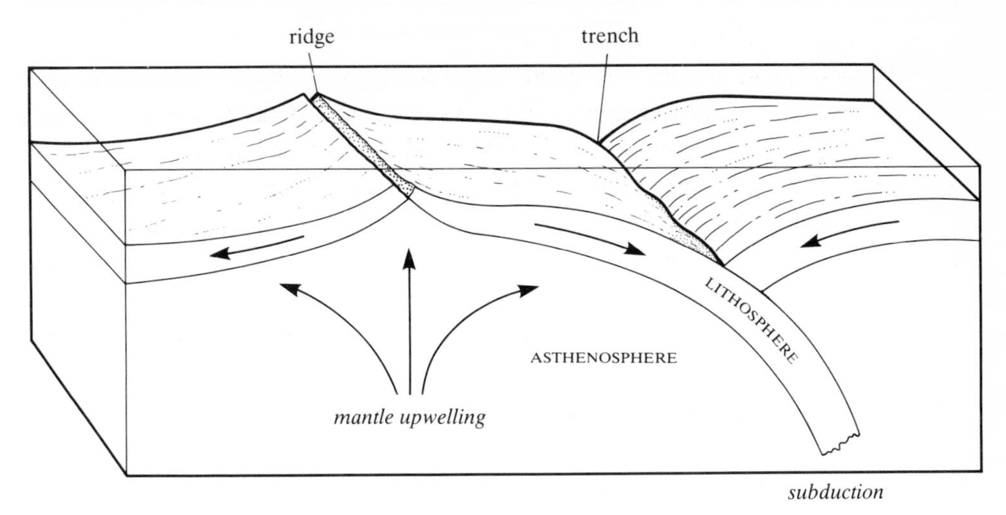

Figure 1.2 Plate tectonics
New oceanic crust is generated above upwelling mantle at mid-oceanic rises; crust is subducted at convergent plate margins, and recycled by melting inside the Earth.

Does recent evidence favour the notion that Venus is geologically active, and has it developed Earth-type plate tectonics? These are just two important questions that scientists hope that Magellan may help to answer, since its radar-mapper can provide images with a resolution ten times better than anything obtained hitherto. Within the space of two Venus years (486 Earth days) geologists have obtained almost complete coverage of the planet at resolutions as good as those obtained of the Earth from Earth-orbiting satellites such as Landsat.

1.4 Unveiling Venus

When, during the 1960s, the giant radio telescopes at Arecibo in Puerto Rico and at Goldstone in California were aimed at Venus, they provided the first reliable data about its diameter (12 104 km) and rotation period (243 days) and also gave scientists the first images of major surface features. Up to that time there had been much speculation about what lay beneath the optically impenetrable cloud mantle: did the planet have high mountains, water oceans, even volcanoes? The first radar images revealed that the planet had continent-sized upland regions, one in particular, Ishtar Terra, being prominent on the images obtained. This success spurred American scientists into building a more sophisticated radar system and putting it on board a spacecraft that entered Venus orbit in December 1978; this was Pioneer-Venus. It was with this craft that NASA scientists successfully achieved the first serious topographic mapping of our sister world.

Figure 1.3 Panoramic view from Venera 9
The probe landed on the plains along the northeastern flank of Beta Regio.

Before Pioneer-Venus, several missions had been sent by the Russians; these constituted the early part of the Venera programme. The first of these probes was launched in 1961, but failed; however, eventually, soft landings were achieved by Veneras 7 and 8, which returned data effectively, despite operating in total for only a few tens of minutes before the inhospitable conditions took their inevitable toll. Then, in June 1975, the very first panoramic pictures of the Venusian surface were sent back by Venera 9, followed by Venera 10, one week later. At this point I think it would be true to say that Venus had been unveiled, once and for all (Fig. 1.3).

1.5 Magellan's voyage of discovery

So aptly named after Ferdinand Magellan, the well known Portuguese seafarer and explorer who circumnavigated the world in the early 16th century, the Magellan Mission was planned to break new ground, not least because it included the despatch of the first interplanetary spacecraft by a launch vehicle with a chequered career, the Space Shuttle. The radar-mapping probe was built and equipped, rather like Magellan the explorer, to provide humankind with a global view of a relatively little-known planetary surface. It was successfully launched from the Space Shuttle Atlantis on 4 May 1989, and it entered orbit around Venus during August 1990.

The alignment of Earth and Venus in 1989 dictated that Magellan had to be launched between 28 April and 29 May 1989. After one false start on 28 April and then a tense period because of inclement weather immediately prior to a second attempt on 4 May, the Space Shuttle Atlantis eventually took the 345 kg spacecraft into the afternoon sky at 2 hr 46 m 59 s eastern daylight time. After a journey of 15 months, the Magellan craft received a command from Earth to brake hard and go into a highly elliptical orbit around Venus. This crucial manoeuvre happened on 10 August 1990. Magellan had successfully arrived at Venus.

Almost immediately, enigmatic problems ensued and at one stage it seemed likely that the mission might be an expensive failure. However, with their usual diligence, both project scientists and technicians at the Jet Propulsion Laboratory (JPL) of the California Institute of Technology (Caltech) wrestled day and night and gradually were able both to calibrate the spacecraft and to send it appropriate commands regarding its orbit and mapping procedures. Most of the identifiable snags were eventually ironed out, and mapping began somewhat later than originally planned, in early September 1990.

On 4 April 1991, a spokesman at JPL revealed that the Magellan spacecraft had completed its primary mission. This, the mapping by radar of 70% of the surface of Venus using left-looking radar, had been achieved just one month short of the first 243-day mapping cycle (equivalent to one complete revolution of the planet beneath the Magellan spacecraft). At that time, NASA's project manager, Tony Spear, announced that, because the returned radar images had been of such excellent quality and certainly had generated tremendous excitement among the planetary community, a second 243-day mapping cycle would commence on 16 May. Since the JPL announcement was made, the second full mapping cycle has been completed (using right-looking radar) and 50% of the surface imaged, including 14% hitherto unmapped. At the time of writing, the third cycle, which commenced somewhat shakily on 21 January 1992, has been completed and has provided stereo coverage of roughly 35% of Venus. Starting on 25 May 1993 the **periapsis** altitude of the probe was lowered from its present 290 km to 140 km, so that Doppler gravity-field measurements can be made of the entire planet. Magellan's orbit was thus modified to a near-circular 94-minute orbit.

The imagery so far processed reveals Venus as a planet of immense complexity and with many geological features unlike anything known on Earth. A new geology is being written and will continue to be amended and extended as scientists gradually come to terms with it.

2 Planet formation

2.1 Birth of the Sun and planets

When viewed dispassionately in the context of the Solar System as a whole, the Earth seems rather insignificant. In the seemingly infinite expanse of what we know as space–time – the known universe – it pales into veritable obscurity. Yet, despite its small size and its attachment to what is a perfectly normal, indeed rather modest, star (the Sun), together with the other planets and the Sun, it is of supreme importance to us. This is because other stars and planets are presumed to have formed by processes similar to those established for our local slice of the universe. It is generally conceded that physical and chemical laws operate throughout space–time in much the same way; therefore, what we have learned about our local bit of space–time applies to it everywhere, at least to a close approximation.

Much of what we "know" about the Solar System is a mixture of observation and theory. It is generally conceded that the chemical elements, other than hydrogen and helium, are produced by nuclear explosions that take place inside some stars, giving rise to supernovae. By this means, the heavier elements are dispersed widely throughout the universe, which otherwise is dominated by the elements hydrogen and helium. There is also a broad consensus that the Solar System formed by gravitational collapse from a cloud composed of dust, ice and gas. After that, however, no single picture of planet formation pleases every expert. One of the most actively debated issues is the way in which the chemical elements initially were segregated within the planets.

Two different models for planet formation have dominated the literature. The first of these has become known as the *heterogeneous accretion model*. According to the former model, which has been supported by workers such as Turekian & Clark (1969), Grossman (1972) and Cameron (1973), selective condensation of different elements and compounds from the solar nebula (which might have been a cloud or a disc of material) occurred primarily in response to the way in which temperature decreased away from the proto-Sun. Thus, Jupiter and the outer planets, because of their much greater distance, accreted in a region where condensation temperatures were below 0°C, while the terrestrial planets and meteorites must have condensed at around 100°C. For this reason, the former had a facility greater than the latter for collecting a far larger proportion of the relatively light, **volatile elements** such as

H, He, C and O. The layered structure of the terrestrial planets would subsequently have arisen during a continuous **accretion** process according to the predicted **condensation** sequence first put forwards by Grossman & Larimer (1974) and shown in Table 2.1.

On the basis of this model, a general scheme for planet segregation would be: **refractory elements** such as aluminium and calcium formed a primitive Ca–Al-rich core, followed by metallic iron and, finally, by magnesium- and iron-rich **silicates** such as **olivine** and **pyroxene**, which eventually produced a mantle layer. Then, owing to gradual warming, further segregation of these elements would have occurred, with iron settling towards the core, and the refractory condensates being displaced outwards into a silicate-rich exterior. Two of the more important implications of this model are that, if significant pre-accretional element segregation had taken place, those terrestrial planets farther from the Sun should contain greater amounts of lower-temperature condensates such as alkali-feldspar, troilite and magnetite than those nearer to the proto-Sun, and that there also should be a trend of decreasing mean density with increasing orbital distance for the inner planets, owing to the predicted falling Fe : Si ratio.

In contrast, *homogeneous accretion models*, such as those proposed by Urey (1962), Kuiper (1952) and Elsasser (1963), perceive the matter that eventually accreted to form the planets as having already condensed into grains and, in the case of the terrestrial planets, being at temperatures below 100°C. They also assume that the chemical elements were fairly well mixed, with a composition approximating to the C1 **carbonaceous chondrites**. Later segregation of elements is seen as a function of heating due to both accretional energy and the decay of radioactive nuclides. Naturally, some condensed material would have been re-volatilized

Table 2.1 Condensation temperatures of solar nebula materials at 10^{-4} Pa. After Grossman & Larimer (1974).

Mineral phase	Composition	Temperature (°C)
Corundum	Al_2O_3	1410
Melilite	$Ca_2Al_2SiO_7$–$Ca_2MgSi_2O_7$	1205
Perovskite	$CaTiO_3$	1200
Spinel	$MgAl_2O_4$	1150
Metallic iron	Fe (Ni)	1130
Forsterite (olivine)	Mg_2SiO_4	1120
Diopside (pyroxene)	$CaMgSi_2O_6$	1100
Enstatite (pyroxene)	$MgSiO_3$	1100
Anorthite (plagioclase)	$CaAlSi_2O_8$	1100
Alkali-feldspar	(Na, K)$AlSi_3O_8$	980
Troilite	FeS	430
Magnetite	Fe_3O_4	135
Ice, methane, etc.	H_2O, CH_4, CO_2, H_2, etc.	<0

(for example oxygen, sulphur, carbon and the alkali elements), that would have been lost to space as oxidized compounds, for instance, H_2O, CO_2 and SO_2.

The resulting oxygen loss would have generated reducing conditions in the remaining silicate materials, and, since both reduction and volatile loss would have increased proportionally with proto-planet dimensions, assuming the same chondritic parent material, the more massive planets should have the least-oxidized compositions. This is largely a function of the greater effectiveness of accretional heating upon the more massive planets. Homogeneous accretion therefore predicts that the more massive of the rocky planets should have the higher densities and should hold the smallest proportion of relatively volatile alkali elements. Subsequ-ent heating due to various radioactive decay sequences would have produced the observed internal layering from the initially homogeneous nebular material. Currently available geochemical and physical data for Venus, the Earth, Moon and Mars are consistent with their having been accreted from the same **primordial** nebular material. Mercury is by way of an exception.

Despite fundamental differences between the two models, there is some common ground, which provides a basis for general discussion. Thus, it is generally agreed (though not universally accepted) that all of the Sun's planets had their origin in an interstellar cloud of dust and gas that was trapped by the rotational energy of a massive and collapsing proto-Sun, and that the angular momentum of the cloud would have forced it to encircle the primary star as a flat disc from which the proto-planets eventually condensed. It is also generally assumed that very large numbers of relatively small proto-planets would have formed, each with its own gravitational field, but that the number gradually diminished as the fragments settled into larger aggregations, owing to gravitational coalescence. Finally, many scientists believe that the chondritic meteorites represent largely unmodified clumps of the primordial particulate matter that condensed from the solar cloud and out of which the planets, including Venus, eventually formed.

2.2 Accretion and planetary heating

There is considerable ignorance about the precise manner in which the initial accretion of what are loosely termed "planetesimals" took place, but it seems clear that a number of distinct phases would ensue as the cloud of dust and gas rotated about the proto-Sun. After an initial stage, during which there was condensation of silicate and ice grains, there would have followed a phase when aggregation of these into larger particles took place, either by gravitational growth or by particle/particle collisions. The larger bodies probably would have developed a **regolith** which effectively inhibited rebound of infalling particles and promoted growth by collision. On the other hand, smaller bodies would tend to shatter as greater infall velocities were achieved and so would disperse. Thus, larger objects in part would grow at the expense of smaller ones and would sweep up much of the finer debris generated by shattering of the smaller fragments. Eventually they would have attained true planet status.

One consequence of the accretionary process is that, each time there was an impact, the kinetic energy of the impacting grain would almost entirely have been converted into heat. The impact record fossilized on the solid surfaces of all of the planets and their satellites indicates that, during the first 500 Ma of Solar System history, all were bombarded heavily by meteoroids, some of very large dimensions. It has been calculated that bombardment on this scale could have raised the surface temperature of the Earth as high as 10 000°C, the exact temperature depending upon how much heat was retained and how much was lost to space. Since there is strong geochemical evidence that Earth's core grew both during and after accretion, the energy required for core generation must have been supplied predominantly by impact energy. A similar amount of heating would have been experienced by Venus.

Radiation and *conduction* are the two processes that would most readily reduce the heat budget of a proto-planet. Both are a function of surface area, so, in terms of heat loss, the smaller planets would have suffered more than their larger companions because of their greater surface-area : volume ratio. Deeper sources of heat would, however, be more effectively retained and should only be lost to the surface by convection. Assuming that the infall time was similar for all bodies, the formation temperature of the proto-planets must have been proportional to their present masses. On this basis, Venus's heat structure should be very similar to Earth's, but with the melted region a trifle closer to the centre.

2.3 Core formation and mantle segregation

Other sources of heat existed early in planetary development. Thus, immediately after accretion, **adiabatic** heating had the potential to raise the Earth's core temperature by as much as 900°C. Venus would have fared only slightly less well in this respect. If there had been no other source of energy, the core temperature of both planets would have fallen almost at once, owing to conduction. However, it can also be shown that the Earth's core could not have been produced simply through adiabatic heating, since the projected increase in melting point with depth for potential core-forming materials is either greater than or at least as great as the **adiabatic gradient**. Estimates of the interior temperature of the Earth at a time immediately after accretion but before core formation show that interior temperatures may have been too low to allow the melting of Mg-rich silicates; nevertheless, metallic iron particles in chondritic parent material may have melted because the melting temperature of the pure metal is somewhat lower than that of the silicate. If this was the case, then droplets of metallic iron may have segregated and begun their descent towards the region of the core. The situation would have been closely paralleled on Venus.

Most thermal models agree that a proto-Venus of approximately chondritic composition with a major part of its radioactivity reservoir locked inside it would have heated up gradually with time. Thermal energy would be generated by the slow decay of long-lived radionuclides such as ^{238}U, ^{235}U, ^{232}Th and ^{40}K. The build-up of radioactive heat would have arisen largely by virtue of the insulating properties of the proto-lithosphere. The segregation of a core is the direct result of such a process, and it is possible that the thermal energy released as the metal phase sank was sufficient to have heated Venus to over 1800°C. At this point the solid mantle may have begun to convect in response to the rising interior temperature. The settling rate of core-forming elements would have been enhanced once the internal temperature had come close to the melting point for the silicates, since at that point the silicate matrix would have become "softened" sufficiently to permit droplets of a dense metallic phase to sink through it. The precipitation of metallic droplets would certainly have preceded the general melting of the silicate phase and there seems little doubt that core formation was a fairly rapid process compared to the generation of the silicate mantle.

The presence of sizeable amounts of FeO and FeS in the Earth's core is now generally accepted and is indicated by the density of the outer core, which is significantly lower than that of an Fe–Ni alloy under the appropriate temperature–pressure conditions. This would have played a significant rôle in depressing the melting point of the metal phase but not the silicate, which would permit core formation prior to any general melting of the mantle. The bulk oxygen content would have played a major rôle in determining the size of **siderophile** cores within the terrestrial planets. Thus, if the Earth was of carbonaceous chondrite composition initially, and no volatile loss had occurred, then it should be almost entirely **lithophile** with just a small **chalcophile** core and certainly no free metal. Because the Earth actually has a 3485 km-radius metallic core that accounts for 32% of the planetary mass (16.4% of the volume), it must be assumed that free iron remained after all of the available oxygen and sulphur had combined with other lithophile and chalcophile elements, and therefore that the Earth must have lost a few per cent of its original oxygen quota.

Since Venus does not have a natural satellite, its moment of inertia cannot be determined. Because it is only slightly smaller and less dense than Earth, it seems reasonable to assume that its internal structure and composition are similar. There is probably a metallic core (somewhat smaller than Earth's) which accounts for about 23% of the planetary mass and there is likely to be more oxidized iron in its mantle; it certainly cannot have a higher Fe : Si ratio than the Earth. The massive CO_2 atmosphere must have been outgassed, like the Earth's, but to **outgas** an oxidized atmosphere of such magnitude means that substantial iron must have existed in the interior. The close similarity in mass and density between Venus and the Earth probably points to a similar early thermal history, controlled by the separation of the core. It is the thermal history, of course, that determines the level of volcanic and tectonic history, the overall pattern being a response to the gradual cooling and thickening of the lithosphere.

The formation of a silicate phase may initially have been localized onlywithin Venus;

however, gradually such pockets as did form must have risen, taking with them entrapped radioactive materials having lithophile affinities. These trapped elements would have slowly caused heating until melting occurred, with the result that small **sialic** bodies were formed. The pattern of surface structures thus far identified on Venus imply the action of compressive stresses in the Venusian crust, and it could be that convective motions were eventually set up within a silicate-rich mantle layer. Mantle melting and segregation were, however, much slower processes than core formation.

Mantle convection probably played a major part in the subsequent development of the larger terrestrial planets. After core formation, cooling dominated the thermal histories of these worlds. The gravitational potential energy released during core infall on the Earth was adequate to melt it completely, but this did not happen; the same is almost certainly true of Venus. The process most likely to have inhibited this is convection within a molten silicate mantle layer, for this would have the potential to remove thermal energy quickly. Eventually the mantle solidified, whereupon convection would have commenced and indeed continues on the Earth to the present day. As yet it is not clear whether this is true for Venus.

Although their thermal histories are not the same, the rate of cooling for all of the terrestrial planets would have fallen with time, as mantle viscosity increased and convective processes became less vigorous. From the time when the first near-surface internal melting occurred, volcanism, in one guise or another, has played a part in planetary development, and, in particular, in surface modification. On Mercury and the Moon, volcanism commenced and terminated early because either thermal energy generation quickly declined or energy was lost to space; by implication such mantle activity as developed must also have diminished relatively rapidly. In contrast, an extended history of volcanism characterized the Earth, Venus and Mars, for which more vigorous mantle activity is indicated. Careful study of the chemistry (including isotope chemistry), mineralogy and morphology of volcanic rocks preserved at the surface of Venus really is the only way to learn about the fundamental internal processes of which volcanic rocks are the surface manifestation. Magellan at least is providing the morphological data.

2.4 Degassing

Geologists are primarily interested in the nature and evolution of a planet's lithosphere. In the case of Venus, Earth and Mars, there are rock types and landforms that are the direct result of the interaction of this lithosphere with the atmosphere (and in the case of the Earth, the hydrosphere). Today, it is generally agreed that all of the planets once had a **primary atmosphere** that consisted of the gases gravitationally trapped by the proto-planets during their formative years. In other words, each had a primordial "inherited" atmosphere. To this very day, the outer planets still have much of this atmosphere; however, the rocky inner planets appear to have lost nearly all of theirs. Instead, **secondary atmospheres** developed by the degassing of their interiors. For this reason the most important components of such atmospheres are carbon, hydrogen and nitrogen – elements that were incorporated during accretion.

Several of the elements that eventually went into the atmospheres of these planets probably started out within the atomic lattices of minerals within rocks (e.g. hydrogen, oxygen and carbon); others (e.g. nitrogen, neon and argon) may have become trapped in the lattices of compounds forming from the nebular gases; still others may have been derived from nebular ices.

All the evidence we have indicates that Venus and the Earth both accreted from materials that were at a temperature of at least 340°C, which is well above the temperature at which water could be incorporated into solid matter condensing from the solar cloud. Yet the Earth is covered in voluminous oceans of water. Then again, Venus has a dense carbon-dioxide-rich atmosphere, yet should consist of matter that condensed at temperatures well above the stability field of carbon compounds. There are a number of possible reasons for these apparent anomalies, some of which will be discussed later. The basic facts suggest that each of the inner planets accreted from a range of different materials with different condensation temperatures, and that, whatever the precise way in which these materials were added to each planet, there is no doubt that the segregation and concentration of some into atmospheres was the result of internal differentiation.

Primary atmospheres are rich in hydrogen and helium; outgassed atmospheres are not. Most hydrogen becomes combined with oxygen to form water or with carbon to form methane. Carbon monoxide (CO), carbon dioxide (CO_2) and nitrogen (N_2) are also generally abundant. The degassing process would be assisted by core formation and mantle segregation, owing to the generation of large amounts of thermal energy. Since these processes can be shown to have taken place early on in planetary evolution, secondary atmospheres undoubtedly are ancient.

2.5 The early history of Venus

The formative period of Venus's evolution probably followed a similar path to that of the Earth. The very similar masses of both worlds suggest that accretion, core formation and mantle segregation took place during the first 1000 Ma or so of their existence and dictated rather similar general patterns to their thermal histories. However, the greater proximity of Venus to the Sun had a profound effect upon the lithosphere–atmosphere reactions that subsequently took place, and also determined that the behaviour of the Venusian lithosphere was significantly different from that of the Earth.

3 Exploring the veiled planet

3.1 Problems of Venus exploration

From Earth we can never see the surface of Venus, for the simple reason that it is permanently veiled by a dense layer of clouds. These are opaque to radiation at visual wavelengths and they prevent astronomers seeing what lies beneath. As a consequence, even the most powerful optical telescopes reveal only the phases, a whitish disc crossed by vague dusky shadings and ephemeral "polar caps". As if the presence alone of such an atmosphere was not sufficient to dampen the resolve of planetary scientists, Nature has contrived to erect another formidable barrier to exploration; this is a direct result of the atmosphere's presence.

The air of Venus is not dominated by nitrogen and oxygen, as is the Earth's, but by carbon dioxide. This gas, although relatively transparent to incoming solar radiation, hampers its escape, owing to re-radiation at wavelengths longer than those which can pass through freely. In consequence, a "greenhouse effect" is in operation and the surface is as hot as molten lead. This hot and chemically hostile surface environment makes it difficult to maintain instruments in an operational mode for all but brief periods; it also relegates the idea of manned landings to the realm of science fiction. The upshot is that scientists have had to develop alternative kinds of imaging systems and to resort to such techniques as dropping instrument packages through the dense atmosphere by parachute.

Luckily, as is often the case with research that was initially a part of one or other of the "great war efforts", help came during the 1940s with the development of a technique known as radio detection and ranging, better known by its acronym – radar. This allowed the military to penetrate through bad-weather clouds and darkness, and, despite either or both, to seek out intelligence on the ground. Subsequently it has become one of the most widely used navigational methods of the modern age, in both civil and military circles. In simplest terms, radar is an imaging system that operates at wavelengths much longer than those which characterize the visual part of the electromagnetic spectrum.

3.2 Radar and radar imaging systems

Whereas optical telescopes utilize that part of the electromagnetic spectrum we know as the visual band (0.4–0.7 μm), radar imaging devices not only provide their own source of energy but also generate radiation at wavelengths 0.8–70.0 cm, much longer than visual ones. At such long wavelengths - which characterize what is termed the microwave region - the radiation is hardly affected by either clouds or darkness, and can penetrate through several metres of solids. Herein lie the obvious advantages of radar to the planetary scientist interested in Venus.

The operation of an "active" radar imaging system (one that produces its own pulse of radiation) is fundamentally different from those "passive" systems that use the Sun's rays for illumination. With radar, an antenna - essentially a reflector whose concave shape focuses the beam into the requisite form - generates a microwave pulse that "illuminates" the surface under scrutiny. Usually the illuminated region is in the form of a long and narrow strip normal to the spacecraft's flight path (Fig. 3.1). Also, in the majority of cases, the "look direction" of the system is downwards and to the side of the antenna. For this reason it is known as **side-looking radar** or SLR. Because this is how radar images are obtained, "pictures" of large regions of planetary surfaces generally are a mosaic made up from individual image strips.

The larger the antennae, the more powerful is the system and therefore the greater is the potential for obtaining high-resolution imagery. In order to optimize on size of aperture, ra-

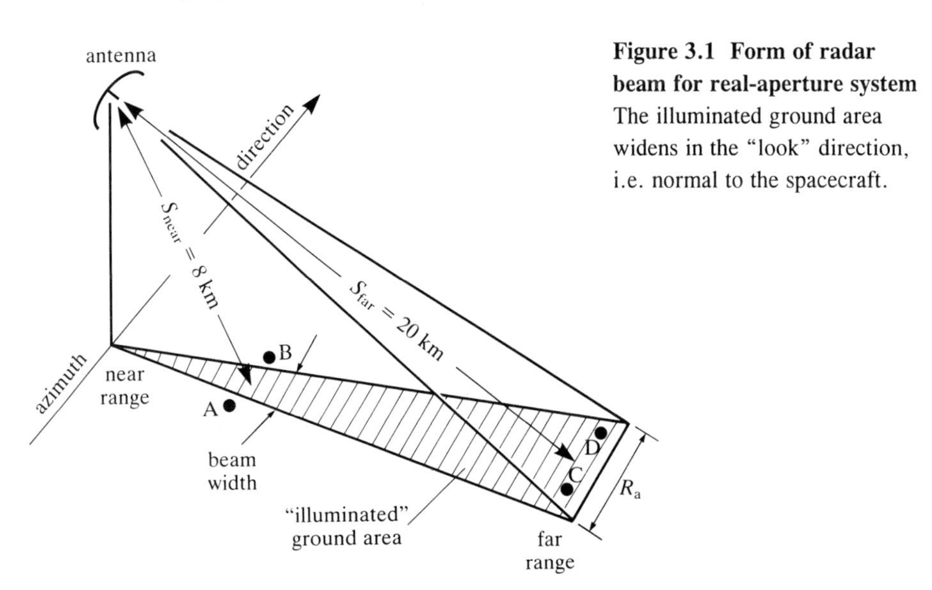

Figure 3.1 Form of radar beam for real-aperture system The illuminated ground area widens in the "look" direction, i.e. normal to the spacecraft.

distance AB = distance CD = 10 m
targets A & B are resolved
targets C & D are not resolved

dar antennae tend to be fixed, so as not to limit their size. In *real aperture imaging systems*, an antenna is directed towards the surface and collects images of as high resolution as it can within the constraint of its own dimensions or "real" aperture. To collect very high-resolution imagery by this method currently is not possible, because very large antennae cannot be carried on board spacecraft.

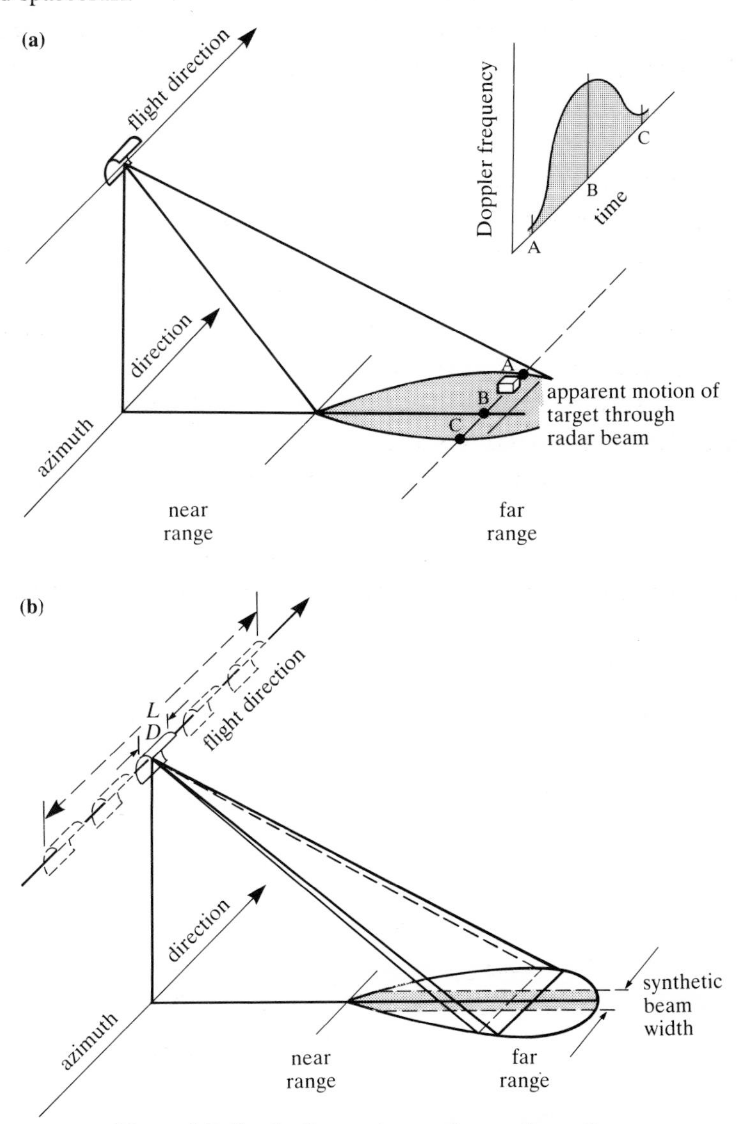

Figure 3.2 Synthetic-aperture radar configuration

In a SAR system there is a Doppler frequency shift due to the relative motion of the ground target through the radar beam (a). The resolution of the SAR system is enhanced due to the synthetically lengthened antenna, L (b). The physical antenna length is D. After Craib (1972, Figs 3 and 5).

Fortunately, there is a way around this obstacle, which utilizes the well known **Doppler effect**: because the antenna is in motion while it is receiving echoes from a planet's surface, special computer processing can simulate a much larger antenna. Thus, the distance a spacecraft moves while a surface feature is within the radar system's field of view effectively determines the functional size of the aperture. This appropriately is called **synthetic aperture radar**, or SAR (Fig. 3.2). The technique demands that the radar system can transmit and receive multiple outgoing pulses and returned echoes, which, when studied on an oscilloscope, are seen to exhibit subtle differences in frequency that are caused by the relative motion of the target and the radar device. Magellan's radar system is of this kind.

The same antenna that sends the radar pulse also receives the returned signal and, in order to avoid interference between the transmitted and returned beams, illumination of the surface is accomplished by sending discrete pulses just a few microseconds in length, separated by pauses. This is achieved by an electronic switching mechanism, which blocks the receiver during transmission periods and prevents interference between the transmitted and received pulses. Then, during each hiatus the returned beam can be received and its contained information stored within a computer. The returned pulse also may be displayed as a line sweep on a cathode ray tube and recorded on film.

Because every radar pulse is electromagnetic radiation, its velocity of transmission is known; it travels at the speed of light in a vacuum. Since it is possible with modern technology to time very accurately the period that elapses between the transmission of a radar pulse and the time of return of the echo, radar may be used to undertake topographic mapping. Therefore, to complement the wide-angle side-looking antenna aboard the Magellan spacecraft, there is a second, narrow-beam, antenna, which points more-or-less vertically downwards. This is providing topographic transects of the Venusian surface.

3.3 Interpreting radar images

Radar images are trickier to interpret than visual ones. For instance, although radar pulses cast shadows when they encounter obstacles, just as visual systems do, highly reflective smooth surfaces that would show as bright regions in a traditional photograph may not appear bright on a radar image. On the contrary, they sometimes appear black, which is highly confusing to the non-expert. The reason for this is that a radar beam hitting a highly reflective surface, rather than being reflected back to the receiver, is reflected away from it, according to the well known law that states that the angle of incidence is equal to the angle of reflection. This being so, no echo is returned to the antenna. On the radar image, therefore, the metallic region would appear black, like a shadow. Because the radar beam hits the target surface at gradually more oblique angles along the strip in the look direction, it is necessary to take account of incidence angle when studying the returned images, as it affects the nature of the return pulse. The frequency of the illuminating pulse also has an effect. Such factors are termed *system properties*.

The radar return is also a function of several *terrain properties*, whereby the transmitted radar pulse interacts with the surface with which it comes in contact and is modified before being sent back to the receiver. Factors that conspire to modify the beam include the material from which the surface is made, its roughness on the centimetre scale, the slope of the ground and the surface's electrical properties. With all these factors having to be borne in mind, not surprisingly the interpretation of radar images is significantly more difficult than that of visual imagery and demands considerable specialist experience. Here is not the place to delve deeply into the complexities of back-scattering functions, dielectric constants and the like; the interested reader can find details of these in Tyler et al. (1991, 1992). Suffice it to say that a great deal can be learned about the properties of the Venusian surface when the full range of echo characteristics are analyzed and integrated.

The specific kinds of radar echoes returned from Venus indicate that there are radar-bright areas, which represent a high degree of back-scattering due to rough surface materials, and radar-dark areas, which generally are a manifestation of smooth surfaces. It is therefore possible to gain some idea about the types of rocks that outcrop from one point to the next, and also of the surface texture. When this information is coupled with altimetric data, scientists can begin to unfold the planet's geological history.

3.4 Pre-Magellan missions

The first US Venus flyby mission was Mariner 2, which, during December 1962, carried on board a microwave radiometer that measured the surface temperature and broad atmospheric structure, and an infrared radiometer that studied the form of the clouds and their temperatures. It also measured the strength of the magnetic field in the neighbourhood of Venus. During this first phase of exploration, it quickly became apparent that the surface temperature was high and the magnetic field was less than one-tenth as strong as the Earth's.

Almost five years elapsed before the next phase began. Then, in October 1972, the Soviet craft Venera 4 – which eventually burned up in the dense atmosphere – sent down a lander probe, which crashed onto the night-side surface. This carried a delicate sensor, put on board to establish the surface pressure. It measured 100 bar, or 100 times the pressure at the Earth's surface at sea level. Venera also found the magnetic field to be even weaker than at first anticipated – a mere 1/100 000 that of the Earth's. Meanwhile, the US Mariner 5 craft flew as low as 4100 km over the surface. By careful radar tracking of the orbiting Mariner probe, scientists were able to measure the radius of Venus more accurately than had been done before. It turned out to be 6054 ± 2 km, which was roughly 66 km less than the radius previously determined telescopically. Mariner also measured the surface temperature; this was very high indeed, almost 430°C. The clouds, whose tops were found to lie 67 ± 10 km above the surface, were confirmed to be composed predominately of CO_2.

With this information to hand, it became clear immediately that Venus was not going to be an easy world to investigate in detail. However, continuing curiosity meant that between late 1970 and early 1982 more Soviet Venera spacecraft were launched. Veneras 7 through 14 managed to soft-land probes on the surface, and it was from these that the first panoramic pictures of the surface were sent back to Earth and the first geochemical analyses of Venusian rocks obtained. Both Venera 9 and 10 touched down near the equator, in the region known as Beta Regio. The Venusian landscape here was shown to be gently sloping and the surface composed of either flat rock slabs or isolated rocky outcrops separated by fine-grained material. During 1982, Veneras 13 and 14 landed farther south, in Phoebe Regio. The former put down between two gently sloping hills on a surface of slabs with intervening fine-grained sediment, the latter in an area generally lacking much fine-grained material (Fig. 3.3a). Because the surface morphology shown at these sites is not dissimilar to the texture found associated with many terrestrial slab **pahoehoe** lava flows, some workers were moved to suggest a volcanic origin for the rocks found at these Venera sites (Fig. 3.3b). The **basalt**-like composition of the rocks analyzed on the surface at the various Venera sites supported such a conclusion.

(a)

(b)

Figure 3.3 Lander views of the Venusian surface

(a) Venera 13 panorama, showing slabby surface with darker fines; (b) Venera 14 panorama, showing less-abundant finer material.

The next landmark was the successful deployment of the US spacecraft, Pioneer-Venus 1 and 2, in December 1978, an achievement that was to cost NASA more than US$175 million. The latter craft, which arrived five days after the former, comprised five separate craft – a hardware carrier known as a "bus", a hefty atmospheric probe and three smaller probes that entered the Venusian atmosphere at different points and determined the structure and composition of the air on both the day and night sides. At much the same time, Pioneer-Venus 1 was placed in orbit; between them, 12 different scientific experiments were included in their payloads. Interestingly, although the periapsis of the orbiter's orbit was originally 378 km, by 1986 it had risen to 2200 km, since which time it has gradually diminished; it had fallen to 150 km by 1992.

The Pioneer-Venus 1 orbiter, as we have seen, carried a varied payload, including instruments to study the surface of the planet, its interaction with the energetic solar wind, the magnetic field, plasma waves and the atmosphere. It also carried a radar-mapping package, which included a modest-resolution mapper and a radar altimeter that allowed the mission team eventually to produce the first accurate topographic map of Venus. It effectively "walked" around the planet once every 243 days. By the end of its scheduled mission, Pioneer-Venus 1 had mapped about 93% of the surface, between latitudes 74°N and 63°S, and had obtained altimetry with a vertical accuracy of about 200 m, integrated over an area 100 km^2. It showed Venus to be quite flat over large areas, but it also showed that the landscape was characterized by extensive regions of upland rolling plains (65%), less extensive areas of lowlands (27%) and much more restricted areas of highlands (8%) that rose between 2 and 11 km above mean datum. The altimetric data and mapper data were combined to give the first general relief map of the planet (see Plate 1).

Of the lowland regions mapped by Pioneer, one – Guinevere Planitia – was particularly prominent and it, together with adjacent lowlands in the equatorial belt, formed a somewhat rectilinear feature shaped like a huge cross. The largest lowland, Atalanta Planitia, lies generally 1.4 km below the mean datum level and has roughly the area of the Gulf of Mexico. Its smoothness and general lack of features resembling impact craters suggested to many workers that it (and, indeed, other of the lowlands) was similar to the lunar maria. Gravity data suggested that the lowland plains were regions of thinner crust and lower density than the upland rolling plains.

In some respects the least extensive of the physiographic regions, the highlands, turned out to be the most interesting and these tend to dominate the Pioneer topographic map. They occur in three principal areas: (a) Ishtar Terra, which is centred at 65°N, 350°E; (b) Aphrodite Terra, on and south of the equator between 60° and 205°E; and (c) Beta Regio, centred at 25°N, 280°E. In terms of scale, Ishtar is about the size of Australia and Aphrodite as large as Africa. The highest point on the planet's surface resides in Ishtar Terra, where a point in Maxwell Montes rises to 11 km above datum. Immediately to its west lies a vast high plateau, named Lakshmi Planum, upon whose surface are two large depressions, Colette and

Sacajawea. Lakshmi is roughly equivalent in area to, but twice as high as, the Earth's Tibetan Plateau. Well developed *ridge-and-trough* systems characterize Ishtar, with lineaments and steep scarps in the higher parts, particularly bounding the high plateau.

No point in Aphrodite Terra rises as high as Maxwell Montes; however, the mean height of this upland region is around 5 km. The highest region lies in the west, while in the east a curved mountainous belt gives rise to the region called Atla Regio, which curves in a way suggestive of a scorpion's tail, which has become its nickname. While exhibiting similar linear ridge-and-trough features to those found on Ishtar, Aphrodite has a more degraded look, which was interpreted by some scientists to mean a greater age. To the south is a deep curvilinear chasm – Artemis Chasma – while further linear valleys cross the central parts in an approximately WSW–ENE direction. One of these was at least 1400 km long and plunged 7 km below mean datum. This firmly places it as being of the same scale as the immense Valles Marineris of Mars.

Pioneer also revealed a range of shallow circular features between 200 and 700 km in diameter, which were interpreted as either impact or volcanic structures. One very prominent radar-bright circular feature, originally named Sappho (7°N, 15°E), has a diameter of 250 km and was surrounded by radiating radar-bright features, believed to be lava flows (Fig. 3.4). The darker central region was suspected by some workers to represent the lava-flooded floor of a volcanic caldera. Similar radar-bright structures were found in Beta Regio. Subsequent work has confirmed their volcanic nature but also has shown that other quasi-circular radar-bright features are more likely to be impact structures.

While details of the Venusian atmosphere will be discussed in Chapter 4, it should be noted that both US Pioneer and Russian Venera spacecraft made a variety of atmospheric measurements. Thus, the zonal flow within the atmosphere was mapped, and its maximum opacity was found to lie 50 km above the surface. Some dark markings observed through ultraviolet (*UV*) filters, from both Earth and Mariner 10, were found to be due to sulphur particles within the lower atmosphere, while the upper cloud layers, generally brighter, were interpreted as due to sulphuric acid droplets. Then again, an important finding from one of the non-imaging experiments concerns levels of the inert gas ^{36}Ar. This gas was most abundant when the Solar System first formed, but since then large amounts of **radiogenic** potassium ^{40}K have decayed, releasing the isotope ^{40}Ar. As a consequence, the ratio of original ^{36}Ar/^{40}Ar gives vital clues relating to planetary evolution. The values obtained by both craft indicate that Venus has much more ^{36}Ar than the Earth. This implies that Venus retained more gas from the primeval solar nebula than did Earth, which came as something of a surprise to atmospheric scientists, who had anticipated the reverse, owing to the stripping away of this early atmosphere by the outward-streaming solar wind.

Pioneer-Venus eventually suffered its demise on 8 October 1992, after nearly 14 years of travelling around Venus. This probably occurred because it exhausted its stock of hydrazine fuel, used to restore its orbit, causing the orbit to decay. The result of a loss of elevation

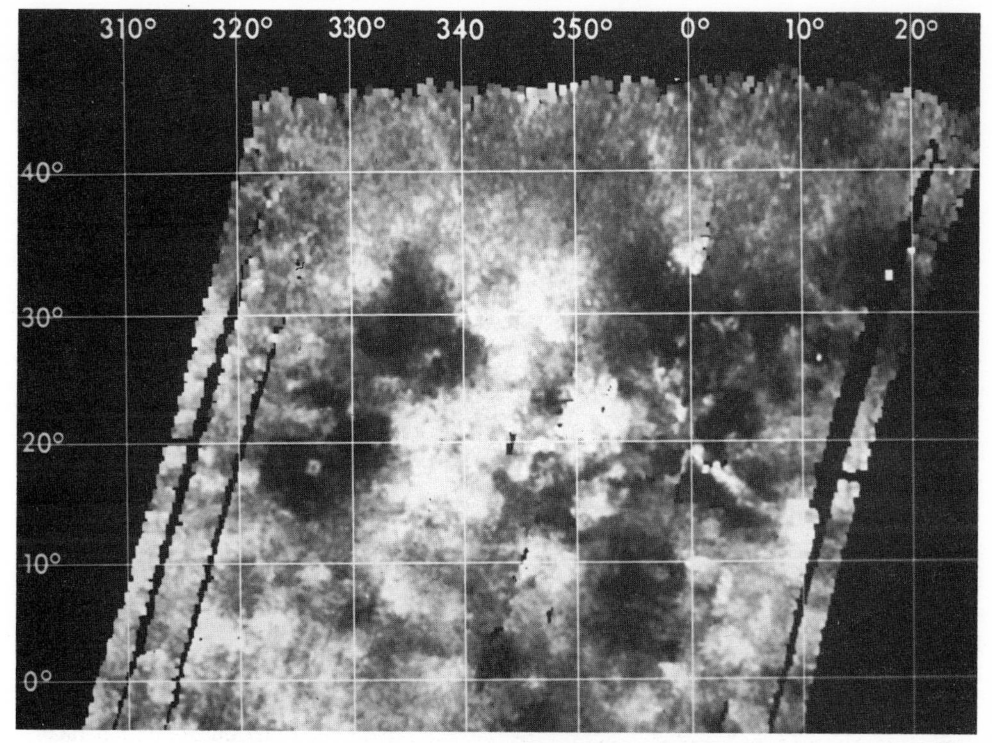

Figure 3.4 Pioneer radar map of western Eistla Regio

Prominent radar-bright features are the volcanoes Sif and Gula Montes with their associated lava flows.

would be severe overheating as the satellite dropped into the denser levels of the Venusian atmosphere. The scientific results of the Pioneer missions can be found in the references.

The stage was now set for a new wave of exploration: from the time of their arrival in October 1983, until July 1984, Veneras 15 and 16 imaged Venus from elliptical orbits with periapsis of 1000 km and **apapsis** of 65 000 km. Venus was imaged by their SLR antennae, using Venera 15 and Venera 16 alternately at a wavelength of 8 cm. The radar survey cycle began when each northward-travelling orbiter reached 80°N, continued over the polar regions and down to between latitudes 30° and 35°N. In this way each swath of strip images 8000 km long and between 120 and 160 km wide was mosaicked to form a large composite image. While the SLR antennae were working, so was the narrow-beam altimetric antenna, acquiring height data with a vertical precision of 50 m, but over an area of between 40 and 50 km, meaning that reliable surface relief data could be obtained only for features that were at least this size.

The greater resolution of these spacecraft revealed all manner of new landforms, including finer details of ridge-and-trough belts not discerned from Pioneer data, complex *parquet terrain*, *fault scarps*, and different types of circular and quasi-circular structures ranging from impact *craters* to volcanic *calderas* and weird *arachnoids*. In one area of 85 $\times$ 10^6 km^2, 140 craters ranging in diameter from 10 to 150 km were shown. About a quarter of these had radar-bright haloes, which were believed to represent impact **ejecta**. The fact that many of these were quite deep indicated that, at least over parts of the surface, the lithosphere was sufficiently rigid (and therefore at least several tens of kilometres thick) to prevent significant relaxation.

The northern polar region of Venus was not imaged by Pioneer-Venus, neither is it accessible to Earth-based radar, so the polar-orbiting Veneras 15 and 16 spacecraft could provide welcome new data. It turned out to be an extensive plain lying near the planet's average elevation. However, the plain is not without interest, since it is crossed by belts of linear ridges and furrows several hundreds of kilometres long and up to 150 km wide. The most prominent of these runs from Atalanta Planitia, through the poles, and then turns south towards the highlands of Ishtar Terra.

The higher-resolution imagery obtained by Veneras 15 and 16 showed that most of the surface of Lakshmi Planum was relatively smooth and 3–5 km higher than the adjacent area.

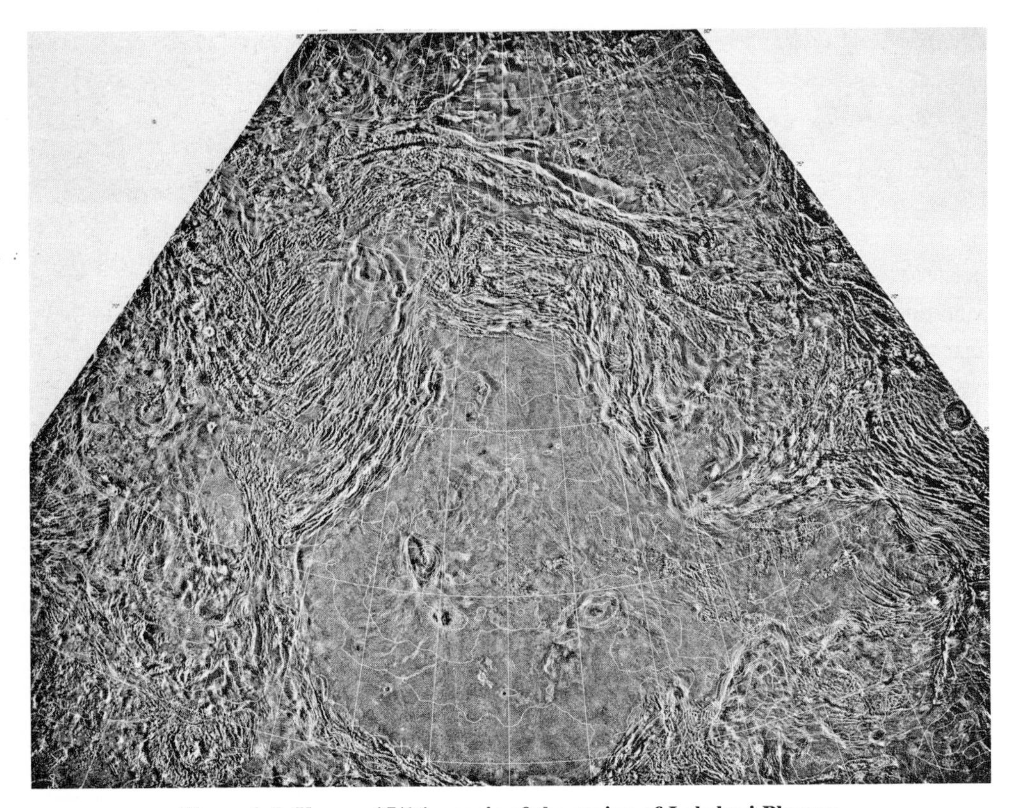

Figure 3.5 Venera 15/16 mosaic of the region of Lakshmi Planum
The relatively smooth surface of Lakshmi Planum is punctuated by the volcanic calderas of Colette and Sacajawea and surrounded by the ridge-and-groove mountains of Freyja, Danu and Maxwell Montes. Image width is 3200 km.

On the plateau surface are two prominent circular depressions – Collette, which measures 120 km × 80 km, and Sacajawea, which is larger (280 km × 140 km). Around the former are radial bands as well as ring-like grooves and ridges, suggestive of volcanic calderas. Lakshmi Planum itself was found to be almost completely surrounded by belts of linear ridges and grooves. The borders of the plateau are ultimately defined by the high ranges of Akna Montes to the west, Freyja Montes to the north and Maxwell Montes in the east, while a narrow belt of less-prominent ridges marks the other margin, which slopes down steeply to a feature called Vesta Rupes (Fig. 3.5). These prominent mountain massifs are the highest on Venus, being 8, 9 and 12 km above datum respectively. Workers who have studied the images of the ridges say that they look rather like stacks of giant plates, suggesting that they could be either folded rocks or complex fractured slices of crust. At the eastern end of the plateau, small regions are traversed by intersecting, closely spaced ridges and grooves, which are often embayed by the plateau surface materials. These relationships seem to suggest that the ridge-and-groove landforms may be relicts of whatever lies beneath the surface skin.

On Lakshmi's southern margin, a giant scarp called Vesta Rupes was discovered, between 3 and 4 km high, dropping off to the south, and changing orientation part-way along its length. Where the scarp meets the plateau top, complex intersecting faults and ridges give rise to a most intricate pattern, which is almost chaotic in its form.

The intersection of ridges and grooves is seen elsewhere on Venus and, because of the distinctive pattern that resembles wood-block flooring, it was originally called parquet terrain but is now more frequently described by the word *tessera*. It is particularly well seen to the east of Maxwell Montes in a zone measuring 1000 km × 2000 km (Fig. 3.6). While there is an unfortunate gap in the radar coverage between the essentially subparallel ridges and furrows characteristic of the eastern slope of Maxwell and the true parquet region, it appears that the subparallel ridges give way gradually to two (or more) intersecting sets. The origin of such landforms is obviously speculative but the fractures probably are developed only in the surface layer(s). This evidence comes from studies of what are considered to be large impact craters. These are distributed quite widely on Venus, even in the size range 100–140 km and, significantly, even the larger and (by implication) older craters appear to have remained substantially deeper than their lunar counterparts. The latter typically become markedly shallower with time, as a result of slow degradation. The implication here is that the Venusian lithosphere must have been sufficiently rigid to allow these large craters to retain almost their pristine profiles and, indeed, the radar-bright haloes surrounding many of the craters suggest that their coarse ejecta deposits have also remained essentially undegraded. Another implication, therefore, is that little major erosion can have affected Venus's surface since the craters were formed. On this basis the parquet terrain cannot represent exhumed, older, fractured crust but must have been formed at the planet's surface and must, therefore, have been generated by stresses in the crust itself, possibly by the drag of the planet's lithosphere over the **asthenosphere** below it.

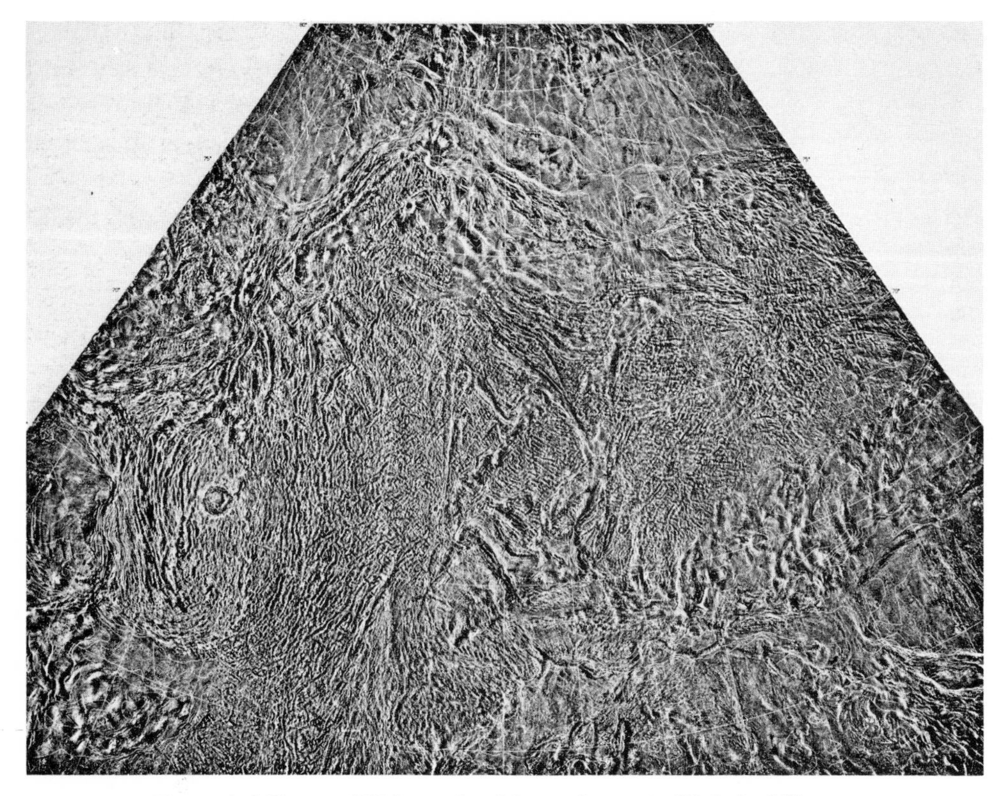

Figure 3.6 Venera 15/16 mosaic of the region east of Lakshmi Planum
On the left is the mountain massif of Maxwell Montes with the distinctive circular impact structure, Cleopatra. Eastwards from here is the complex tessera massif of Fortuna Tessera.

East of Ishtar is a region that is predominantly a plain but which has several upland massifs; this is Tethus Regio. In many ways it is similar to the southern part of Ishtar Terra, but Venera images reveal that it differs in having areas covered with groups of dome-shaped hills and also several 300–500 km diameter elliptical structures. The latter typically have a core of somewhat irregular relief, which may be surrounded by up to 12 rows of concentrically arranged ridges and grooves. Soviet workers have coined the term *coronas* for these. Similar structures occur elsewhere on Venus, but most are confined to the latitude zone 55° to 80°N. One corona, situated at 64°N, 129°E, has parts of its interior lower than the surrounding region and other parts higher. The surrounding ridge belt is 100–200 km wide and up to 2 km high and, in places, is encircled by a huge moat; furthermore, large areas on the flanks are corrugated by sinuous wrinkles that have a roughly concentric arrangement, while on the inner north slope there appears to be a massive landslide (Fig. 3.7).

At the time of their discovery, ideas concerning the origin of these peculiar landforms were considered entirely speculative, although radar-bright flow-like markings surrounding some coronas were thought to represent lava flows. Furthermore, their concentration, especially

Figure 3.7 The corona Nightingale This is located at 63°N, 130°E and measures 560 km × 480 km. Venera 15/16 image.

at the western and eastern margins of Ishtar Terra, suggested some underlying structural control on both their formation and location. Since volcanism and faulting usually go hand-in-glove, the most favoured explanation was that they had an endogenic origin and were in some way a manifestation of peculiarly Venusian volcanic processes.

Southwest of Ishtar Terra and separated from it by plains is Beta Regio, an upland region in which reside two prominent 4.5 km high mountains, Theia and Rhea Montes. Venera revealed that both of these have low flank slopes and show radar-bright radiating streaks, which are presumed to represent lava flows. Veneras 9 and 10 both landed on the edge of Beta Regio and showed the surface rocks to have a basalt-like chemistry. Considerable interest had been shown in this region owing to somewhat circumstantial evidence that pointed to geological activity in very recent times. US scientists analyzing both terrestrial and Pioneer data noted that a spacecraft antenna had recorded low-frequency radio emissions clustering around both Beta Regio and Phoebe Regio (to its south) and also over Atla Regio, another region considered to have volcanic structures. These were most readily interpretable as lightning discharges, which frequently have been observed in the plumes of erupting terrestrial volcanoes. The other piece of evidence came from observations that levels of sulphur dioxide and sulphuric acid hazes in the atmosphere of Venus fell by about 90% during the period 1978–83. Since the 1978 levels had been well above those predicted after studies of the planet from Earth, Larry

Esposito conjectured that large-scale volcanic eruptions must have occurred during the late 1970s, sending sulphur dioxide high into the atmosphere. All this evidence, when put together, strongly suggested that Beta Regio is a major and currently active volcanic region. One of the objectives of the Magellan Mission was to search for current activity of this type.

Only the northern part of this region was surveyed by the last two Veneras, which showed the transition from the adjacent plains to be gradual rather than marked. In passing up slope, numbers of linear and often arcuate scarps appear, which were hundreds of kilometres long and 5–10 km in width. In places these were found to pass into graben-like depressions akin to minor rift valleys, which appear to be a continuation of further faulting to the southeast of the region. The association of volcanic structures and faults is quite marked and it bears close resemblance to the relationships they show on both the Earth and Mars.

3.5 Earth-based studies

Radar observations have been made from the Earth for over two decades; indeed, they gave us our first information about the planet's surface. Such observations have been made principally from three US facilities: the 43 m antenna at Haystack in Massachusetts, which operates at 3.8 cm wavelength; the 64 m antennae based in the Californian Mojave Desert at Goldstone, operating principally at 12.5 cm; and the 300 m system in Puerto Rico at Arecibo, which operates mainly at 70 cm wavelength. Because of the geometry of the Earth with respect to Venus, high-resolution images can be obtained only around inferior conjunction – once every 19 months.

The initial work using this technique was undertaken purely for distance measurement and these data were collected during the late 1960s. However, by the year 1970, radar images of the Venusian surface had been collected by each of the observatories. At this time the best horizontal resolution that could be achieved was around 50 km; thus it could only be shown that Venus had some fixed features and that there was a regional variation in radar brightness. The first actual features to be resolved, as techniques improved, were circular ones believed at the time to be craters. The highest resolution achieved at Goldstone is around 10 km. The most powerful of the three radar "telescopes" is that at Arecibo, where the best imagery can resolve features as small as 3 km. At the forefront of research here is the team of Campbell & Burns, who were the first to build up a mosaic of the surface, which covered about 25% of the globe at a resolution of 10–20 km. The images show considerable detail, which mainly reflects differences in roughness from one area to the next (Campbell & Burns 1980, Campbell et al. 1989). The radar-bright signatures from regions such as Freyja and Akna Montes, bordering Lakshmi Planum, from Maxwell Montes and from the two massifs of Rhea and Theia Montes in Beta Regio imply rough topography at these radar wavelengths (Fig. 3.8). Several

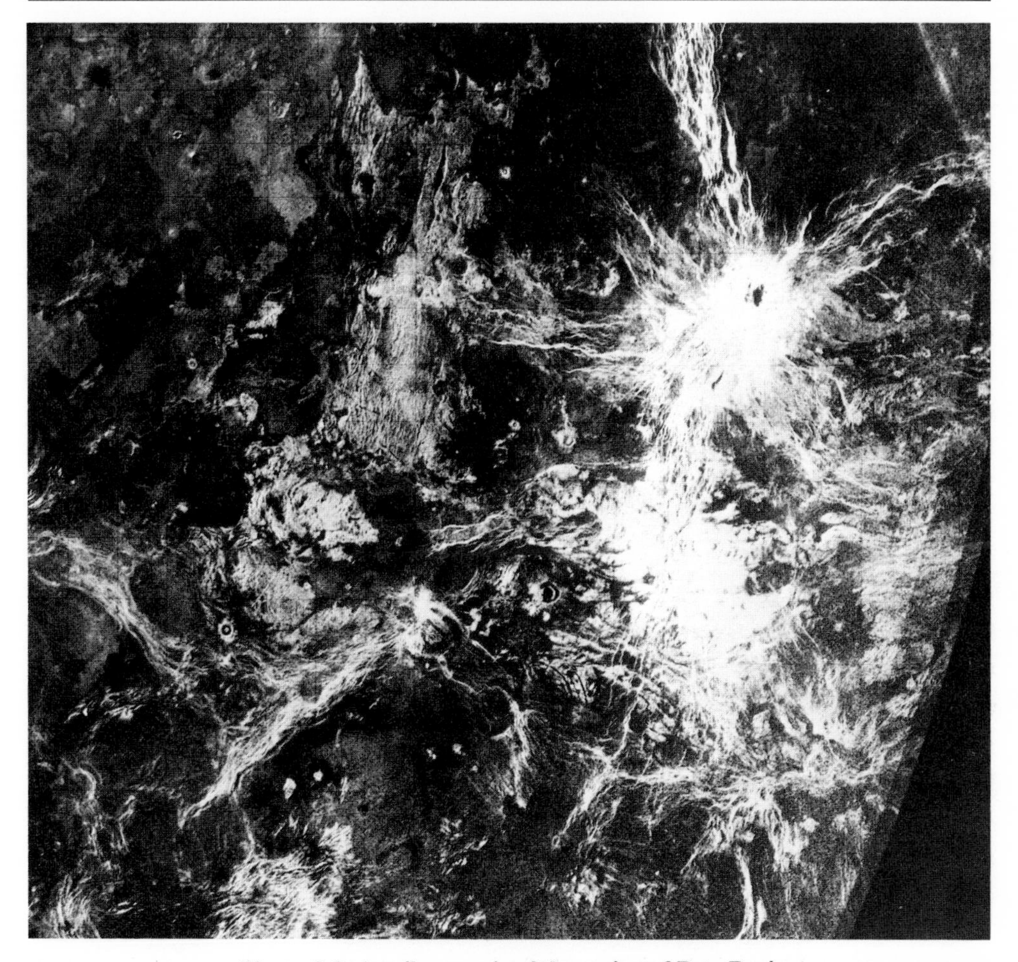

Figure 3.8 Arecibo mosaic of the region of Beta Regio
The distinctive radar-bright signatures of the volcano Rhea Mons, with its dark caldera floor, lies to the north of the dome-shaped mountain of Theia Mons. Fracture belts radiate out from this region. The radar-bright upland of Phoebe Regio lies to the south of Beta, while to the west is Asteria Regio. Note several impact craters with bright rims and dark floors with bright central peaks.

large circular structures with radar-bright rims are seen, together with linear zones of banded terrain, only in the highest-resolution imagery, and are strongly suggestive of folded rocks. The existence of global-scale belts of radar-bright terrain, in places cross-cut by linear fault zones, was clearly established by the radar mapping undertaken from Puerto Rico.

3.6 The Magellan spacecraft and radar system

During the early 1970s, plans for a new and very sophisticated mission to Venus were drawn up by NASA. This was named Venus Orbiting Imaging Radar, or VOIR. It was to have a circular orbit around Venus and a complex array of experiments. Its principal science investigators were selected in 1979. However, after detailed studies by technicians and scientists, the mission proved too costly to be viable and was scrapped during 1982. However, the notion for such a mission did not die, and in October 1983 a much-reduced mission objective saw the inception of the Venus Radar Mapper, or VRM. Its sole objectives were to undertake radar imaging, altimetry and radiometry and gravity-field measurements. Also, instead of the circular orbit planned for VOIR, VRM was to have an elliptical polar one. Because there was a need to reduce the budgetary costs by one-half, the VRM mission was to utilize proven technologies and spare parts from other programmes, such as Voyager, Ulysses and Galileo. In 1986 VRM was officially renamed Magellan.

Magellan was scheduled as the first spacecraft to be launched by the Space Shuttle, and a launch date of May 1988 was fixed. Then, disastrously, came the Challenger explosion in 1986. This set back the launch schedules of all missions owing to a 32-month suspension of Shuttle activity. Furthermore, because of planetary configurations, it also meant that the Galileo mission to Jupiter now had to go in October 1989 (the planned new launch date for Magellan), and it was eventually decided that Magellan should be launched in May 1989, but, instead of the original trajectory planned to take it to Venus over a period of four months, in what is called a type-IV trajectory. This meant that Magellan would spend 15 months in space and would go one-and-a-half times around the Sun before arriving at its objective. Nevertheless, at last a date had been fixed for the launch of this US$551 million mission.

The Magellan spacecraft can best be described as a low-cost, high-performance vehicle. Its main function was to transfer to Venus orbit an array of radar antennae, instruments, propulsion modules, solar panels and computers that would allow scientists on the Earth gradually to build up a detailed picture of the surface physiography and geology of Venus backed up by gravity data. To this end Magellan carried four antennae: a parabolic high-gain antenna (HGA), medium- and low-gain antennae (MGA, LGA) and an altimeter antenna (ALTA). It also carried a radiometer. These are mounted as shown in Plate 2.

The main antenna, the HGA, is the primary antenna for radar operations, the transmission of radar data, the receipt of radio signals from the Earth, and the sending of information regarding spacecraft engineering "health" status. The MGA and LGA essentially are back-up antennae that can be utilized when the HGA cannot be turned towards the Earth. Each operates in sideways-looking mode and as part of a synthetic-aperture system. The altimeter antenna, on the other hand, points directly downwards, in one-dimensional mode, to give height readings for surface features as the spacecraft passes overhead. The solar panels can generate 1200 W of power to enable the instruments to function.

During each orbit – and there will be 1852 in the first 243-day phase of the mission – the craft has to be manoeuvred four times: (1) away from Venus to aim the HGA towards Earth for transmission of data, (2) towards space to scan stars, which allow for precisely determining the craft's orientation, (3) towards Earth again to resume data transmission, and (4) towards Venus for radar mapping. Thus, in the first mapping cycle the probe is required to make major adjustments to its attitude 7408 times! Clearly, if such frequent adjustments relied upon rocket motors, inordinately large volumes of propellant would have to be carried. Instead, technicians rely upon a spinning, electrically driven *reaction wheel* which, when rotated in the direction opposite to the one in which Magellan needs to move, relies upon Newton's famous third law: "for every action there is an equal and opposite reaction" to achieve the required changes.

The launch took place on 4 May 1989 and the Magellan spacecraft entered a highly elliptical orbit around Venus in August 1990. Almost immediately there were several extremely puzzling problems, and at one stage it looked as though the whole mission could become a very expensive failure. Fortunately, with their usual skill and resourcefulness, NASA technicians were able to calibrate the spacecraft and send it appropriate commands to allow it to enter the correct orbit and begin mapping. This it commenced, somewhat later than originally planned, in September.

3.7 Mapping Venus with Magellan

Using its array of data-collecting instruments Magellan completed a total of 1852 data-collection and playback passes of the planet during the first 243-day mapping cycle. Each of the mapping passes collected images over a swath of surface about 25 km wide and 16 000 km long. Magellan also imaged a new area of the planet at each pass, as the planet rotated slowly beneath the spacecraft (which remained essentially fixed in inertial space). Because of the orbital configuration, the HGA had to point to the left as the mapping proceeded; as a result it could not image the planet's south pole during the first mapping cycle. However, this was accomplished during the second mapping cycle (in right-looking mode) and consequently extends the radar coverage to the entire surface.

During mapping, the Magellan spacecraft approaches as close as 294 km and recedes as far as 8743 km from Venus during each orbit. The imaging system is able to resolve objects as small as about 120 m – a major improvement on previous spacecraft-mounted radar instruments and also the Arecibo radar telescope. Each 3 h 15 min orbital period is divided up into a number of shorter periods of mapping, data playback and idling time. This is because the complexities of a radar system mean that data cannot be collected and transmitted back to receiving stations on the Earth at one and the same time. Furthermore, an alternate mapping strategy has been pursued, whereby on one orbit both high and low latitudes were imaged, while on the next only the lower latitudes were covered. This avoided "doubling up" coverage of the polar regions, which present a much smaller percentage of new ground during successive passes than do the lower latitudes. Thus, the probe alternated between mapping from the north pole down to latitude 56.9°S on one orbit to mapping only from 66.9°N to 74.2°S on the next. During each of the two playback periods allocated for each pass, data are sent back to Earth at the rate of 268.8 kilobits per second, which is only one-third of the rate at which new data are recorded. There is thus a considerable lag between the actual acquisition of new data and their transmission to receivers on the Earth. For this reason Magellan has to have a considerable memory storage capacity, putting the backlog of data into memory until it becomes possible to return them to Earth.

The slow retrograde motion of Venus means that major topographic features, such as Beta Regio, Ishtar Terra and Aphrodite Terra, have been imaged in sequence as a function of days after Venus orbit insertion. The initial images were of the highland region of Beta Regio, which already had excited geologists after having been beautifully imaged at relatively high resolution by the Earth-based Arecibo facility. During each pass, scientists have been continually amazed at the amount and type of detail revealed by Magellan. The initial impression suggests Venus has a geologically active surface that shows evidence of widespread volcanism, tectonism and impact cratering, as well as erosion. The relatively small number of impact craters compared with those counted on the Moon, Mercury and Mars suggests that Venus's surface is comparatively young.

Magellan began by mapping near Beta Regio, a volcanic massif cut by rift faults, and then the highland region of Lakshmi Planum, a massive volcanic plateau larger than that in Tibet. This is nearly 5 km above the mean datum and is surrounded by extensive belts of mountains with a ridge-and-trough morphology. Subsequently, Magellan imaged the fascinating regions of Guinevere and Sedna Planitiae, low-lying areas that exhibit long lava flows and many small volcanic features. It was near here that the Soviet probes of the Venera series of missions landed, photographed the surrounding area and analyzed some of the surface materials. It is amazing to think that it was over 15 years ago that the first of these – Venera 9 – sent back our first close-up glimpse of the surface of this hot, hostile world. Passing over the southern hemisphere, Magellan passed over a group of peaks called Ushas, Innini and Hathor Montes. Prior to Magellan, these were envisaged as being large shield volcanoes, since Earth-based radar images revealed peripheral lava flows adjacent to gently sloping rises in the Cytherean surface. In many ways they appear to resemble the line of large volcanoes that we know as the Hawaiian Chain.

As the first cycle progressed, images of Fortuna Tesserae, east of Maxwell Montes, were received. Magellan began transmitting images of a confusion of distorted and fractured blocks that look like wood-block flooring on a massive scale. These proved to be even more complex than Arecibo and Venera images had suggested, and they have the structural experts stretched to their intellectual limits. Later in the cycle, clusters of strange spider-shaped structures, called arachnoids, came into view; their origin is still an enigma. Then the dark-floored volcanic caldera of Sif Mons, the complex highland region of Alpha and Bell Regiones, and, during late November, the fascinating continent-sized region known as Aphrodite Terra, successively came under scrutiny. At the close of the first mapping mission, Magellan found itself again over Beta Regio with its huge shield volcanoes and long rift valleys.

At the same time as images are being studied, other teams will be analyzing data concerned with the radio thermal emissivity of the Venusian surface, the dielectric permittivity and detailed scattering properties (Pettengill et al. 1992). In this way they will endeavour to characterize the surface roughness, rock density and (possibly) chemical composition of the Venusian crust.

Bearing in mind that Magellan Mission scientists are involved in the detailed mapping of an Earth-sized planet, at a resolution equivalent to the best Earth-orbiting probes, it comes as no surprise to appreciate that there is an unprecedented amount of new imagery to be processed and inspected. Nevertheless, what the images show is a staggering improvement on anything we have hitherto seen and has meant the rewriting of many earlier ideas about the planet. This is going to be the new geology of Venus.

At the time of writing, NASA has informed the scientific community that Magellan will terminate its mission after the fifth cycle is complete; that is, in March 1994. By this time high-resolution gravity data of the equatorial and polar regions should have been collected.

3.8 Vital questions regarding Venus geology

Plate tectonics theory now welds together various disciplines within the broad field of terrestrial geology and it provides a model for explaining the internal workings of our own dynamic planet. Of very high priority is the finding of a definitive answer to the question: "Is there plate tectonics on Venus?" This was formulated first by Solomon & Head (1982), and is not as simple to answer as it is to pose, since the major geological structures so far identified on Venus are a mixture of linear features that possibly are driven by plate tectonics, and focused features that by some have been appositely described as "spotty" and which may possibly have formed around hotspot activity. Then there is the matter of whether features potentially due to such plate movements are currently active or formed in response to motions that have long since ceased.

Naturally an understanding of Venusian stratigraphy is also vital. While there is little likelihood that fieldwork can ever be undertaken on the veiled planet, as it can on Earth, or that extensive lander missions that provide absolute ages for different units can be funded, careful analysis of imagery may elucidate temporal relationships between different units.

There is also a need to extend the geochemical databank, so that different types of crustal material can be characterized. The basic question that needs to be answered is whether Venus, like Earth, has both **mafic** and **felsic** crustal material, or whether it is predominantly basalt-like and therefore more closely similar to Mars. If the geochemical data can be expanded, this will lead to a better understanding of the interior of Venus, a vital issue in terms of the planet's geology. That there is one significant internal difference is clear: Venus has negligible magnetic field, and therefore its core must be in a rather different state to Earth's. It is also important to get a feel for lithospheric thickness variations, and more refined gravity data are therefore needed. Magellan should provide them.

The high-resolution imagery being sent back from Magellan should also help to define the parameters of another important issue, namely, the extent, character and intensity of surficial processes. Certainly there must be active chemical weathering of the surface rocks, aeolian processes and gravity-aided mass movements. There is a need to understand the importance and distribution of these modifying forces.

4 The Venusian atmosphere

4.1 Introduction

Astronomers have observed transits of the disc of Venus across the Sun since the 18th century; the next one will occur in the year 2004. Significantly, it was during the transit of 1761 that M. V. Lomonosov noticed that, as Venus passed between the Sun and Earth, the disc was obviously blurred. He attributed this to the presence of an atmosphere – an observation that was largely ignored by astronomers for many years. Then, in the late 18th century, Johannes Schröter confirmed, once and for all, that an atmosphere did exist and that the changing face of the planet was due to changes in the clouds, not the solid surface.

In 1923 the first really good photographs of the planet were obtained by Ross at the Mount Wilson Observatory in California; better ones soon were obtained by Kuiper at the MacDonald Observatory in Texas, Dollfus at Pic du Midi and Richardson at Mount Wilson. Both Kuiper's and Richardson's photographs showed that the planet had a banded appearance, due to unknown atmospheric effects. Subsequently, two French astronomers, Boyer and Guèrin, observed a dark Y-shaped feature in the atmosphere and calculated a retrograde rotation period of 4 days. We now know that this is true for Venus's upper clouds.

4.2 Atmospheric composition of the terrestrial planets

The atmospheres of the inner planets exhibit a great diversity of characteristics, yet there are pervasive strands that link them all. For example, the volume of carbon dioxide endowed upon the Earth – most of which is now trapped within carbonate rocks – is comparable with the amount of CO_2 found within the Venusian atmosphere. Then again, most of Mars's water is locked within the regolith and much of the CO_2 in the polar ice caps. The current diversity is a reflection of the different ways in which their atmospheres have evolved over geological time, which, in turn, is a function of several things, including their position with respect to the Sun, their mass and the effects upon them of the solar wind and other energetic particles travelling through local space.

The gaseous and particulate characteristics of the atmospheres of the terrestrial planets have now been measured by spacecraft. A selection of their data are shown in Tables 4.1 & 4.2.

A small proportion of the water vapour found in the atmospheres of the terrestrial planets is dissociated by short-wave solar radiation into hydrogen atoms and hydroxyl (OH) radicals. The latter are very reactive and are important in determining atmospheric chemistry, since they assist in the conversion of gaseous sulphur into sulphuric acid vapour and also behave as catalysts for the annihilation of ozone gas. The UV radiation is also important since it converts oxygenated gases into other oxygen-bearing species. Whereas on Earth molecular oxygen is the principal oxygen-bearing gas, on Venus (and, indeed, on Mars) CO_2 takes its place. Nevertheless, small amounts of molecular oxygen and minute amounts of atomic oxygen and ozone exist in the Venusian atmosphere owing to chemical reactions induced by UV dissociation of carbon dioxide (CO_2) into carbon monoxide (CO) and atomic oxygen.

On Earth, sulphur gases are present only in trace amounts. Most either dissolve easily in the hydrated cloud droplets that exist in the terrestrial air, or where there are no clouds, they react with oxygen-bearing radicals, water vapour and other gases to form particles that eventually find their way to the planet's surface. In contrast, the very high temperatures characteristic at the Venusian surface mean that sulphur is not removed from the atmosphere. The consequence of this is that SO_2 is abundant in the lower levels of the atmosphere and it provides a source for the dense layers of H_2SO_4 particles that are found 50–80 km above the surface. No sulphur-bearing gases have hitherto been detected in the Martian atmosphere, but these may in the past have been released from volcanoes and converted to S-bearing particles that went to form sulphate minerals. The latter were found in the regolith at both Viking lander sites.

Table 4.1 Properties of the atmospheres of the terrestrial planets.

Planet	*P* (bar)	*T* (°C)	Major gases	Particulate composition
Mercury	10–15	167	He, Na, O	None
Venus	90	457	CO_2, N_2	Conc. H_2SO_4
Earth	1	15	N_2, O_2, H_2O, Ar	H_2SO_4, sulphates, silicates, NaCl
Mars	0.007	-55	CO_2, N_2, Ar	Silicate, dust, water and CO_2 ices

Table 4.2 Concentrations of gases within the atmospheres of the terrestrial planets (vol%).

Earth		Venus		Mars	
N_2	76.1	CO_2	97.0	CO_2	95.32
O_2	20.9	H_2O vapour	0.1	N_2	2.7
Ar	0.9	CO	0.005	Ar	1.6
CO_2	0.03			O_2	0.13
Ne	0.00182			CO	0.007

The presence of reflective clouds has a twofold effect on temperature. On the one hand, they allow less sunlight to penetrate to a planet's surface, and on the other they absorb a proportion of the re-emitted heat of which a part is re-radiated back to the surface, thereby warming it. It is the latter process that helps to generate the notorious greenhouse effect and elevates the surface temperatures of Mars, Earth and Venus by 3°, 29° and 227°C respectively. Both the Earth and Mars experience only a modest increase in greenhouse warming (although there is every likelihood that we are changing that rather rapidly on Earth!) because the air of both worlds is transparent at some IR wavelengths; the Venusian air, on the other hand, is not.

The opacity of the terrestrial planet atmospheres at IR wavelengths is a function of different gases and particulate materials. On Earth, carbon dioxide, water vapour and water-bearing clouds are the predominant ones; on Mars, carbon dioxide and dust particles are most important; while for Venus – where there is complete absorption of thermal energy radiated from its surface – carbon dioxide, water vapour, sulphur dioxide and sulphuric acid clouds are the major contributors to the extreme surface heating that Venus experiences. Of these, CO_2 accounts for about 55% of the trapped heat, H_2O vapour for a further 25%, and SO_2 absorbs another 5%; the residual 15% of the greenhouse effect is due to hazes and clouds.

As mentioned in Chapter 3, the amounts of ^{36}Ar in the Venusian atmosphere are significantly higher than those in the Earth's; this is also true of ^{38}Ar (Von Zahn et al. 1983). However, Venus has less ^{40}Ar than the Earth. Now ^{40}Ar is generated by the radioactive decay of ^{40}K, and its lesser abundance on Venus implies either that Venus has less potassium than Earth or that the release process for this element is less efficient on Venus.

Venus also appears to have three times more krypton than Earth; again, this is not what had been expected. One possibility that has been suggested to explain this apparent anomaly is that, during the early stages of Solar System history when the solar wind was denser than it now is, Venus received large amounts of various gases from the Sun. If this had happened, then solar gases would have impacted onto the primordial matter from which Venus was accreting. Also, the proto-Venus might have blocked off the Earth and Mars from the enriched solar wind, preventing it from transferring such gases to these more distant worlds. The argon/krypton ratio for Venus is 700:1, which compares with a 30:1 ratio for both Mars and the Earth and of 2000:1 for the Sun. This fact certainly suggests that these noble gases may have emanated from the Sun, forming a veneer on the primordial dust and ice grains that were collecting at the orbit of proto-Venus.

4.3 Atmospheric temperature and pressure on Venus

The temperature and density of a planetary atmosphere vary considerably with altitude. Density, in particular, diminishes with height so that, at any given altitude, a close balance exists

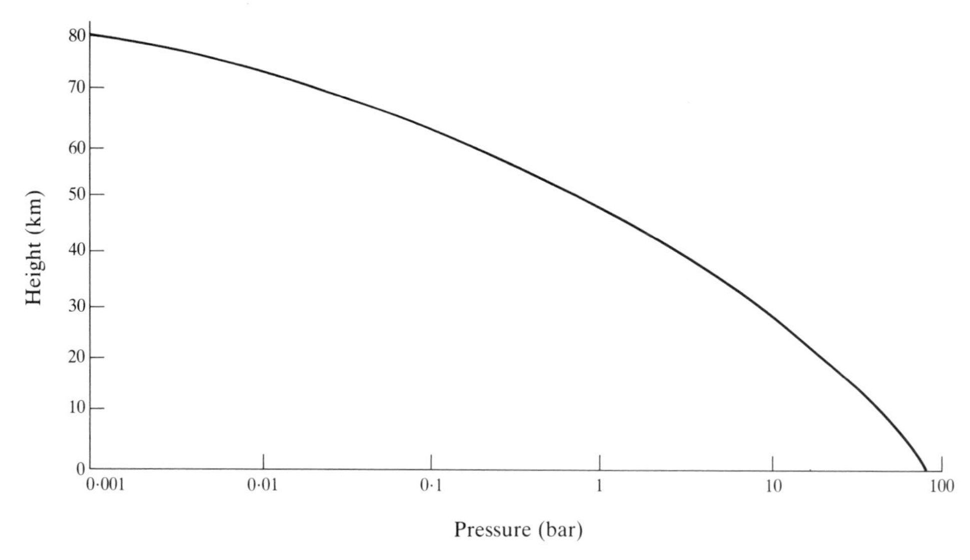

Figure 4.1 Pressure data for the atmosphere of Venus
Pressures were measured by the Veneras 8–12 descent probes. After Seiff (1983).

between gravity, which is pulling inwards, and gas pressure, which is pushing outwards. For all three of the terrestrial planets that have atmospheres, air density changes by a factor of 3 for every 10 km increase in altitude. For Venus the pressure ranges from 90 bar at the surface to 0.005 bar at a height of 80 km (Fig. 4.1).

The temperature profile of the terrestrial atmosphere has been used to subdivide it into four principal regions: the *troposphere*, where the temperature falls steadily with height; the *stratosphere*, where the reverse occurs; the *mesosphere*, where the temperature falls initially, but then rises; and the *thermosphere*, where there is a steady increase with height. In the case of the Earth, the existence of the characteristic temperature profile within the stratosphere is attributable to the abundant presence of ozone, which absorbs UV wavelengths; this more than compensates for the cooling that is brought about by the emission of heat by molecules of ozone and carbon dioxide.

On Venus, up to a height of about 35 km, there is a gradual decrease in temperature; but, above this point and up to 48 km (at the cloud base), the air does not cool off with altitude as rapidly (Fig. 4.2). This is because in this region the gas molecules are not so densely packed and in consequence are more transparent to thermal radiation (Tomasko 1963), so that heat is more readily exchanged between adjacent layers, producing a shallower temperature gradient than would otherwise occur. In effect, there is no true stratosphere above Venus.

On the other hand, there is an Earth-type thermosphere on the day side of Venus, where temperatures range from -93°C at a height of 100 km to about 27°C at an altitude of 150 km. On the night side, however, a true thermosphere does not exist, temperatures falling from about -93°C at a height of 100 km to as low as −173°C at a height of 150 km. The transition across the terminator is very abrupt.

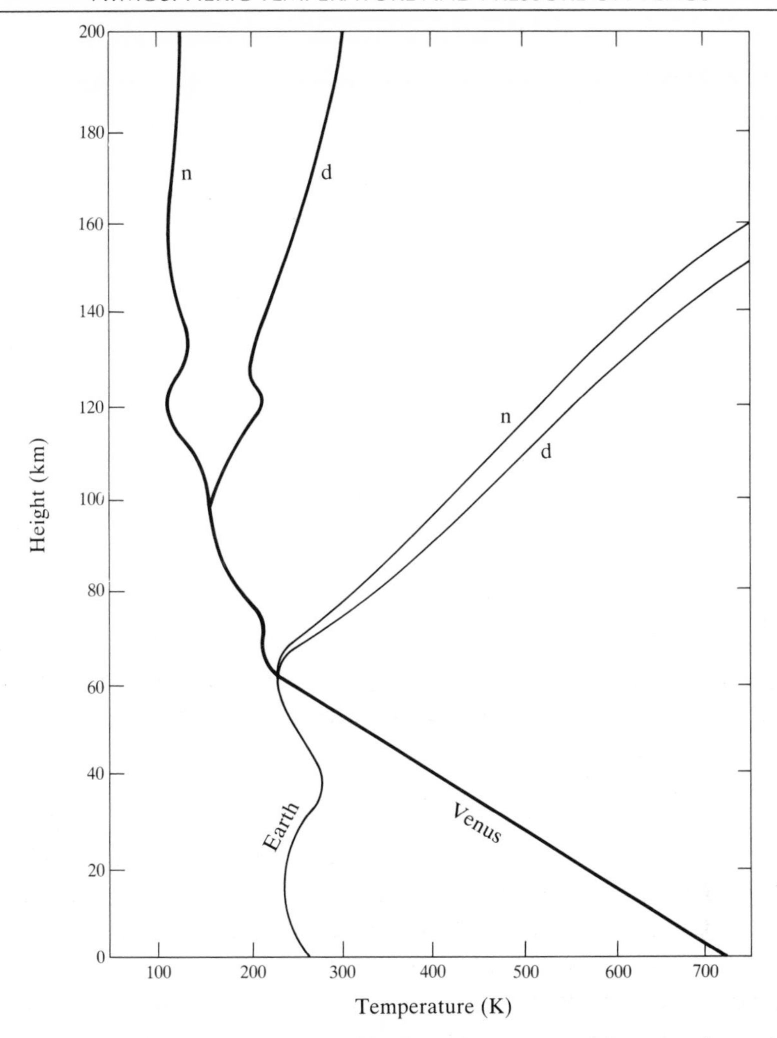

Figure 4.2 Comparison of the vertical thermal structures of the atmospheres of Venus and the Earth

The day and night hemisphere profiles are shown as (d) and (n) respectively. After Hunt & Moore (1982).

At higher levels within the terrestrial atmosphere, the presence of ozone again influences the temperature profile. Thus, in the Earth's mesosphere there is less ozone than in the stratosphere, and hence the gradient shows a reversal, radiative cooling being a function of the presence of carbon dioxide. In the uppermost layer – the thermosphere – there is a continual rise in temperature with height because there are too few molecules and atoms present to cool the atmosphere by emitting heat. Since the thermosphere of Venus is much more densely packed with gas atoms and molecules (principally of carbon dioxide), it is much better equipped to radiate away energy into space. In the very outermost regions of the Venusian atmosphere (the *ionosphere*), carbon dioxide atoms are stripped of electrons by sunlight. The most abundant ions at this level then become those of oxygen, as they do for both the Earth and Mars (Taylor et al. 1983).

4.4 Venusian meteorology and atmospheric circulation

Perhaps the most surprising characteristic of Venus's atmosphere is its very rapid rotation at cloud-top level. It takes a mere 4 days to complete one revolution. The solid-body/ cloud-top rotation ratio for Venus is 60:1, over eight times that of Saturn's moon Titan. Additionally, the factors that produce meteorological phenomena and very strong winds around Venus cause the atmospheric features to flow in a direction opposite to, and 20 times faster than, the overhead movement of the Sun with respect to any point on the Venusian surface (Schubert 1983).

Most solar energy is absorbed at cloud-top level on Venus. It is here that the heating is provided that drives the atmospheric motions observed by spacecraft. Since, during a yearly cycle, more solar radiation falls upon the equatorial regions of a planet spinning on an axis at right angles to the Sun than upon the poles, the equatorial belt is warmer than the polar. Such a latitude-dependent temperature gradient causes winds that transfer heat from the warmer regions to cooler ones. By this process the equator/pole temperature differences on Venus are reduced to as little as a few degrees, because of the air's incredible mass.

The kind of motion set up where air rises over warm regions and sinks over cool ones is known as a **Hadley cell** (Fig. 4.3). On the Earth this is modified by the *Coriolis force*, which imparts a horizontal component to the movement at right angles to the circulatory cell. For each of the terrestrial hemispheres there are three such cells; on Venus a single cell probably stretches from pole to equator at any one level; however, several cells probably exist, but at different altitudes. Theoretically, the planet's very slow rotation should mean that a near-ideal north–south Hadley pattern should exist; however, spacecraft measurements have shown that, in the lower and middle levels of the atmosphere, there are strong east–west winds (Schubert 1983). At ground level these have a velocity of around 1 ms^{-1}, rising to 100 ms^{-1} at the cloud tops. These are probably generated by centrifugal force. In the case of the Earth and Mars, such high wind velocities are encountered solely in jet streams. On Venus, the high-velocity winds that blow around the entire globe give rise to the *super-rotation* of the atmosphere.

While Hadley circulation probably is the dominant mechanism whereby momentum is transferred from the denser to the more rarefied regions of the atmosphere, high-level *eddies* may transport it towards the equator, thus spreading momentum over a wide range of latitudes. That such eddies exist is shown by the small-scale cloud structures revealed in high-resolution images of Venus. Such swirls and wisps are generated by a variety of processes, such as small-scale gravity waves, planetary-scale waves and convection cells. The larger-scale ed-

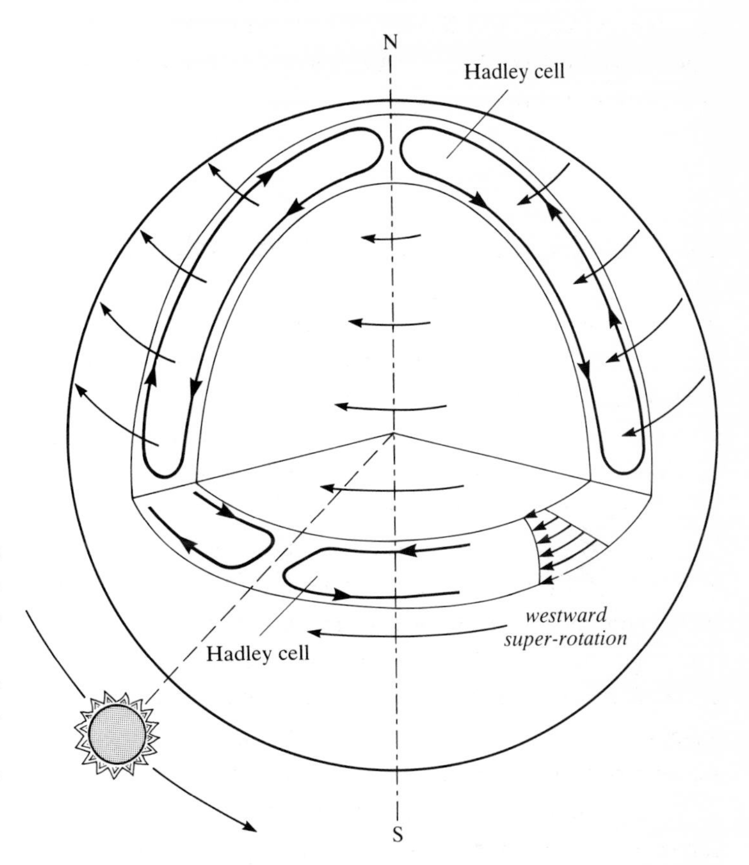

Figure 4.3 Hadley cells A Hadley circulation pattern is a response to differential latitudinal heating. Thus, air rises in warm regions, flows towards cooler zones, sinks and returns to warm latitudes again.

dies are manifested in atmospheric structures such as the well known dusky "Y", which often is to be seen in images obtained through ultraviolet filters (Fig. 4.4). This feature more-or-less encircles the planet and may result from the superposition of two large eddies whose individual components drift in and out of phase before dissipating.

Ultraviolet observations of the planet have enhanced our understanding of cloud structures. It is clear that the weather does change, for cloud brightness and detailed eddy structures were observed to have changed between the Mariner 10 and Pioneer fly-bys. Furthermore, infrared observations show quite clearly that there is a dipole structure around the north pole, where two clearings in the cloud layer rotate about it once every 2.7 days (Apt & Goody 1979) (Fig. 4.5). It is generally assumed that this provides clear evidence of atmospheric subsidence at the core of a polar *vortex*. Because similar downward motion has not been found elsewhere, it also supports the notion that a single Hadley cell may fill the entire northern hemisphere, at least at a level near to the cloud tops. Another feature observed from Earth – the much-observed crescentic dusky "collar" – has been found to be a region of anomalous and variable temperature and cloud structures that encircle the pole at 70°N; it also rises some 15 km above the adjacent cloud tops.

Figure 4.4 Venus as seen by Pioneer-Venus in 1979
The Y-shaped cloud pattern is formed in response to eddies in the Venusian atmosphere. JPL 400-344C.

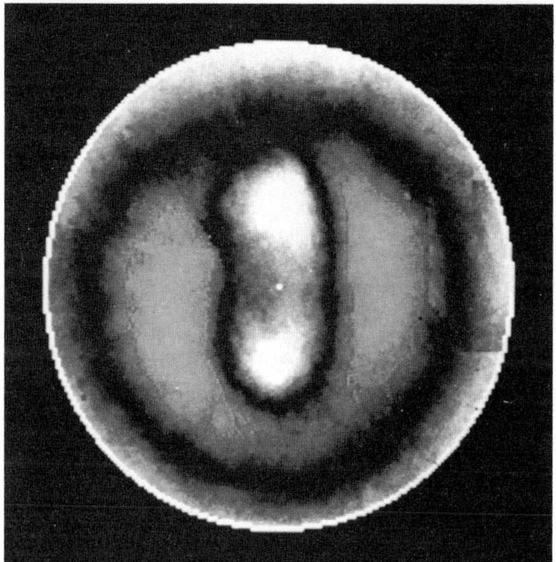

Figure 4.5 The polar dipole feature
Pioneer-Venus Orbiter infrared data were used, at a wavelength of 11.5 μm.

4.5 Evolution and origin of Venus's atmosphere

Since it is now generally accepted that only minute amounts of the Sun's primordial atmosphere have been retained by any of the terrestrial planets, any volatiles that do exist must have originated from the particles from which they accreted. Either these could have come from original "rocky" grains and planetesimals that formed close to the current position of the planets, or they were brought in from more distant parts of the Solar System, probably after most of the silicate matter had aggregated. Interestingly, whichever way they were supplied, they would have ended up with more-or-less the same amounts of volatiles.

If the source had been local, then the composition of the particles would have been similar to meteorites, in which volatiles such as water would have bound up in the atomic structure of silicates such as **serpentine**. Nitrogen and oxygen, on the other hand, together with carbon and hydrogen, would probably have been tied up in non-biological organic compounds. Because there was a thermal gradient away from the Sun, volatile-rich planetesimals would have been more abundant near Mars than near either Venus or Mercury; however, as time went on, Earth and Venus grew in mass much more quickly than either Mars or Mercury, with the result that any original inequality in volatile content (in favour of Mars) would have been more than compensated for as Earth and Venus became more competent at attracting a greater proportion of the volatile-bearing planetesimals. In other words, during the latter stages of growth, all of the inner planets would have accreted comparable amounts of volatile constituents.

Had volatiles been brought into the inner Solar System from, say, beyond the orbits of Jupiter or Saturn, the chief bearers of carbon and nitrogen again would have been organic compounds, while water ice would have been the principal H_2O-bearing phase. If they had originated farther in, within the asteroid belt, then hydrated silicates would have borne the water instead. The massive giant planets, in either case, would undoubtedly have perturbed them and, indeed, possibly held them up on their inward journey, such that they would have eventually arrived in the vicinity of the inner planets late on in the accretion process. Earth and Venus would have received somewhat more of the volatiles than either Mars or Mercury and ended up with roughly comparable proportions to those arrived at from a purely local source.

Early-impacting planetesimals would have arrived at much lower velocities than later ones, partly because initially they came from nearby, and partly because the growing planets were not yet large enough to accelerate them to high impact velocities. As time progressed, however, and the planetary masses grew, impact velocities significantly increased. In consequence, arriving volatiles would have been vaporized upon impact and added to the planets' atmospheres. It has been calculated that the impacting planetesimals began to *degas* when the protoplanets had masses about 0.3% of the Earth's modern mass; they would have become totally degassed when the growing planetary masses had reached around 5%. The result of the mass differences between the inner group of planets is that, for Earth and Venus, all of the gas locked inside the arriving planetesimals would have been added to the atmospheres, but, in the case of both Mars and Mercury, much less degassing would have taken place.

Of course, the released gases did not remain forever in the primitive atmospheres. On Earth, where surface temperatures were not too high, much of the water condensed to form oceans. In the case of Venus, because the total amount of incident solar energy was only about three-quarters of the modern value, there may once have been oceans present. However, as time went by, surface temperatures became significantly higher than on Earth, and more of the water vapour would have been returned to the atmosphere. This having occurred, there would have been enough water vapour in the air to prevent any of the re-radiated IR radiation escaping back into space through "IR windows". Greenhouse heating would have begun.

Solar energy was not the only factor in contributing heat to the proto-planets; the impact process itself provided enormous quantities of thermal energy. In fact it probably more than counterbalanced the generally smaller amounts of solar energy being generated during the early and middle stages of accretion. Later in the process, it is likely that a runaway greenhouse situation prevailed on both Venus and the Earth, in which case an atmosphere of steam may have gathered above their molten silicate surfaces. Gradually, however, the supply of planetesimals diminished and solar heating became the dominant heating source. This favoured the Earth, as less solar radiation reached it than Venus. Earth cooled perhaps to around 30°C, but Venus retained much higher temperatures, and what is termed a *moist greenhouse* developed; that is, Venus had hot oceans, as opposed to no oceans at all.

That there once was a much greater volume of water in the Venusian atmosphere is suggested by studies of the behaviour of rapidly flowing hydrogen during hydrodynamic escape, and of its heavier atmospheric constituents, particularly its **isotope** deuterium (Watson et al. 1982). Measurements of the hydrogen (H) to deuterium (D) ratios indicate a 100-fold enrichment in deuterium, which led Watson et al. (1982) to suggest that there was a stage when H_2O accounted for approximately 2% of the atmospheric volume. This would have been sufficient to provide water to fill an ocean approximately 1/500th the volume of Earth's (Donahue & Pollack 1983). More recent analysis of Pioneer-Venus data (Donahue 1993) suggests that ancient Venus may have had three and a half times more water than once was thought – enough to cover the entire surface to a depth of between 8 and 25 m. This appears to rule out the possibility that Venus was initially dry and that the present meagre water inventory came from occasional cometary impacts.

The situation for Venus did not improve from this point on as water was continuously broken down by UV light, and hydrogen was remorselessly lost to space. The hot oceans would have responded to this by replenishing the atmosphere with what water vapour they still held and thereby promoting intense chemical weathering of the crustal rocks by reactions involving oxygen and carbon dioxide. The latter phenomenon played a crucial rôle in what transpired, for it inhibited oxygen and carbon dioxide dominating the chemistry of the air and diluting the concentration of water vapour. Very likely, when there was only 0.1–1% of the original water remaining, it could not remain as a liquid at the surface of Venus at all. In consequence, all the CO_2 released by the intense heating of the Venusian surface would have remained within the atmosphere. From this point onwards, the evolution of Venus's atmosphere went along the road towards an exceedingly hot surface and a dense atmosphere dominated by carbon dioxide. This is the situation at the present day. A full discussion of the evolutionary process can be found in Donahue & Pollack (1983).

5 Topography and gravity

5.1 Global altimetric and gravity mapping

The first altimetric data for the planet were collected by Earth-based imaging systems during the 1960s. The earliest topographic profiles were drawn from radar measurements made with the 64 m Goldstone antenna in California (Campbell et al. 1972) and the 43 m antenna at Haystacks in Massachusetts (Rogers & Ingalls 1969, 1970). The 300 m antenna of the Arecibo observatory subsequently joined with the other two in forming a powerful trio of Earth-based radar ranging and imaging facilities. During the 1977 inferior conjunction of Venus, three separate receivers were used together to apply multiple interferometry techniques to Venus for the first time (Jurgens et al. 1980). These early observations indicated the existence of three isolated mountain massifs, which rise at least 2 km above their surroundings; furthermore, several large craters, small mountains and one prominent 800 km long double ridge system were recorded. Then, in the late 1970s, the orbiting Pioneer-Venus spacecraft sent back an extended series of altimetric and imaging data that covered 93% of the surface, enabling scientists to achieve the first truly global topographic mapping of Venus. During this programme, elevations were measured to a vertical accuracy of about 200 m, although the dimensions of each areal resolution box varied widely – between 7 km × 23 km and 101 km × 101 km – depending upon the spacecraft's altitude (Pettengill et al. 1979, 1980). The resultant maps, published at a scale of 1:50 000 000, were generated by applying digital relief shading techniques to a gridded array of the radii of Venus as constructed from the same Pioneer altimeter data (Fig. 5.1).

The Pioneer-Venus data, combined with those from Earth-based antennae, established that 60% of the surface lay within 500 m of the mean radius of 6051.4 km, the predominant physiographic unit being a huge area of *rolling plains*. Of the remaining terrain, a lesser part (8%) lies above this level than below (27%); that is, *highlands* are less extensive than *lowlands* on Venus. The major areas of highland terrain are Aphrodite Terra, which extends along the equator between 60°E and 210°E; Ishtar Terra, extending between 300°E and 40°E, 55°N and 75°N; the smaller massifs of Beta and Phoebe Regiones, centred upon 280°E, 30°N and 280°E, 10°S respectively; and Alpha Regio, whose situation is farther east at 5°E, 25°S. Aphrodite is roughly the same size as Africa, while Ishtar is comparable with Australia.

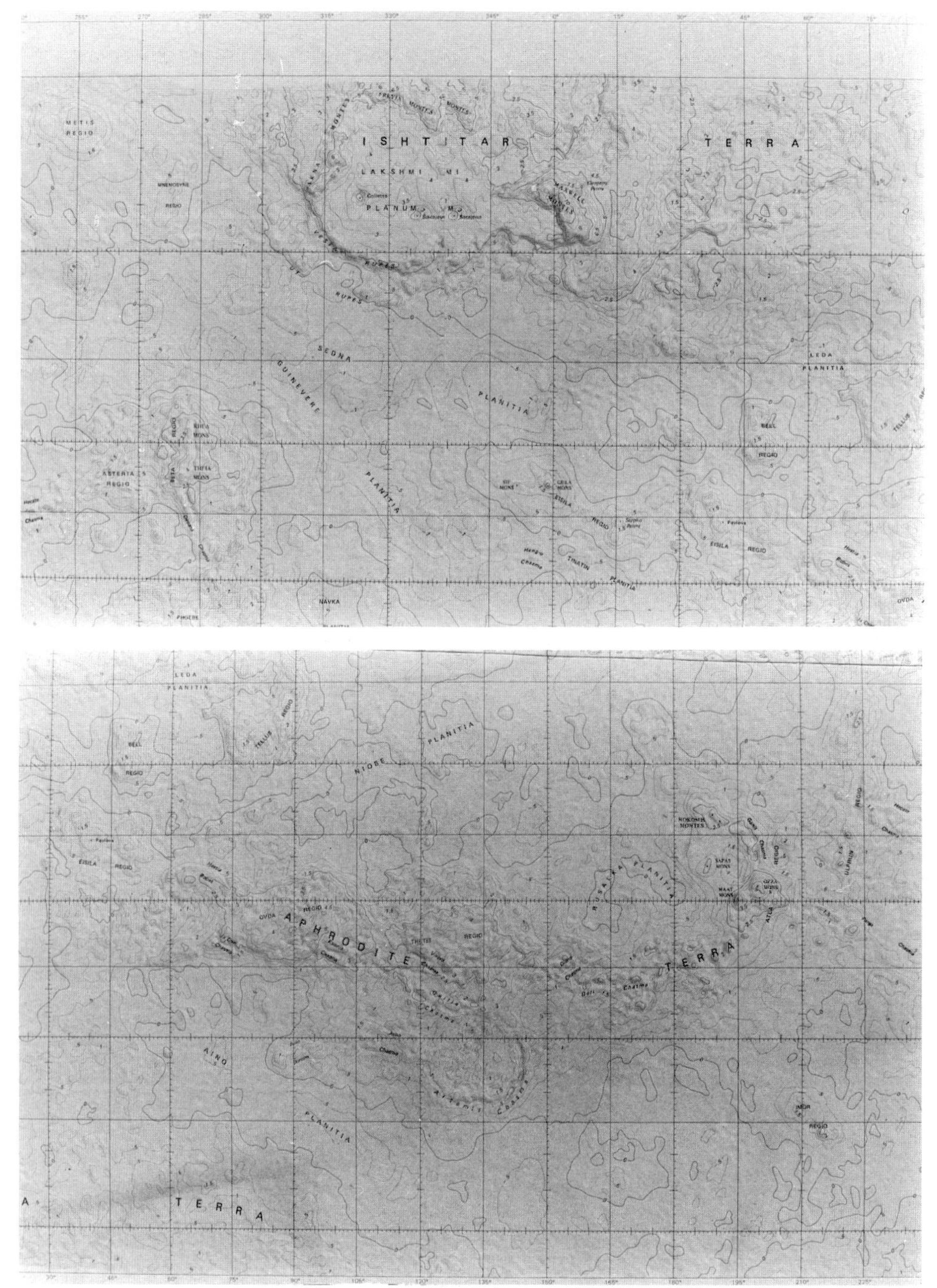

Figure 5.1 Altimetric map of Venus

These maps were generated by applying digital relief shading to a gridded array of the radii of Venus from Pioneer altimetric data.

Lowlands are located primarily in Atalanta Planitia (centred at 65°N, 165°E), Leda Planitia (45°N, 65°E) and Sedna Planitia, located south of Ishtar Terra. The lowest point on the surface was found within a deep rift valley known as Diana Chasma, 4 km below datum, which makes it deeper, relatively, than Earth's Red Sea.

In contrast to the Earth, where the continents and oceans give a bimodal distribution to the elevation histogram, the histogram for Venus is *unimodal* and the topographic peak narrow (Fig. 5.2). This latter is also in complete contrast to the curves for both the Moon and Mars, and this is a reflection of the extreme average smoothness of the Venusian globe. In fact, 90% of the surface lies within a 3 km height interval. Slope measurements made during the same exercise indicated that the steeper slopes (on the metre scale) were characteristic of the more elevated regions. There is also a broad tendency for the roughest surfaces to coincide with the higher terrain, which is why the radar-bright regions on the imagery generally delimit the highland massifs or the rims of raised circular features such as craters, or the tops of large domes. While Earth and Venus share the general property of having much the same

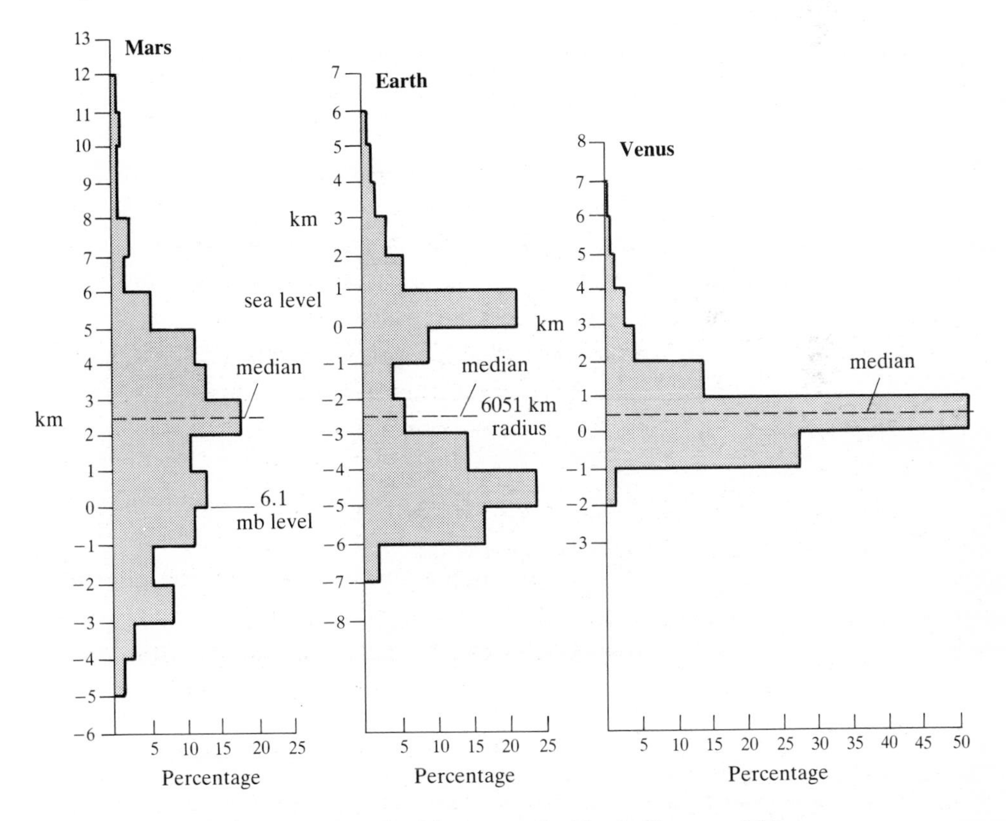

Figure 5.2 Elevation histograms for Earth, Venus and Mars

Mars shows a trimodal distribution reflecting the northern lowlands, Tharsis volcanoes and cratered plateau; Venus's distribution is unimodal. The Earth's bimodal distribution reflects continents and ocean basins.

proportion of rugged mountains, this is not, of course, to say that they formed by the same process, as will be discussed in a later section.

That there is a significant difference in the distribution of slope characteristics between Earth and Venus was revealed by an analysis of Pioneer and other data by Sharpton & Head (1986), who noted that, although regional slopes on both planets span the same range (0–2.4°), the slope frequency distributions are quite different, the Earth being typified by an excess of very low slopes, owing to the effects of planation and sedimentary deposition, driven by vigorous hydrospheric activity (see Plate 3). Roughly twice as much of the terrestrial surface as that of Venus is characterized by slopes > 0.24°, a difference principally attributable to the existence of continental margins on Earth.

A vertically looking radar altimeter was also on board both the Russian Venera 15 and Venera 16 orbiting spacecraft, which reached Venus during 1983. These achieved a vertical precision of ± 50 m, within a nominal altimeter box diameter of about 50 km; however, special processing of the data enhanced the spatial resolution to 5 km (Barsukov et al. 1986). In addition to accumulating a higher-resolution data set, the Russian results extended the altimetric and geological data to the polar regions not imaged by Pioneer-Venus owing to orbital constraints. High-resolution imagery acquired by Venera also provided insights into the geological characterization of the surface layer on a global scale, since this new information could be combined with the Pioneer-Venus data to derive surface roughness and reflectivity data (Bindschadler & Head 1986a,b). These studies suggested that over 70% of the surface imaged by Veneras 15 and 16 consists of plains, and about 25% of tectonically-deformed units. Florensky and his colleagues showed that the Venusian surface is dominated by blocks and a laminated pavement and has only small amounts of regolith or soil (Florensky et al. 1977).

During the later 1980s, the Arecibo radar antenna obtained images with even higher resolution than those sent back by Pioneer-Venus and it continues to do so, enabling scientists to build up mosaics that cover around 25% of the planet's surface at a horizontal resolution of between 10 and 20 km (Campbell & Burns 1980, Head & Campbell 1982).

Pioneer-Venus provided the first really useful gravity data, the movements of the orbiting Pioneer spacecraft being tracked by radio from the Earth over a period of 220 days. Using the well established Doppler tracking method, features between 300 and 1000 km were resolved within a region bounded by parallels 10°S and 40°N and by 70°E and 130°E. A large number of relatively mild **gravity anomalies** were recorded (Sjogren et al. 1980). The Venusian gravity field appears to correlate strongly with topography, such that major gravity highs coincide with large positive topographic features.

Disappointingly, although Magellan gravity data have higher signal-to-noise ratios, they have not provided higher resolution than Pioneer data. This may improve during the fourth and final mapping cycle, but even if it does, the enhanced resolution will be available only for low latitudes, as the spacecraft is too high as it approaches higher latitudes, and its perception is "clouded". It is interesting, therefore, that Reasenberg & Goldberg (1992) recently have extracted gravity models from the Doppler tracking of Pioneer-Venus orbiter, using a new technique that removes the computational artefacts that introduce uncertainties into traditional methods of gravity data interpretation. In earlier methods, topographic information is incorporated into the calculations that allow construction of gravity maps, and this has led some scientists to hold reservations about the reality of the close correlation between gravity and topography described above. The gravity and smoothed topography maps resulting from their studies are reproduced as Plate 4. The clear correlation between high gravity and high elevation over the regions of Ishtar Terra, Aphrodite Terra and Beta Regio is shown. This is in contrast to the Earth, where there is very poor correlation between the two. While 100 mgal anomalies are common on Mars and the Moon, those of Venus generally are smaller (40–50 mgal), although the largest – that associated with Beta Regio – reaches 135 mgal according to Esposito et al. (1982).

Modelling of the anomalies that would be produced by the observed topography implies that considerable compensation must have occurred on Venus. Initial studies by Phillips et al. (1981) suggested that depths of compensation of the order of 100 km or more were necessary to explain the support of the topography solely by density variations. They also concluded that the topography can only be supported passively if it was very young, i.e. of the order of 10^7 years. On the other hand, it could be that the lithosphere is supported by dynamical motions of the mantle, for instance, convection (McGill et al. 1981), a subject to which we shall later return.

5.2 Regional topographic variations

The global distribution of elevations from pre-Magellan data is depicted in Figure 5.3, derived from data shown on the United States Geological Survey map I-2041 (USGS 1989). It will be seen that the northern polar region is by nature a large plain crossed by ridge belts and has an average elevation close to the average elevation of the planet, i.e. 0.5 km above datum. South of latitude 80°N and extending between 300°W and 72°W is the continent-size upland of Ishtar Terra, where elevations 2 km above datum are typical and where an altitude of over 11 km is reached in the mountain massif known as Maxwell Montes. The highland massifs that constitute Ishtar Terra extend as far south as latitude 52°N, close to where the Pioneer-Venus probes landed. At the western end of Ishtar is the particularly striking plateau of Lakshmi Planum, which covers an area roughly twice that of Earth's Tibetan Plateau and has an average elevation of around 3 km. This is surrounded by narrow belts of rugged terrain that attain heights of 4 km.

Rolling plains, with elevations between 0 and 1.5 km, occupy extensive regions between 60°N and 60°S and may be separated from highland massifs by transitional, intermediate-

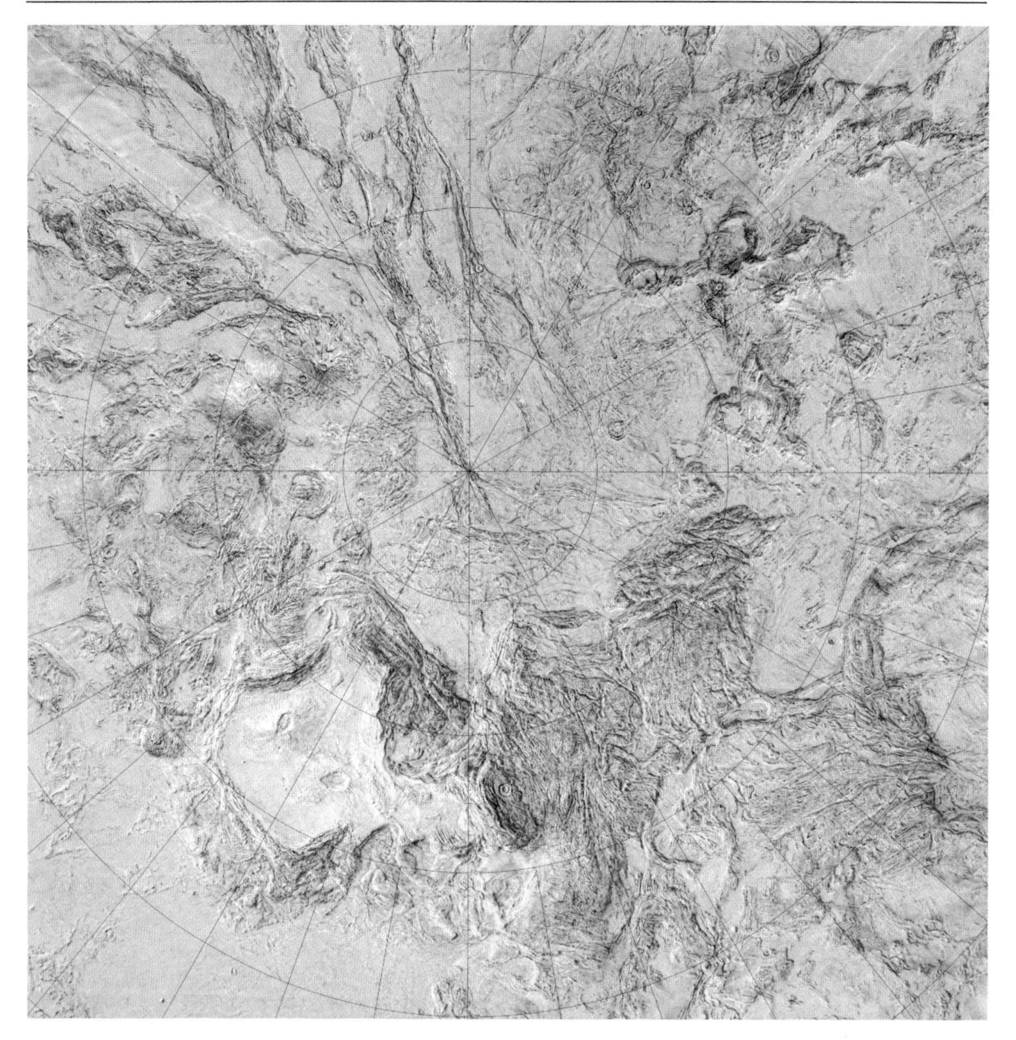

Figure 5.3 Topographic maps of Venus
This map of the northern hemisphere was prepared by the Joint US/USSR Working Group on Solar System Exploration. Map I-2041, part 2. From USGS (1989).

height areas of ridge-and-groove topography known as *tesserae*. The large tessera block, Tellus Tessera, which lies southeast of Ishtar Terra, in general stands 1–2 km above datum, and covers an area approximately that of peninsular India. Between longitude 55°E and 210°E and within the latitude range 10°N–20°S lies another major highland region – Aphrodite Terra – where the elevations range between 1.5 and 4.5 km above datum. The distribution of relief within Aphrodite is, however, complicated by many, often deep *rift valleys*, one of which, Diana Chasma, descends to more than 3 km below datum. Smaller highland blocks are located west of Aphrodite, in Eistla Regio, where elevations above 1.5 km are found. Further highland massifs, typified by elevations above 2 km, are located at Beta and Phoebe Regiones (centred on 30°N), Alpha Regio (30°S), west of Ishtar Terra, at Metis Regio (72°N), and east of Ishtar, in Tethus Regio.

The major regions of low elevation are located in Atalanta Planitia (centred at 65°N, 165°E), which descends to 1.5 km below datum and is roughly the size of the Gulf of Mexico, and in the broad X-shaped area that includes Guinevere, Sedna, Niobe, Lavinia and Aino Planitiae, which extend between 60°N and 60°S. Smaller lowlands, with elevations below zero datum, are found in Navka Planitia (centred on 310°E, 7°S). Around the south pole is a further extensive region of intermediate-level rolling plains.

5.3 Venera and Vega on the surface of Venus

The Russian spacecraft of the early 1970s gave the scientific community the first chemical data for Venusian surface rocks and a limited glimpse of what the hitherto mysterious surface of the planet looked like. The first soft-lander was Venera 8, which landed on 22 July 1972, near Navka Planitia (Fig. 5.4). It was able to measure the radioactivity levels within

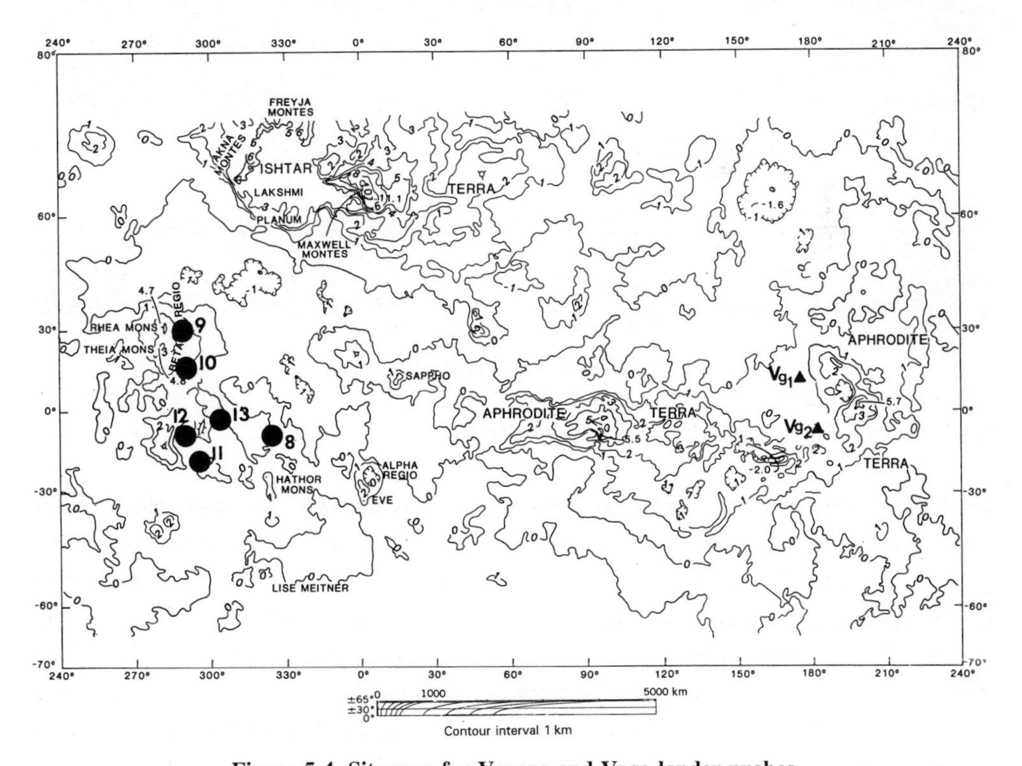

Figure 5.4 Site map for Venera and Vega lander probes

the surface rocks with a gamma-ray spectrometer and somewhat surprised the planetary community by recording levels of U, Th and K similar to those found in some terrestrial continental volcanic rocks (Vinogradov et al. 1973.). It also established the density of the rocks to be 1.5 g cm^{-3}.

Both the Veneras 9 and 10 soft-landers arrived safely, the former on the northeast flank of Beta Regio and the latter about 2500 km farther south. Both obtained excellent pictures of the actual Venusian surface, and successfully sent these back to Earth. Venera 9 landed on a sloping rock-strewn surface, the rocks being slabby, the individual slabs ranging in size from

(a)

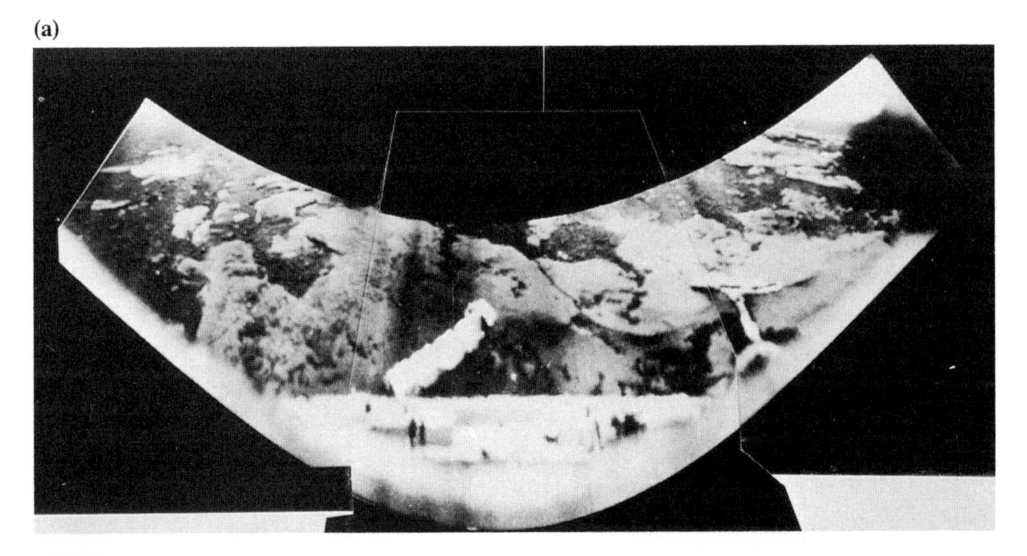

(b)

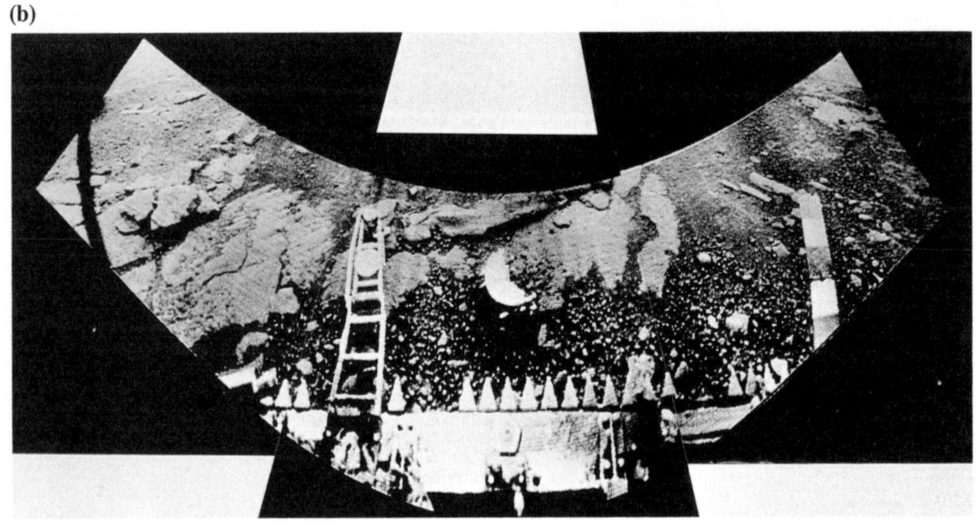

Figure 5.5 Venera Lander views

(a) Panorama from the Venera 10 landing site. (b) Panorama from the Venera 13 landing site.

20 to 70 cm across. The rocks were separated by darker fine-grained material. Venera 10, which landed three days after its companion, took photos of a flattish plain composed of scattered flat 1–3 m outcrops with intervening dark fine-grained debris with an albedo of less than 3% (Fig. 5.5a). The probes recorded rock densities of the order of 2.8 ± 1 g cm^{-3} and also measured abundances of the same trio of elements as did their predecessor (Surkov et al. 1977), this time finding levels more typical of terrestrial basaltic rocks (Table 5.1).

Each lander had on-board a simple anemometer, which recorded low windspeeds of between 0.4–0.7 m s^{-1} at the Venera 9 site and 0.8–1.3 m s^{-1} at the Venera 10 site (Keldysh 1977).

The next pair of successful Russian spacecraft, Veneras 13 and 14, landed during October 1975; each transmitted data for one hour and sent back one excellent panoramic picture of the surrounding surface. Both set down on the eastern slope of Phoebe Regio, the former at an elevation somewhat higher than the latter (see Fig. 5.4). In general terms, both sites resembled that of Venera 10, but a distinctive feature of the Venera 13 site was a large area of dark regolith in the vicinity of the lander probe (Fig. 5.5b). Some workers found the surface appearance rather reminiscent of some terrestrial volcanic slab pahoehoe surfaces. Veneras 13 and 14 also obtained more detailed geochemical data by XRF analysis than did their predecessors. The analyses are comparable broadly with terrestrial oceanic basalts (Table 5.2).

The last probes to land on Venus were Vegas 1 and 2, which landed in June of 1985, having been dropped through the dense Venusian atmosphere on balloons, as a spin-off from the Giotto mission to Halley's comet. Both landed in the region known as Rusalka Planitia, lo-

Table 5.1 Venera measurements of certain radioactive elements.

Spacecraft	U ($\times 10^{-4}$ wt%)	Th ($\times 10^{-4}$ wt%)	K (wt%)
Venera 8	2.2 ± 0.7	6.5 ± 0.2	4.0 ± 1.2
Venera 9	0.60 ± 0.16	3.65 ± 0.42	0.47 ± 0.08
Venera 10	0.46 ± 0.26	0.70 ± 0.34	0.30 ± 0.16

Table 5.2 Veneras 13 and 14 rock analyses and selected terrestrial igneous rocks. After Barsukov (1982).

Oxide	Venera 13	Venera 14	Oceanic basalt	Continental crust
SiO_2	45.0 ± 3.0	49.0 ± 4.0	51.4	63.3
Al_2O_3	16.0 ± 4.0	18.0 ± 4.0	16.5	16.0
MgO	10.0 ± 6.0	8.0 ± 4.0	7.56	2.2
CaO	7.0 ± 1.5	10.0 ± 5.0	9.4	4.1
TiO_2	1.5 ± 0.6	1.2 ± 0.4	1.5	0.6
MnO	0.2 ± 0.1	0.16 ± 0.08	0.26	0.08
FeO	9.0 ± 3.0	9.0 ± 2.0	12.24	3.5
K_2O	4.0 ± 0.8	0.2 ± 0.1	1.0	2.9

cated near the northern flank of Aphrodite Terra. Vega 1 carried a gamma-ray spectrometer, while Vega 2 also carried an XRF spectrometer, the two probes providing an extension of the dataset for Venus's surface rocks. The contents of natural radioactive elements in these Venusian rocks as determined by Vega were interpreted by Surkov and his colleagues (Surkov et al. 1986a,b) as being representative of gabbroic rocks. Comparisons between the surface properties of the Vega landing-site rocks, including surface roughness and RMS slopes, and of those imaged by the Veneras 15 and 16 spacecraft, led Bindschadler et al. (1986) to conclude that the Rusalka rocks were more like the rocks characteristic of the northern plains of Venus that the deformed rocks of the highlands.

5.4 Magellan altimetric and gravity data

As it orbited Venus, the Magellan spacecraft collected altimetric data in short bursts, about 25 ms long, via a horn antenna. These are interleaved between somewhat longer bursts during which the synthetic-aperture radar (250 ms bursts) and radiometer (50 ms bursts) are operational. The bulk of the processing of the altimetric data is undertaken by a team at the Massachusetts Institute of Technology (MIT) who derive surface height, metre-scale roughness and normal-incidence power reflection coefficient values along the ground track. The footprint dimensions of the area resolved by the Magellan altimeter varies with the position of the spacecraft in its orbit; thus, at periapsis, the ground resolution is 2 km, whereas at high latitudes over the polar regions, it is only 20 km. Owing to uncertainties associated with determining the probe's orbit, although the theoretical accuracy of the height determinations is as little as 5 m, the actual error is of the order of ten times this.

Magellan data have refined the mean radius of Venus to 6051.84 km and confirmed the unimodality of the *hypsographic* curve of Venus, 80% of the surface lying within 1 km of mean datum. They also show that average kilometre-scale slopes with gradients $<30°$ are not uncommon on the southwest flanks of Maxwell Montes, the southern slope of Danu Montes and in the walls of chasmata to the east of Thetis Regio (Ford & Petengill 1992).

A first analysis of Magellan gravity data by McNamee et al. (1993) shows that the data are consistent with previous results. Using the spherical harmonic representations of the topography and of the gravity field of Venus, it has been possible to constrain models for the interior structure of the planet. For instance, using the product of the gravitational constant and the mass of Venus ($324\ 858 \pm 0.03$ km^3 s^{-2}) the mean density was found to be 5243.83 ± 0.4 kg m^{-3}. It was also found that if a triaxial ellipsoid was, by least squares, fitted to the topography, the centre is displaced considerably less than it is for any of the Earth, Mars or the Moon. This may be a function of the very slow rotation of Venus.

A striking resemblance is observed between the topography contours and the radial accelerations associated with the gravity field near the surface of the planet. Furthermore, a very strong correlation between topography and the gravity field is matched by an equal strong anti-correlation between the latter and topography. This is taken to indicate that isostatic compensation, whether passive or dynamic, is very prominent at the wavelengths used (degrees 3 and 11).

Interestingly, the gravity–topography analysis indicates that there are two kinds of highland regions on Venus. The first type shows the following features:

- the theoretical gravity calculated from the topography is much larger than the observed gravity;
- the Bouguer anomalies are negative and very large in absolute terms; and
- the isostatic anomalies are negative.

Such properties are consistent with passive isostasy associated with crustal thickening, and are exemplified by western Ishtar Terra, Ovda and Thetis Regiones, each of which (as will be suggested later) may be thickened crustal blocks repeatedly tectonized. The second type of highland region is characterized by the following:

- the theoretical gravity computed from topography is larger than the observed gravity, but not nearly as large as for the first type above;
- Bouguer anomalies are negative and quite small in absolute terms; and
- isostatic anomalies are positive.

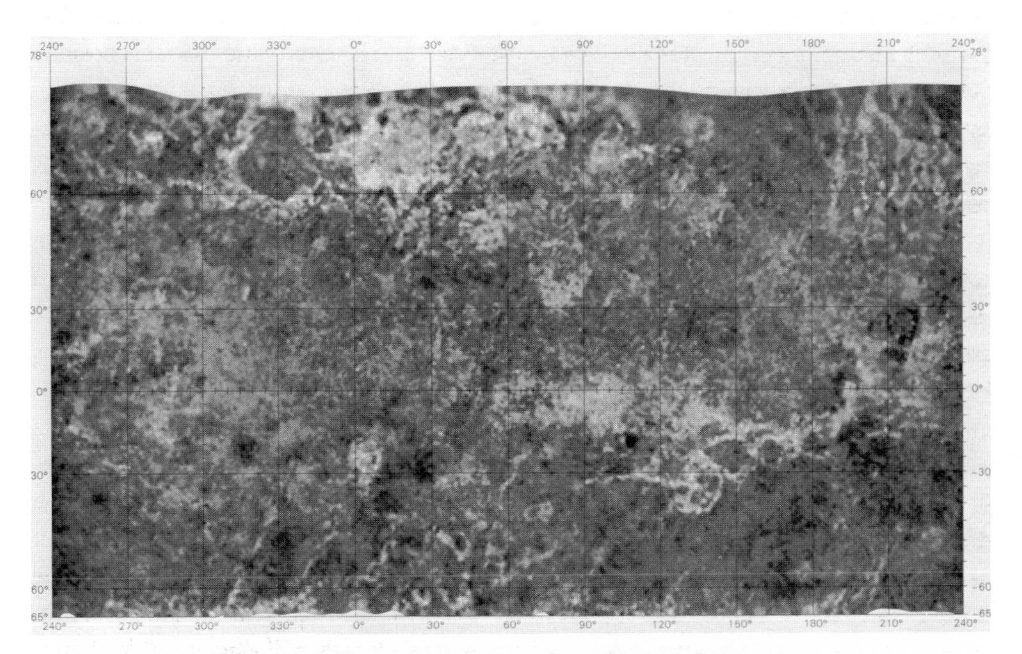

Figure 5.6 Venus RMS metre-scale surface roughness map

This map is derived from the Hagfors scattering model. Lighter tones represent rougher surface elements.

Examples of this type are Beta and Atla Regiones, and the properties they show are consistent with dynamic support of topography. It will later be shown that such areas lie above hotspots.

At the time of writing, Magellan data are still being compiled; already, however, the first few of a new generation of altimetric, topographic and physiographic charts have been published. Subsequently, the entire Magellan data set will be used to generate new state-of-the-art global topographical and geological maps of the planet and more detailed charts of specific regions. The planet's gross topography – with those regions not imaged by Magellan during the first mapping cycle picked out in grey – is shown in Plate 5, while Figure 5.6 is a chart of planet-wide RMS metre-scale surface roughness, both based on data from the first Magellan mapping cycle. The correlation between the major highland regions and the higher values of RMS metre-scale slope is clear.

5.5 Venus, Earth, Moon and Mars: a comparison

Venus has highlands and lowlands, as do both the Moon and Mars; Earth has these too, but is unique in also having oceans. In §5.1 it was suggested that the Venusian highlands appear to be (in part, at least) isostatically compensated, probably at relatively shallow depths. Therefore, Venus appears, in this respect, to be more Earth-like than the other two worlds in this respect. However, strong positive correlation between gravity and topography is not a feature of the Earth, where gravity lows underlie the continents (equivalent to highlands) owing to complete, or near-complete, isostatic compensation. Strong correlation between gravity and topography is a characteristic of Mars; however, although there is a large free-air anomaly associated with the Tharsis rise, there are also large free-air anomalies associated with the adjacent lowlands of Amazonis and Chryse, which there are not beneath the lowlands on Venus. This can be satisfactorily explained only by assuming that there is incomplete isostatic compensation beneath both the highlands and lowlands of Mars, or that the depth of compensation is very great (i.e. >1000 km). This apparently contrasts with the situation on Venus and confirms a widely held suspicion that both the rheology of the crustal rocks, and the means whereby heat is lost from inside Venus, are rather different from those of its neighbours.

Analysis of slope and surface roughness data is also instructive in terms of comparative geology. The distribution of regional 3° by 3° slopes on Venus is very different from that on both the Earth and Mars. Pioneer-Venus data reveal there to be no extensive regions of zero slope within the interiors of either the Venusian highlands or plains, which is quite unlike the situation on the Earth, where extensive gradation and deposition have modified the continental surfaces to produce widespread smooth regions with zero slopes. The range of RMS slopes on Venus (on the metre scale) at the Pioneer-Venus radar wavelength of 17 cm is 1° to 3° –

low compared with those measured by Apollo on the volcanic maria of the Moon (range 3° to 4°, radar wavelength 13 cm) and both the very rough volcanic plains of Tharsis–Memnonia–Amazonis on Mars (more than 10°–15° at 12.6 cm wavelength) and also the cratered and smooth Martian plains (4°–6°). There are a few isolated regions on Mars, for instance Syria Planum, parts of Syrtis Major Planitia and Elysium Planum, that do have 1°–2° slopes like those of Venus. These are, however, atypical.

The actual extremes of elevation of Venus and Earth are much the same, as has been noted above, but Venus is much the smoother of the two. The highest point on Venus, Maxwell Montes, is relatively greater than the height of Everest above sea level. Diana Chasma, the lowest point on the Venusian surface, is 4 km below the adjacent ridged terrain, but a mere 2 km below datum. Compared with terrestrial oceanic trenches such as the Marianas Trench, which descends 11 km below sea level, and the Valles Marineris on Mars, whose floor in places sits 7 km below datum, Diana Chasma is a mere one-fifth as deep as the terrestrial trench and only one-half as deep as the Martian canyon system. Total relief on Venus is approximately two-thirds that of the Earth and one-half that of Mars.

As was discussed in §5.1, the hypsographic curve (topographic "spectrum") for Venus is unimodal, contrasting markedly with a strongly bimodal distribution for the Earth. This latter reflects the two significant physiographic levels, namely the oceans and the continents. As for the Moon, the mean levels of the cratered highlands and maria are separated by nearly 1.5 km difference, giving the lunar hypsogram a weakly bimodal characteristic. It also contrasts strongly with Mars, whose curve is weakly trimodal owing to the tripartite distribution of levels, representing the cratered plains, the Tharsis uplands and the northern plains.

The distinctive nature of the Venusian elevation spectrum must have some explanation. Two possibilities come to mind: first, Venus originally had a crust of uniform thickness and/ or density; or secondly, its lithosphere is readily deformable owing to the high ambient temperatures that prevail. If the former was the case, then the development of lowlands – presumably overlying a thinned crust – and highlands – where the crust had become thickened – may have been a response to convective motions in the Venusian mantle.

Finally, what of the composition of Venus's crust? The limited data set shows that basalt-like lavas are found on the flanks of Beta–Phoebe Regio, but also indicate more differentiated types, e.g. **syenitic** or **granite**-like, over parts of the rolling plains. The smooth nature and low elevation of the Venusian lowlands certainly indicate the likelihood of their being floored by basaltic lavas, like the lunar maria. However, some of the Venera data point to considerable differentiation of the Venusian crust to form *continental* material akin to that so typical of the Earth. The sending of Magellan has gone a long way towards elucidating this issue.

6 Physiographic regions

6.1 Physiographic map of Venus

Prior to the analysis of Magellan data, available physiographic maps were constructed from a combination of Venera 15/16, Goldstone and Arecibo high-resolution imagery, and, of course, Pioneer-Venus data. The acquisition of Venera 15/16 images during the mid-1980s enabled scientists to characterize the major physiographic units in considerably greater detail than hitherto had been possible; however, it should be remembered that Venera 15/16 imaged only 30% of the surface, between latitude 78°N and approximately 30°N. Despite its restricted coverage, the improved quality of the new data marked a major step forward in the understanding of Venusian geology. Analyses of the Venera 15/16 data by various teams of (mainly) Russian scientists during the mid-1980s (Barsukov et al. 1986, Basilevsky et al. 1986, Basilevsky & Head 1988, Ivanov et al 1986) allowed recognition of several different types of complex plains, interpreted to be mainly of volcanic origin, and a diverse array of tectonic features.

According to the largely pre-Magellan work of Head (1990a), the major geomorphological units recognized, in decreasing order of abundance, are: plains, tesserae, ridge belts, coronas, volcanoes and mountain belts. Their global distribution was found to be distinctly nonrandom. Subsequent studies by Saunders et al. (1991) and members of the Magellan Mission team (Saunders et al. 1992) using the new improved-resolution Magellan imagery achieved a more detailed understanding of the nature and global distribution of Venusian land-forms. Features first revealed by Magellan include *wind streaks* – familiar from the plains of Mars – extensive exposures of outflow materials associated with *impact ejectamenta*, *lava flow lobes* and different types of *channels*, *aeolian dunes*, and *impact craters* as small as 3 km in diameter. Magellan has largely confirmed the earlier work, but has also extended it as other features have been discovered and the enhanced resolution has refined earlier data. Initial results reaffirm that the most widespread terrain type on the planet is lowland volcanic plains, but Magellan has also imaged very large numbers of previously unsuspected volcanic structures. It is now apparent that Venus can be thought of in terms of two principal landscape types: *volcanic plains* upon which are superimposed thousands of individual volcanic constructs and which cover 85% of the globe, and *tectonically deformed highlands*, which cover the remaining 15%.

The latest physiographic maps reveal in its entirety the southern highland, Lada Terra, and the massifs of Beta–Phoebe–Themis Regiones, which lie astride the 285° meridian, together with a wealth of other detail. One Magellan topographic product is shown here (Fig. 6.1). Magellan revealed in its entirety the southern highland region, Lada Terra, centred on 8°E, 67°S, for the first time. Each of the major highland areas is separated from the others by deep basins within which are located smaller uplands, such as Alpha Regio (6°E, 25°S), Bell Regio (49°E, 33°N), Eistla Regio (39°E, 18°N) and Tellus Regio (82°E, 39°N). Some of these are joined by deep rifts, termed *chasmata,* while straighter but shallower rifts, arranged in parallel sets, cut obliquely across the equatorial zone. Lithospheric deformation is manifested in global-scale mountain and ridge belts of various styles and in the occurrence of large numbers of coronas. In general terms the distribution of these areas of deformation suggests strain over rather broad areas, rather than a restriction to the narrow zones so typical of Earth's lithospheric plate margins.

While the structure of the highland areas is undeniably quite complex, the regions of what are now termed *complex ridged terrain* (CRT) (also known as *tesserae,* and previously called parquet terrain) are uniquely Venusian and arguably the most tectonically complex. Comprising a complicated network of intersecting troughs and ridges, they tend to occur as elevated plateaux with steep sides, which typically are very rough at metre and smaller scales. After study of Venera 15/16 imagery (Sukhanov 1986), tessera terrain was shown to cover at least 10% of the planet north of latitude 30°N and was suggested to cover large areas within several of the highland massifs. Magellan images essentially tripled the known extent of the tesserae, which can now be identified in Ovda, Thetis, Phoebe, Beta and Asteria Regiones, and within the regions of Lada Terra and Nokomis Montes (Solomon et al. 1992). The largest contiguous area of such terrain covers 10 million square kilometres and stretches between Ovda Regio and the westward extension of Aphrodite Terra.

The greatest area (85%) of Venus is covered by plains deposits, which have been studied by several individuals and groups. Thus, Barsukov et al. (1986) recognized five morphological types: ridge-and-band plains, band-and-ring plains, patchy rolling plains, dome-and-butte plains and smooth plains. Distinctions between the types, as can be deduced from the names, were based largely on the superimposed landform types. Head (1990a) argues that the distribution and characteristics of landform assemblages demonstrate that vertical and horizontal stresses are operating on the Venusian crust and lithosphere in different ways in specific regions. We shall return to this when discussing the regional geology.

The improved resolution of Magellan images has enhanced our knowledge of the fine structure of the plains, a beautiful example of which is shown in Figure 6.2, an area of Niobe Planitia showing the typical landforms of the plains regions: coronae, impact craters, volcanic shields, domes and volcanic flows. Details of over 1600 volcanic land-forms, of coronas, and of complex strain structures and deformational belts were revealed for the first time. The latter are of more than one type and were first noted on Arecibo high-resolution images.

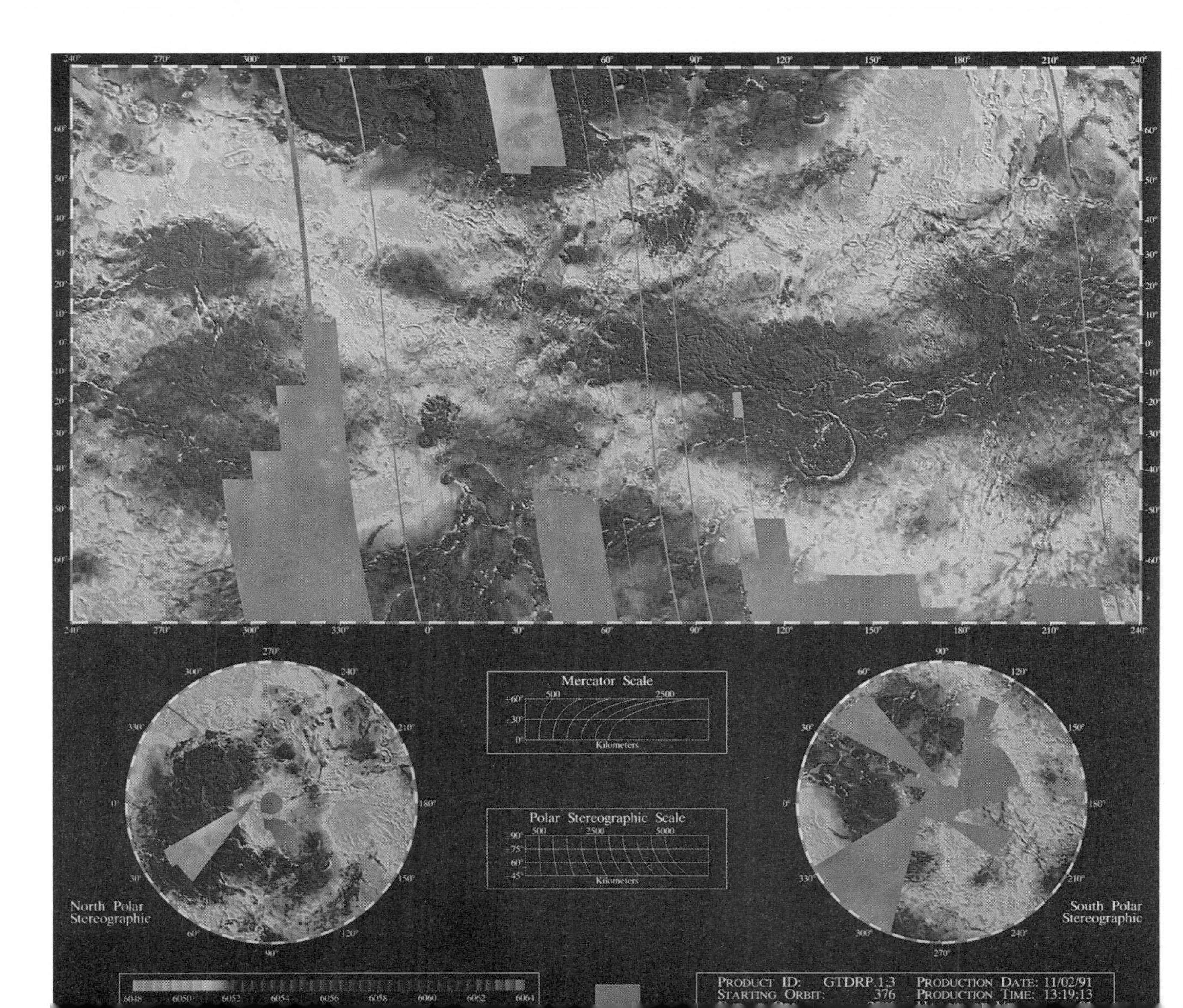

Mercator Scale
Kilometers
Polar Stereographic Scale
Kilometers
North Polar Stereographic
South Polar Stereographic
PRODUCT ID: GTDRP.1;3
STARTING ORBIT: 376
PRODUCTION DATE: 11/02/91
PRODUCTION TIME: 13:19:13

Figure 6.2 Magellan image of plains in Niobe Planitia

The typical landforms of Venusian plains are seen: a 25 km diameter impact crater is located amongst ridged plains towards the top left corner of the image. To the south lie more strongly textured units upon which can be seen several volcanic shields, shield fields, pancake domes and coronas. From Magellan F-MIDRP 15N112;201.

They tend to concentrate in or near broad topographic lows, particularly in Lavinia, Atalanta and Vinmaria Planitiae. In contrast, the uniquely Venusian structures – coronas – tend to avoid the lowlands, but occur in large numbers in the rolling plains. However, this difference may be more apparent than real, since it is possible that any coronas that formed on the lowland surfaces may subsequently have been buried by flood lavas. Those recent detailed studies which have been undertaken indicate a complex depositional and subsequent deformational history for the plains of Venus, but have also begun to reveal that there is a first-order similarity in the structural evolution of widely separated plains provinces.

Figure 6.1 Topographic map of Venus (opposite)

Mercator and polar projections reveal the gross topography of the planet. Magellan GTDRP.1;3.

To the south of Ishtar are the extensive lowlands of Sedna and Guinevere Planitiae, whose deepest levels descend 1 km below datum. These plains are punctuated by the isolated highland massifs of Eistla, Alpha and Bell Regiones. The former is dominated by the twin peaks of Sif and Gula Montes, 3.5 and 4.4 km high respectively, which are located astride southeast-striking chasmata, and associated with which are extensive volcanic flows. Alpha Regio has a more complex structure and is by nature a plateau surrounded by clusters of high volcanic domes. Bell Regio, on the other hand, is a broad dome approximately 1000 km in diameter, elongated in a north–south direction. The twin volcanic peaks of Tepev Mons and a variety of tectonic features are located here. On the plains also are many circular coronas,

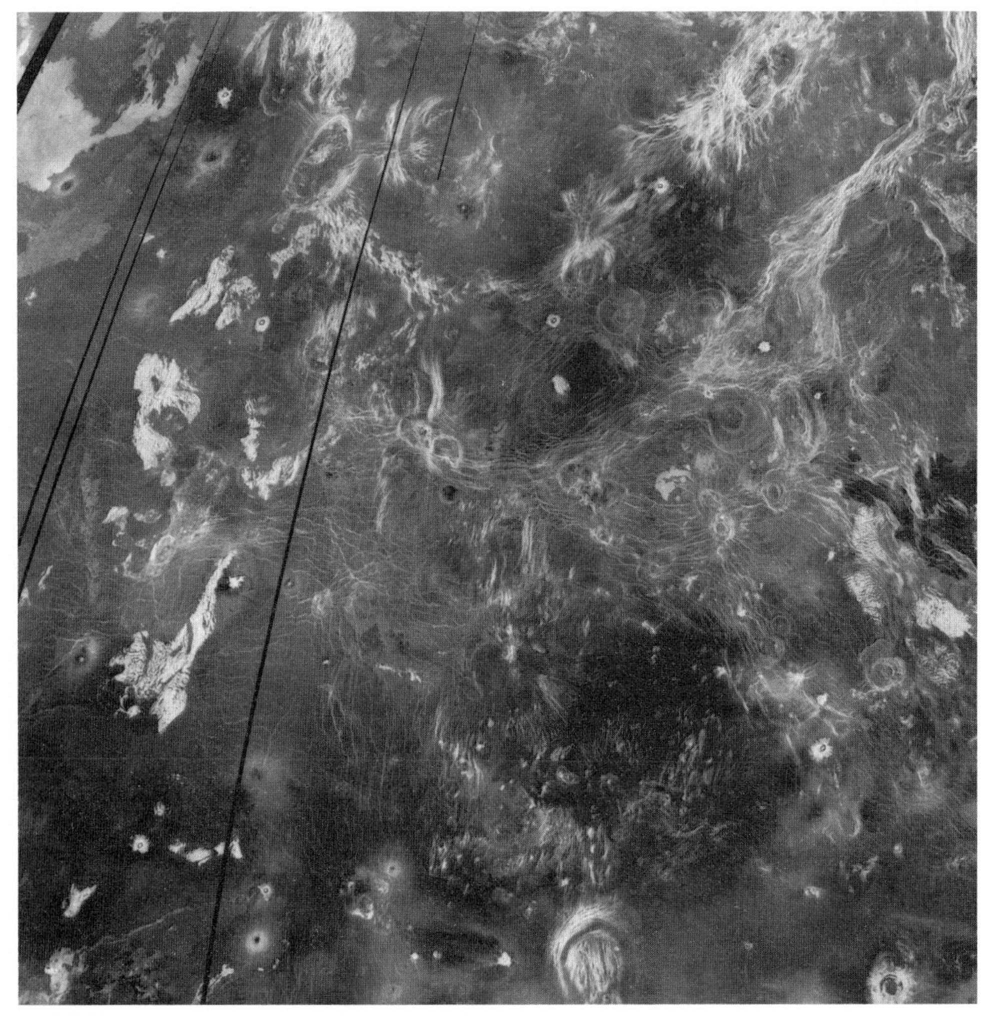

Figure 6.3 Plains of Bereghinya Planitia

Landforms seen here include several large coronas, together with volcanic edifices and parts of ridge belts. From Magellan image C2-MIDRP 30N026;1.

peculiarly Venusian deformational structures apparently formed above mantle plumes or hotspots (Fig. 6.3). Saunders et al. (1992) record 175 such features. One in particular, the 750 km diameter Heng-O (355°E, 2°N), is prominent on radar maps generated from Magellan imagery.

East from Guinevere Planitia the surface rises along a belt of equatorial highlands, which together comprise Aphrodite Terra, the most extensive highland region on the planet, depicted in false colour in Plate 6. There are several individual massifs, beginning in the west with the steep-sided scarps of Hestia Rupes (70°E, 5°N), and passing through Ovda Regio (95°E,

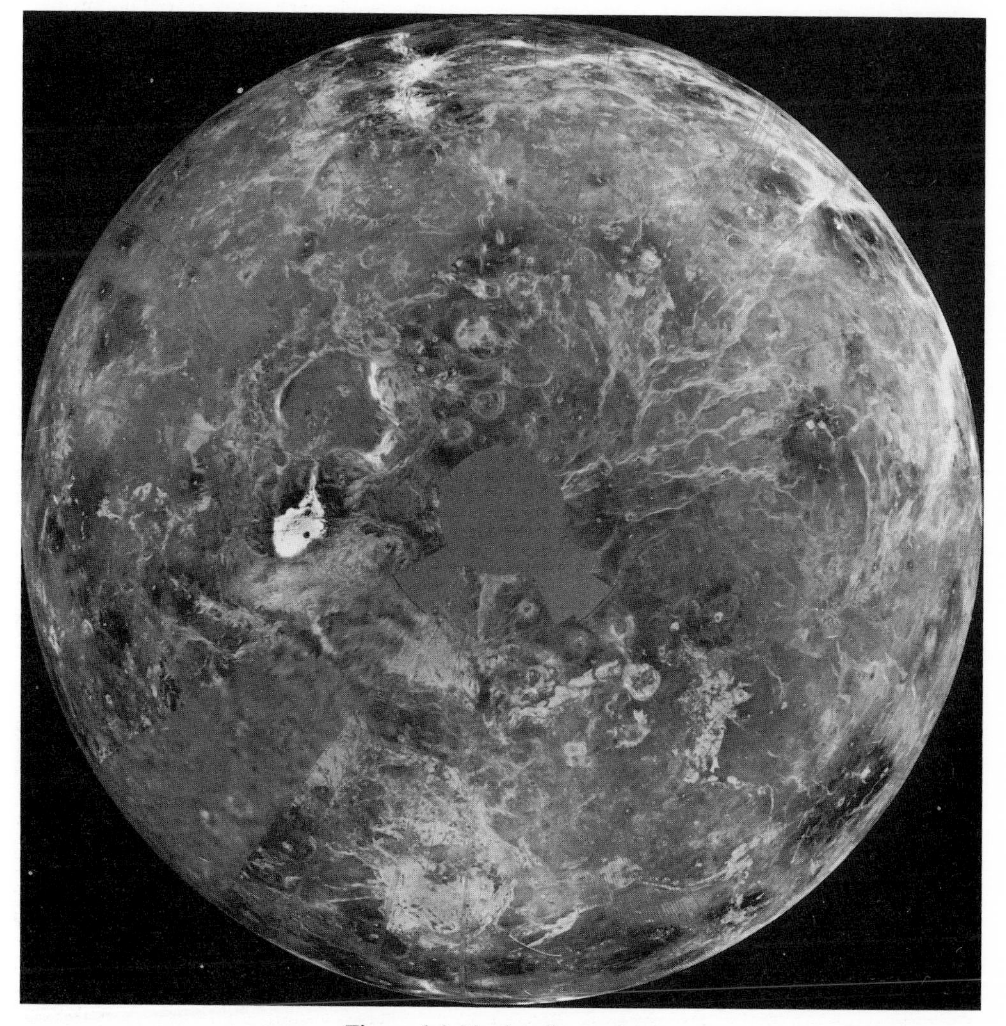

Figure 6.4 North polar region.
This polar projection of Magellan imagery shows the ridge belts fanning out from the north pole, while the radar-bright rim of Lakshmi Planum and Maxwell Montes are both clearly visible. Physiography from Venera 15/16 data where Magellan imagery non available. From Magellan image P-39224.

6°S) and Thetis Regio (130°E, 10°S) to Atla Regio (200°E, 3°N), separated by deep rifts, such as Diana, Dali and Artemis Chasmata.

To the north of Aphrodite are the plains of Leda, Niobe and Rusalka Planitiae; the isolated upland massifs of Bell and Tellus Regiones rise on either side of the first of these. To the south of Aphrodite lie the extensive plains of Aino Planitia. Eastwards from Aphrodite a further series of chasms extend towards the north–south trending highland complex of Beta–Phoebe–Themis Regiones, massive volcanic complexes that have grown along a north–south system of chasmata. Two of the larger volcanic structures – Rhea and Theia Montes (located in Beta Regio) – show evidence of having collapsed, the former into the deep rift on its eastern flank, and the latter into a 3 km deep summit caldera. Theia is also cut by an east–west rift (Hecate Chasma), which joins Asteria and Ulfran Regiones, both west of Beta Regio. The more northerly massif is joined to neighbouring Phoebe Regio by the 2500 km long, 90 km wide rift of Devana Chasma. At its deepest point this descends 2.5 km below datum.

The most recently discovered highland region, Lada Terra, is situated largely south of latitude 50°S and is bordered on the north by the lowland plains of Helen, Lavinia and Aino Planitiae. Alpha Regio, some 3500 km to the north, is joined to Lada by a complex of deep rifts with raised rims.

The northern polar region is essentially a broad plain cut by tectonized zones and surrounded by highland and tessera terrain (Fig. 6.4). The ridge belts that traverse it are several hundreds of kilometres in length and tens of kilometres in width. The longest continuous belt runs along the 200° meridian from Atalanta Planitia into the polar region and then turns south along 80°E, where it abuts against Ishtar Terra. Much of the southern polar region still remains to be imaged.

6.2 Physiography of highland regions

The Earth's continents cover approximately 30% of the globe; in contrast, the most continent-like massifs on Venus – the highlands – which rise on average 4 to 5 km high above datum, cover a mere 10–15% of the planet's surface. Despite their relatively small extent these complex regions are particularly intriguing and it is unfortunate that, to date, rock analyses from highland locations are restricted to basaltic plains units from eastern Aphrodite Terra, which were analyzed by Vegas 1 and 2 (Crumpler 1990). However, it is generally assumed that, like terrestrial continental rocks, their composition is more evolved chemically than those of the basaltic lowlands analyzed by Venera and Vega probes. Significant perhaps is the observation that at least one of the continent-sized highlands is encircled by steep escarpments, which are not dissimilar to terrestrial continental slopes.

6.2.1 Ishtar Terra

The highest point on Venus, situated within Maxwell Montes, rises 12 km above mean datum and is situated towards the eastern end of the Australia-sized highland of Ishtar Terra (Fig. 6.5). It forms the core to a massive elevated region with a particularly steep western flank. The unusually high degree of polarization of radar echoes received from this area at the Arecibo Observatory is doubtless attributable to this steepness. The most recent Magellan data indicate that, in a horizontal distance of only 10 km, the southwestern flank of Maxwell rises 7 km, a slope of 35° (Fig. 6.6). However, it falls away more gently eastwards towards the highly deformed area known as Fortuna Tessera, which extends eastwards from the 30° meridian for a distance of nearly 2000 km.

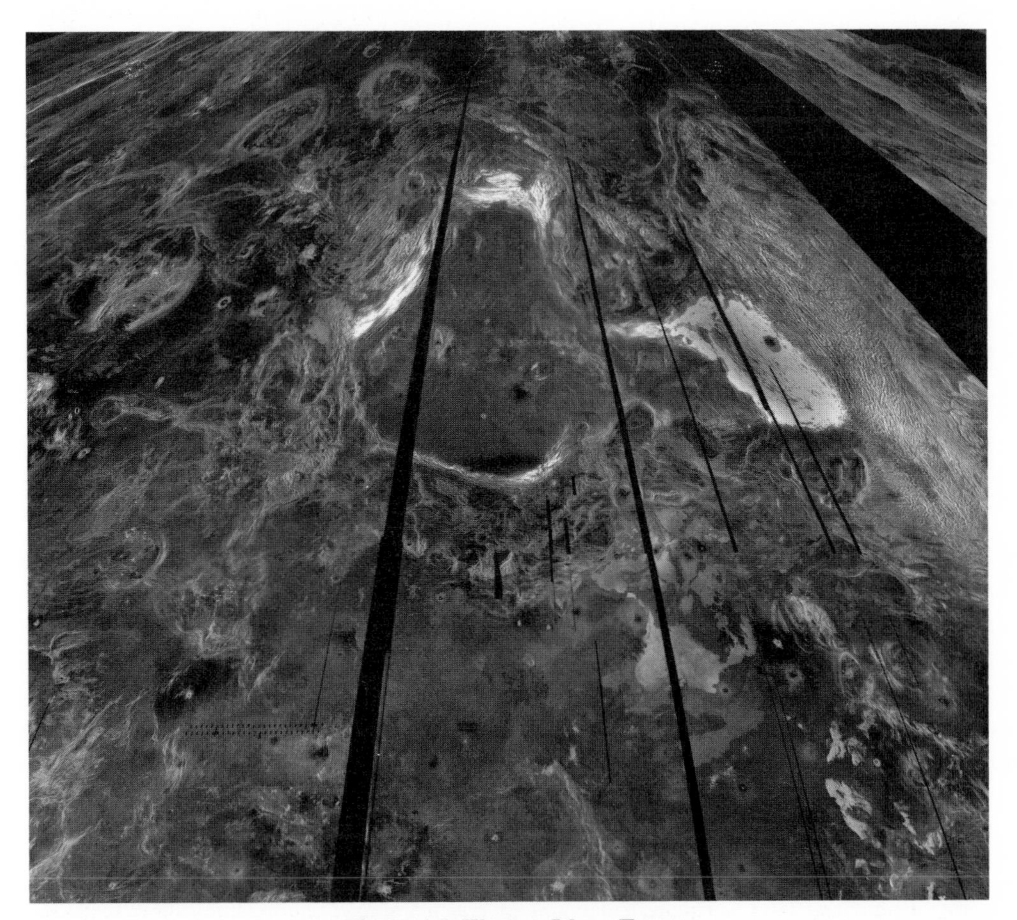

Figure 6.5 Western Ishtar Terra

The radar-dark upland plateau of Lakshmi Planum has a radar-dark signature. The surrounding radar-bright mountain belts and Maxwell Montes (right side of image) are prominent. From Magellan image C2-MIDRP.60N333E.

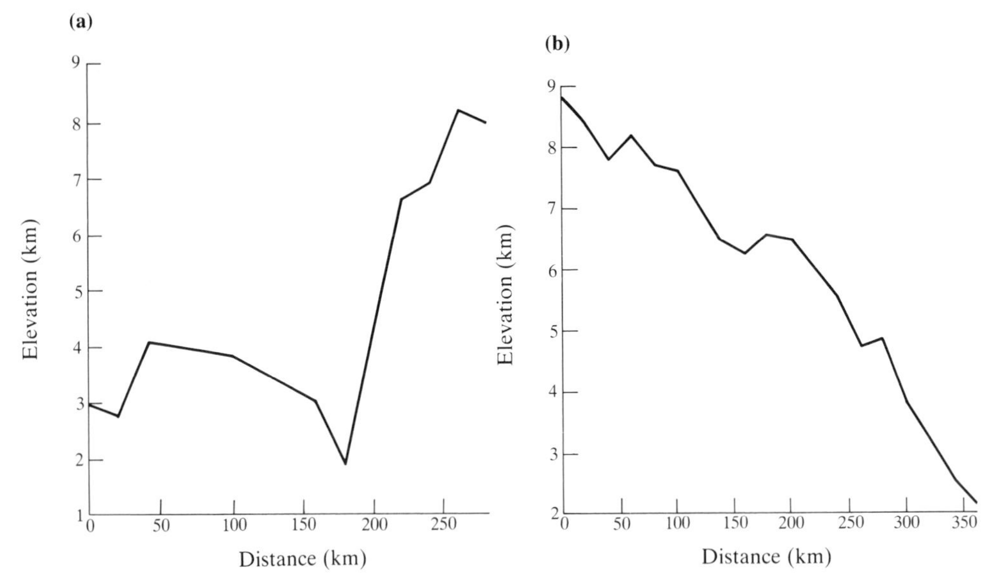

Figure 6.6. Altimetric profiles across Maxwell Montes
(a) northwest flank; (b) southeast flank. After Smrekar & Solomon (1992).

The western side of Ishtar is rather different, comprising the elevated plateau, Lakshmi Planum, which is bounded by curvilinear belts of high mountains. The plateau surface rises 4 km above the adjacent plains and covers an area of two million km^2, approximately twice that of the Earth's Tibetan Plateau. The surface is rather featureless at medium resolution, and is broken by two major shield volcanoes with deep calderas, Colette and Sacajawea Paterae, whose floors lie 3 and 2.5 km below the plateau level and which have diameters of 100 and 200 km respectively. These dominate the volcanic and deformational history of central Lakshmi Planum. The encircling mountain belts – Danu, Akna, Freyja and Maxwell Montes – form the north, west, south and east boundaries respectively, while the steep escarpments of Vesta Rupes, to the southwest, and un-named scarps to the southeast and northeast sharply define its limits. External to the encircling mountain belts are elevated regions of tessera that extend down slope over distances of between 100 and 1000 km.

6.2.2 Aphrodite Terra – the Equatorial Highlands

Aphrodite Terra is the most extensive highland region on Venus, spanning the equator between the 45° and 210° meridians and exhibiting considerable variation in physiography along its length (Fig. 6.7). This variation is paralleled by gravity data that indicate that apparent depths of compensation increase from west to east (Kiefer & Hager 1992). Western Aphrodite is occupied by the highland massifs of Ovda and Thetis Regiones, which rise between 3 and

(a)

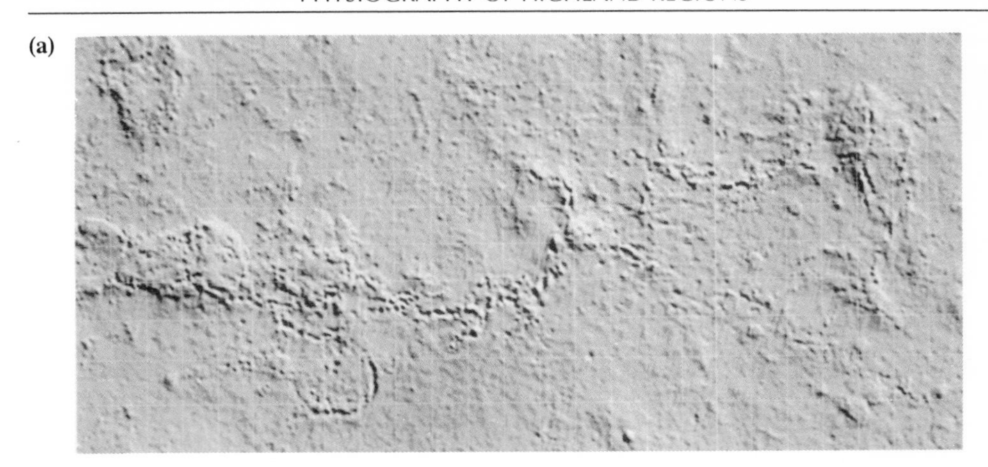

(b)

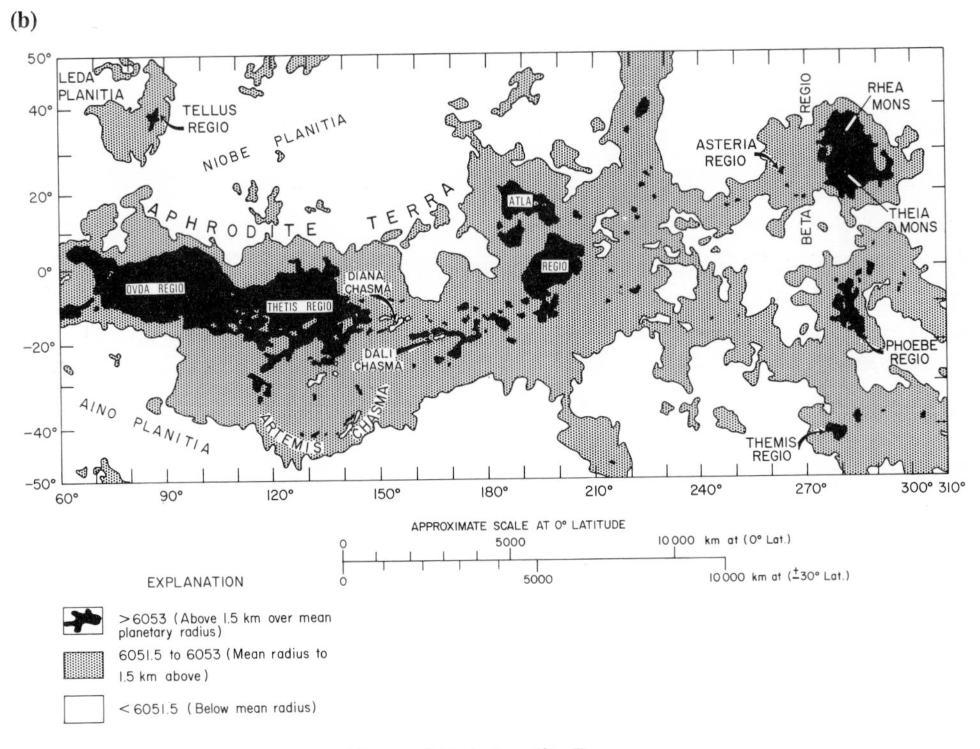

Figure 6.7 Aphrodite Terra

(a) Shaded relief map from Pioneer-Venus altimetry. (b) General altimetric map. After Schaber (1982).

4 km above the datum, and extend from 45° to 140°E, a distance of about 6000 km. The uplands are traversed by linear WNW-striking rifts, represented by Ix Chel, Kuanja and Virava Chasmata. Magellan has confirmed that both massifs are complexly deformed, a suspicion

Figure 6.8 Ovda Regio

Compressed Magellan image of Ovda Regio and the complex region to its west. The extreme brightness is a function both of metre-scale surface roughness and highly reflective surface materials. From Magellan C2-MIDRP 00N080;1.

founded on Pioneer-Venus imagery, while their margins are often highly fractured. Their interiors are a complex of blocks, broad domes, linear ridges and troughs, which generate predominantly radar-bright signatures indicative of rough surfaces. However, the floors of many grabens are covered by radar-dark materials presumed to be volcanic flows.

Ovda Regio is a high plateau 3000 $\times$ 2000 km in extent and has a rugged surface where elevation changes of over 1 km may occur over horizontal distances as small as a few tens of kilometres (Fig. 6.8). Centred on the highest point is a 35 mgal gravity anomaly – considerably less than those observed at Beta and Atla Regiones – which suggests an apparent compensation depth of 70 $\pm$ 7 km (Smrekar & Phillips 1991). The predominant terrain type is complex ridged terrain, i.e. Ovda is a region of tessera structure. Indeed, the same terrain extends to the west of Ovda for at least 1500 km into the region that the escarpment of Hestia

Rupes bounds on the north, and may well stretch for a further 500 km (Bindschadler et al. 1990b), making it even more extensive than Fortuna Tessera.

Thetis Regio is also a highland plateau but has somewhat less steep margins than Ovda and, furthermore, is highest not at the centre but close to its margins; it more closely resembles Ishtar Terra than nearby Ovda. The complex ridged terrain of the interior is embayed by radar-dark plains in several places; furthermore, a large embayment of this type exists to the north, extending for nearly 1000 km in that direction (Fig. 6.9). However, the gravity data are significantly more complex for Thetis than Ovda and no single interpretation is yet accepted by the scientific community. Solomon et al. (1992) are of the opinion that the tectonic framework exposed here is consistent with an early phase of crustal shortening, which was followed by more modest extensional deformation. On the other hand, Bindschadler et al. (1992a) favour the idea that the Venusian mantle undergoes downward circulation beneath these plateau-like highland regions.

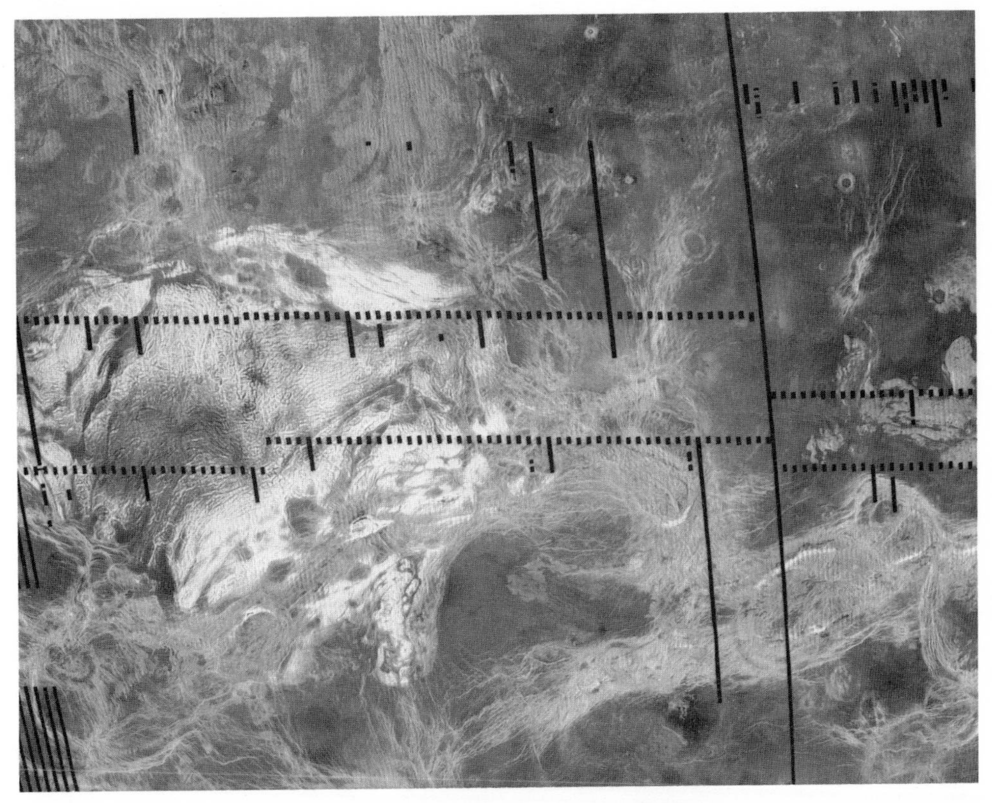

Figure 6.9 Magellan image of Thetis Regio

The image measures 2850 km × 2250 km, and is centred at 130°E, 12°S. The complex ridged terrain is partly embayed by dark plains-forming materials in the southwest sector. The radar-bright linear features towards the southeast corner of the image extend eastward into Dali and Diana Chasmata. From Magellan C2-MIDRP 00N131;1.

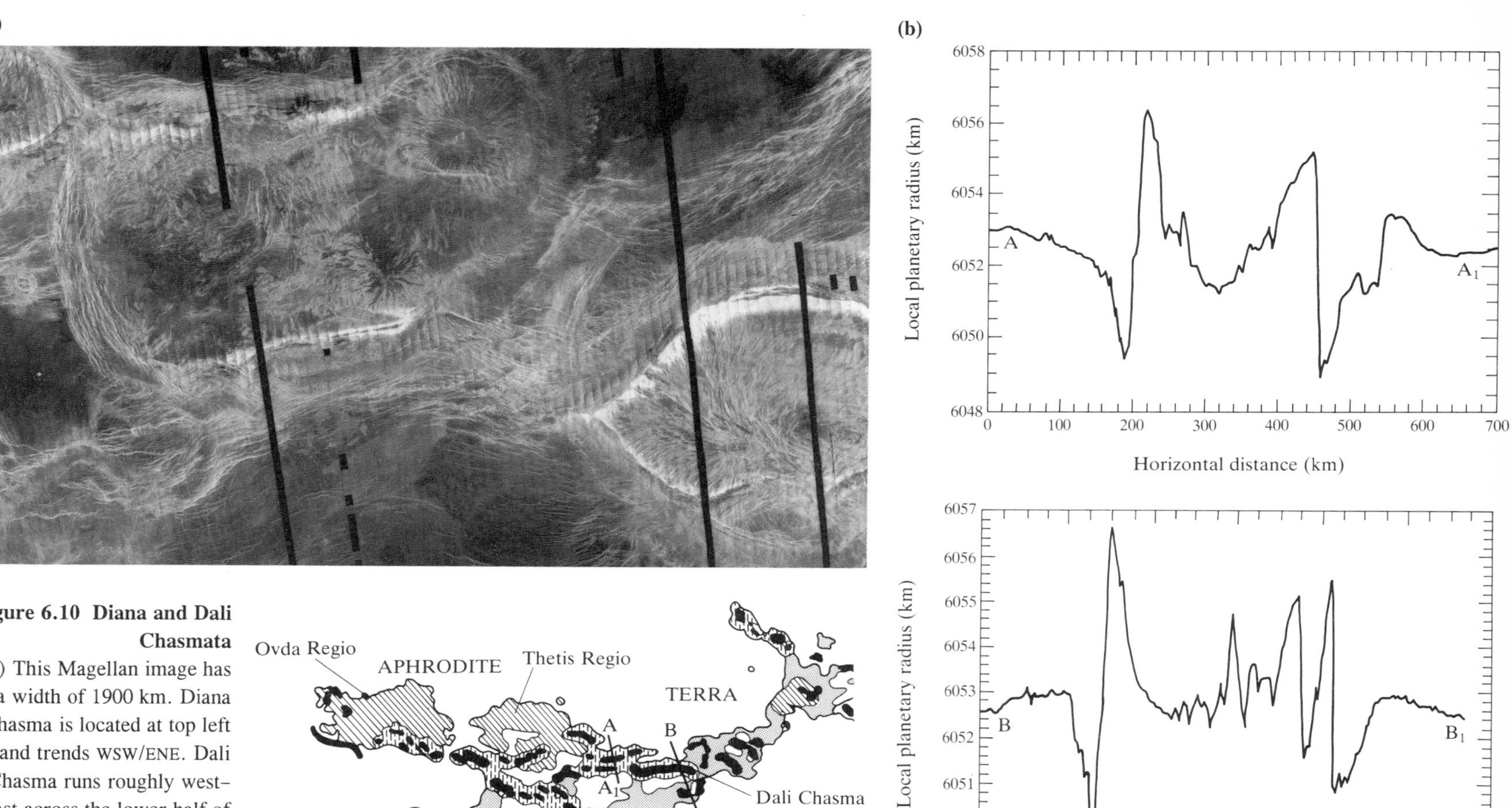

Figure 6.10 Diana and Dali Chasmata
(a) This Magellan image has a width of 1900 km. Diana Chasma is located at top left and trends WSW/ENE. Dali Chasma runs roughly west–east across the lower half of the image. Note the two coronas: Miralaidji (upper centre) and Latona (lower right). From Magellan C1-MIDRP 15S163:1. (b) Topographic profiles. After Ford & Pettengill (1992).

To the east of Thetis Regio, elevations seldom exceed 1.5 km and this is so for a further 3000 km. Aphrodite's central region is characterized by a series of ENE-trending troughs and ridges. Particularly imposing topographically are the troughs of Diana and Dali Chasmata, the former striking ENE, the latter roughly west–east (Fig. 6.10). As is typical of such troughs, one side is flanked by a high ridge comparable in height to the trough depth. Within these rifted zones, elevation differences of 7 km are not uncommon over horizontal distances of only 30 km, meaning average slopes of the order of 13°.

The eastern part of Aphrodite is occupied by Atla Regio with an average elevation of 3 km and which is a broad upland dome on whose surface are many rifts and volcanic centres. Above its surface rise the imposing volcanic peaks of Ozza and Maat Montes. Ozza Mons (6 km high and 300 km across) has a central area of radar-dark material 100 km across on whose surface are many pits, domes and volcanic flows; Maat Mons (9 km high and 200 km across) has several summit calderas (Fig. 6.11, and see Plate 8). A plethora of flows and radial graben are associated with the former, while the latter has the flows but lacks the radial faults. A third similar but smaller volcanic centre, Sapas Mons, is situated farther to the west. Not surprisingly, the surface of Atla is dominated by volcanic deposits and therefore differs significantly from the highland regions of western Aphrodite. Associated with Ozza Mons is a 130 mgal gravity anomaly from which Smrekar & Phillips (1991) have deduced a depth of compensation of around 200 km.

Figure 6.11 Perspective view of Maat Mons

Maat Mons is a 5 km high volcano with a summit caldera complex. This view, looking south, shows radar-bright volcanic flows extending northwards and embaying ejecta from a 23 km diameter impact crater. Magellan image P-40700.

While the volcanoes are undoubtedly imposing, the geological characteristic of greatest prominence is the presence of major rifting. Of particular interest is the 1000 km long Ganis Chasma, which connects Maat Mons to Nokomis Montes, located to its northeast. This fracture system, composed of many individual graben 1–10 km wide, has a maximum width of 300 km and shows a relief difference of between 1.0 and 1.5 km with the surrounding surface. The individual faults cross-cut one another, forming a branching network. Ganis is, however, just one of many topographically prominent discontinuities with a general NW strike that are visible in both altimetric and radar-imaging data of eastern Aphrodite (Crumpler 1990). Interestingly, the morphological similarity that these features show to the faults along terrestrial divergent plate boundaries have led some workers to the view that here we have a manifestation of terrestrial-style crustal spreading (Schaber 1982, Crumpler 1990).

South of central Aphrodite Terra lies the curvilinear Artemis Chasma. This appears to be a massive corona structure rather than a part of a regional fracture system.

6.2.3 Tellus and Bell Regiones

Three thousand kilometres to the north of Ovda Regio is the elevated plateau of Tellus Regio, which rises to around 2 km. With fractured margins and a radar-rough surface, it is similar to both the Ovda and Thetis highlands, even in having embayments of smoother radar-dark plains within its complexly ridged terrain. However, although there is a weak positive gravity anomaly of 15 mgal associated with the eastern part of Tellus, the gravity data differ significantly from other such highlands in that the major anomalies in its vicinity are strongly negative (–30 mgal over the lowest parts of Leda Planitia immediately to the north and west). Bell Regio, which sits west of Tellus at 32°N, 50°E, rises to between 1.5 and 2.0 km, and hosts a 500 km diameter volcanic shield (Tepev Mons) and the 370 × 295 km corona structure, Nefertiti. Like Atla Regio, this is more by nature a volcanic rise than a plateau structure, and both have been interpreted by Bindschadler et al. (1992a) to lie above mantle plumes.

6.2.4 Beta Regio

This 2000 × 3000 km highland is traversed by the north–south trending trough of Devana Chasma, characterized by at least 6 km of relief in places, and generally interpreted as being an extensional fault complex comparable with the East African Rift (Fig. 6.12). There is a large free-air positive gravity anomaly beneath the region, suggesting an **apparent depth of compensation** of the order of 300 km.

Theia Mons lies at the southern end of Beta Regio, and is a huge volcano that Magellan imagery reveals to be superimposed on the rift; however, the structure is also cut by later

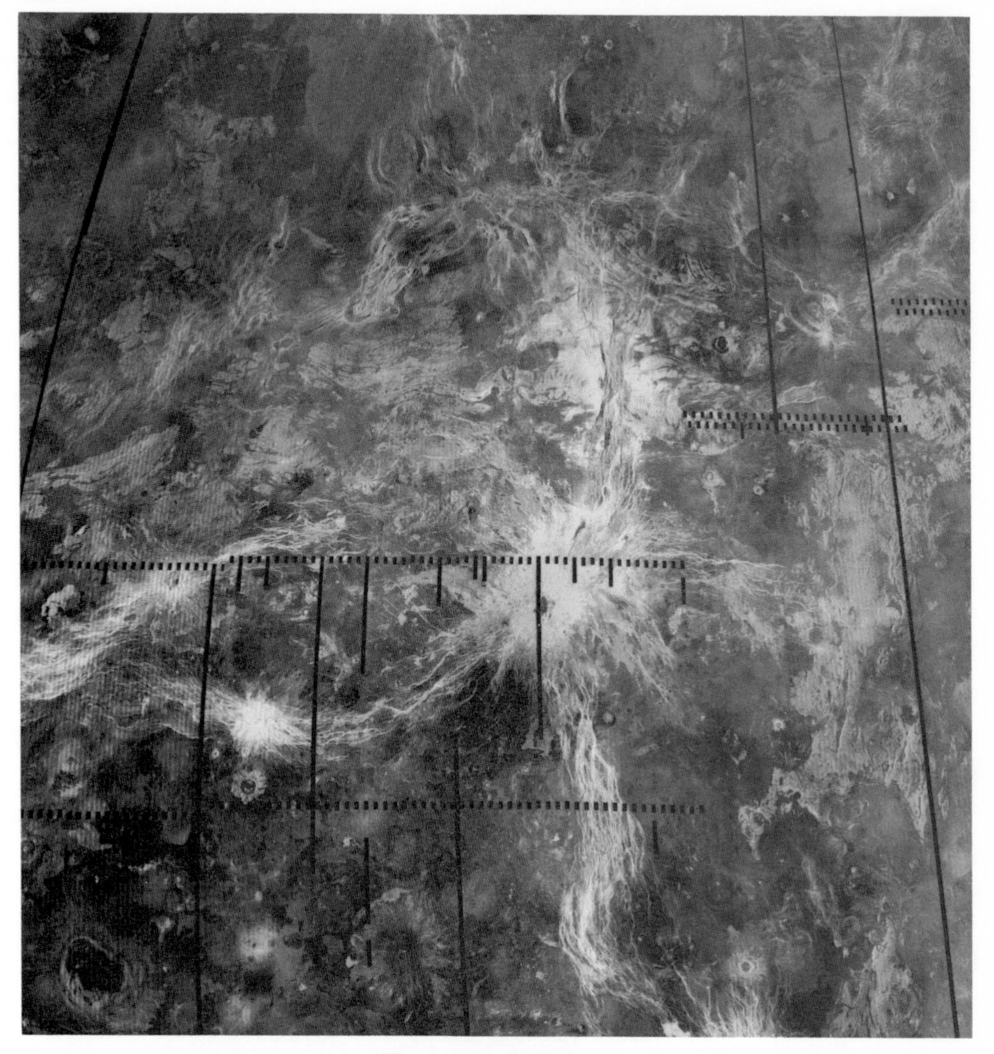

Figure 6.12 Beta Regio

Compressed Magellan image of Beta Regio. Theia Mons is the radar-bright feature located below centre at the left. Devana Chasma runs to the north and south of it. Note the other rifts and also the tessera blocks to the east. Magellan C2-MIDRP 30N264;1.

faults. With a diameter of 350 km and an altitude of 5 km, Theia is the focus of a plethora of long lava flows that partly infill the fault trough (Stofan et al. 1989). Theia also sits at the junction of features with different tectonic trends, the most obvious being the west–east trending Hecate Chasma. Not surprisingly, this has led some workers to draw parallels with terrestrial rifts, where rift faults are propagated and may eventually link up between individual regions of hotspot-related magmatism. Rhea Mons, located in the north of the highland region, formerly was interpreted from Arecibo and Venera images also to be a volcanic structure, hav-

(a)

(b)

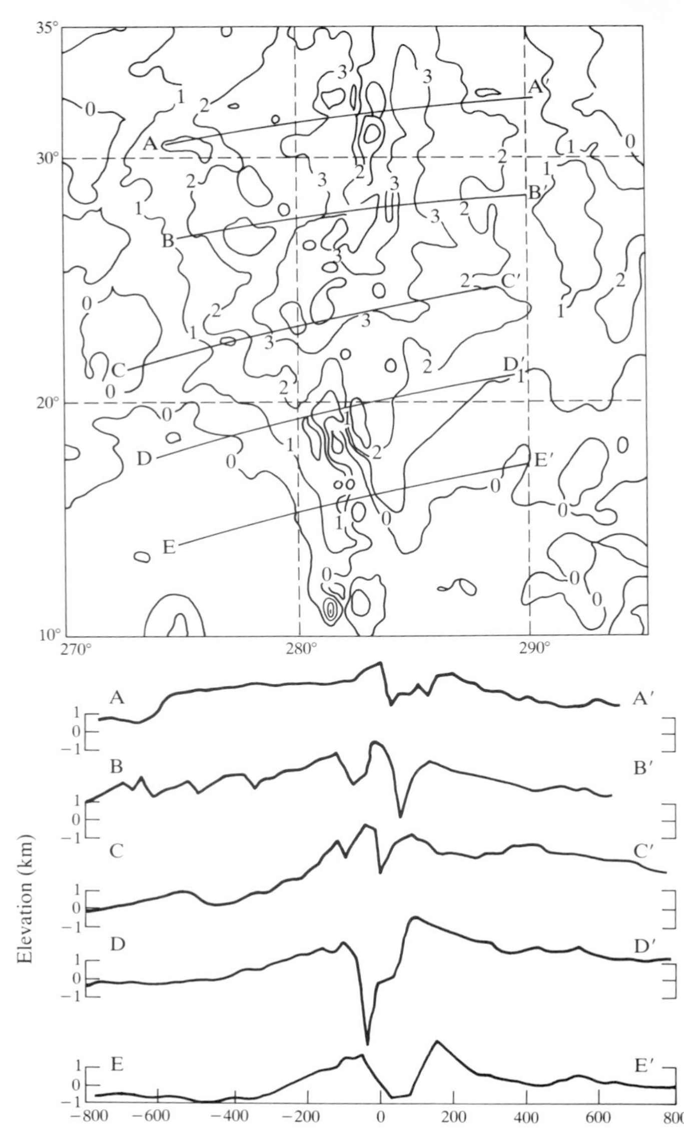

Figure 6.13 Magellan image and cross sections of Devana Chasma, Beta Regio

(a) Mosaic of the region. (b) Profiles constructed from altimetry measurements located within 15– 25 km of evenly spaced points along a great circle. After Solomon et al. (1992).

ing a shield-like form with similar radar back-scattering signatures to Theia. However, Magellan reveals the most elevated part to be an uplifted area of tessera, which, at the summit, is mantled by smoother volcanic deposits. What previously had been envisaged as volcanic flows associated with Rhea now seem more likely to be lobate areas of highly deformed tesserae whose similarity of form understandably confused earlier interpretations.

Between the two topographic highs, faulting occupies a broader region than across their flanks. Near Rhea the fault trough is relatively deep ($>$ 2 km) and narrow (80 km), whereas in northern Beta Regio the trough is broader (130 km wide) and there are higher-standing horsts in the rift centre. South of Rhea, that is between Beta and Phoebe Regiones, Devana Chasma has as much as 6 km of relief and is flanked by well defined high shoulders (Fig. 6.13).

There are other differences within the region: for instance, in the central and southern parts, the main rift is surrounded by fractured volcanic plains; to the north, however, complex ridged topography is encountered, that is, tessera terrain. To the northwest of Rhea Mons, the tessera region is dominated by NE-trending grooves and ridges spaced 1– 2 km apart, but the trend of these structures swings to become more nearly north–south to the east of Devana Chasma. It is also cut by normal faults and graben, which fan out from the main rift, so that, in places, tessera blocks can be identified on the floor of the down-faulted region. Furthermore, in several localities the tessera is embayed by radar-dark volcanic plains. To the north of Beta Regio is a chain of large coronas, which are linked by graben faults. These are younger than fractures splaying out northwards from Devana Chasma, and may represent a later phase of volcano-tectonic activity in this region. To the south of Beta Regio the roughly north–south tectonic line is continued through first Phoebe, then Themis Regio and Tefnut Mons, which, as more Magellan imagery is returned, may well be revealed to have volcanic origins too.

6.2.5 Types of highland terrain

Current knowledge suggests that we can view the Venusian highlands in terms of three kinds:

- *Beta-type highlands* – broad topographic rises showing moderate extensional deformation, shield-style volcanism and deep levels of compensation (e.g. Beta, Eistla, Atla and Bell Regiones);
- *Ovda-type highlands* – topographic plateaux distinguished by intense deformation, more limited volcanism and shallower depths of compensation (i.e. $<$ 100 km) (e.g. Ovda and Thetis Regiones); and
- *Lakshmi-type highlands* – elevated plateau with surrounding narrow mountain belts, moderate volcanism and deep compensation depths (Ishtar Terra is the only one).

The reasons for these differences will be explored at a later stage.

6.3 Venusian plains

In broad terms, two non-highland topographic units can be recognized: *rolling plains* and *lowlands*. The more extensive of the two – rolling plains – typically lie between 2 km and zero datum, the lowland areas below 0 km. In the early 1980s, when the only available imagery was that from Pioneer-Venus and Earth-based facilities, it was already clear that the main lowlands – Sedna, Guinevere, Aino, Niobe and Atalanta Planitiae – form an X-shaped region of low country that crosses the equator and whose eastern arms are bisected by Aphrodite Terra (see Plate 1). It was also realized that to the north of this belt of low ground lay a zone where the greatest relative elevation differences on the planet were to be found.

Besides elevation differences, rolling plains were shown to be characterized by surface scattering properties intermediate between the highlands and lowlands, and RMS slopes in the range 1.5° to 3° on the metre to dekametre scale. Goldstone and Arecibo images also showed many large (20–300 km diameter) circular structures with low rims, many of which were suspected to be impact craters (Rumsey et al. 1974, Goldstein et al. 1976, 1978). Several large radar-dark circular features seen on the same imagery were interpreted to be impact basins (Saunders & Malin 1977). All were characterized by low depth : diameter ratios.

With the advent of Veneras 15/16, it was discovered that plains also covered the north polar region and that long and relatively narrow deformed zones traversed many of the rolling plains units (Barsukov et al. 1986). It was also possible to refine data on Venusian impact craters superposed on the plains units (Ivanov et al. 1986). The improved resolution revealed a host of new detail, and showed that many of the circular features imaged from Arecibo and believed to be of impact origin were actually volcano-tectonic structures; these were termed coronas. Their discovery prompted considerable interest in Venusian plains deposits, since they were a landform type found nowhere else in the Solar System (Sukhanov et al. 1989, Head 1990a, Frank & Head 1990).

Lowlands, with their darker and smoother signatures, have been likened to lunar maria and interpreted as lava-flooded basins (Masursky et al. 1980). Because they also lack the significant numbers of impact craters recognized on the rolling plains and had small gravity lows recorded over them, at this time they were widely accepted as being regions of thinned low-density crust covered in basalt-like lava flows. It also transpired that not only were impact craters present, but also volcanic flows, domes and shields, and extensive systems of ridge belts (Barsukov et al. 1986). What the improved resolution of the Venera data had shown was that the lowlands were much more complex than hitherto had been realized. Atalanta Planitia – the deepest regional depression on the planet – rather than being simply a featureless lava basin, is an extensive depressed plains area, the greater part of which is traversed by ridge belts that continue across the north polar region (Fig. 6.14). However, the western part, adjacent to Tellus Regio, is rather different and is host to clusters of dome-like hills and a network of radar-bright ridges. Indeed, the ridge belt crossing Atalanta turns out to be the

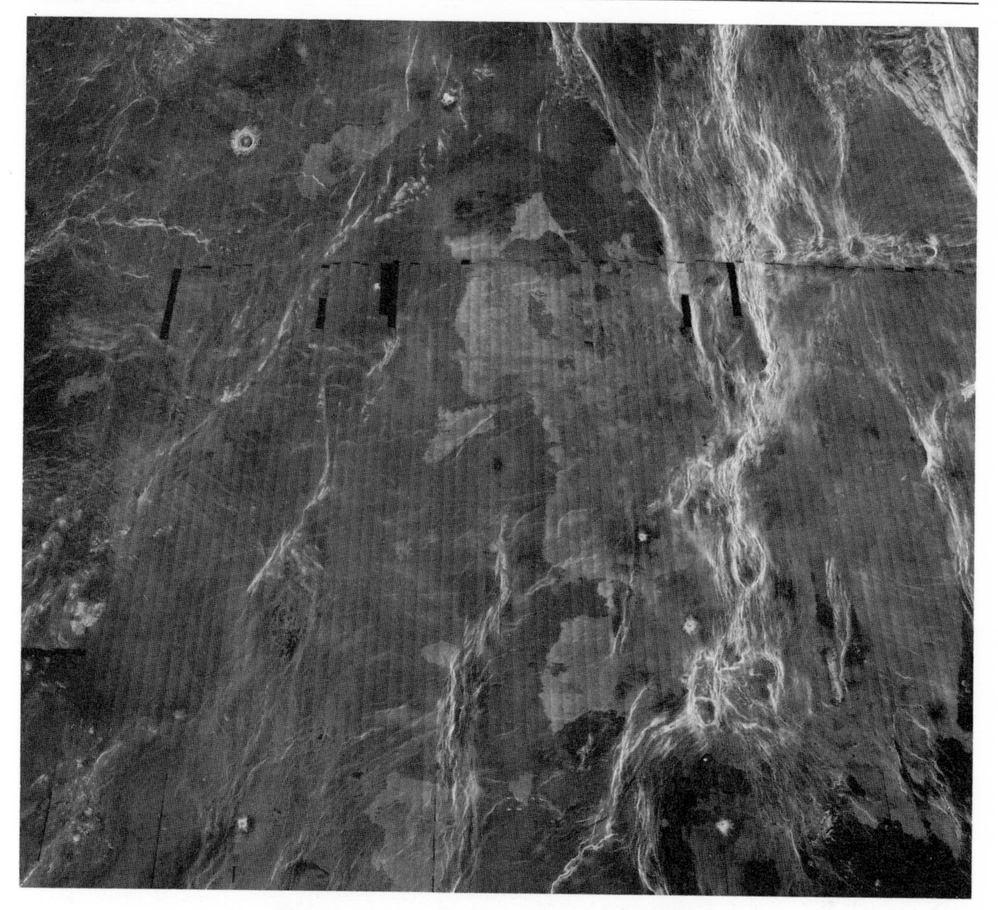

Figure 6.14 Atalanta and Vinmara Planitiae
A compressed Magellan image 2550 km × 3150 km across. The lowland is dominated by radar-dark plains units with wrinkle ridges. Magellan C1-MIDRP 60N180;1.

longest and most prominent on Venus. Guinevere Planitia is in many respects similar to Atalanta but also has several large coronas on its surface; additionally, in the south, Guinevere becomes rather hummocky, as though older units are modified or buried by plains materials. Similar developments occur widely within the plains elsewhere. Where this occurs, the elevation difference between what clearly are volcanic plains and hummocky plains is generally in the 0.5 to 1 km range. This has led several workers to surmise that, where hummocky plains development is found, we are seeing the effects of mantling of older tessera units by younger plains materials (Nikishin 1990). This, in turn, raises the interesting question of whether or not the volcanic plains of Venus, in general, rest on an older basement of fragmented tessera material.

A dramatic increase in the study of the plains accompanied the arrival of Magellan data (Head et al. 1991, Saunders et al. 1991, Solomon et al. 1991). Fine detail on the plains' sur-

(a)

Figure 6.15 Types of volcanic plains on Venus
(a) Smooth plains near Artemis Chasma; (b) reticulate plains; (c) gridded plains; and (d) lobate plains. Magellan images (a) F-MIDRP 40S145;1; (b) F-MIDRP 25N119;1; (c) P-36699; (d) F-MIDRP 05S177;1.

(b)

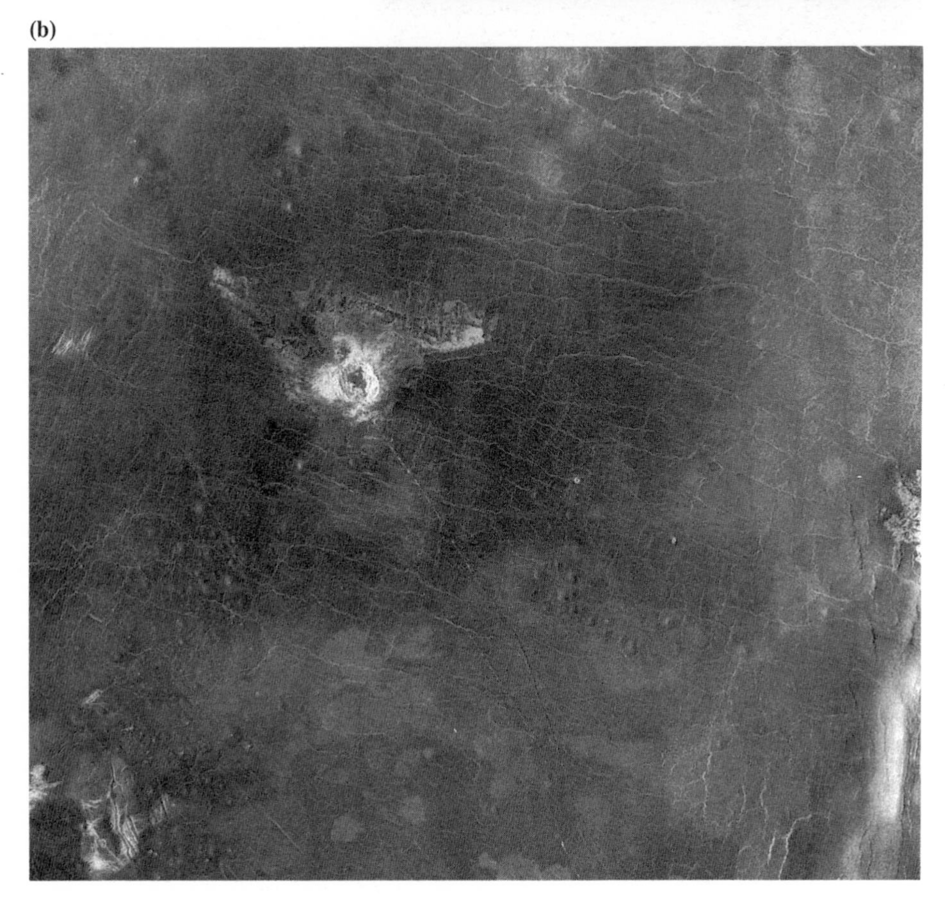

(c)

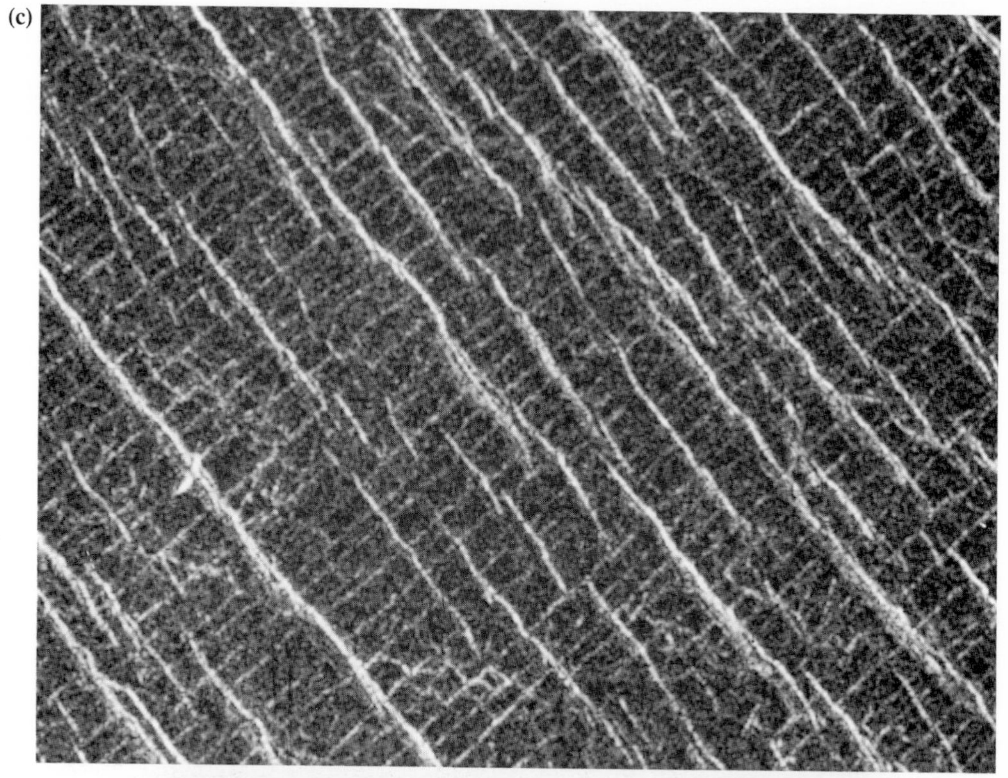

(d)

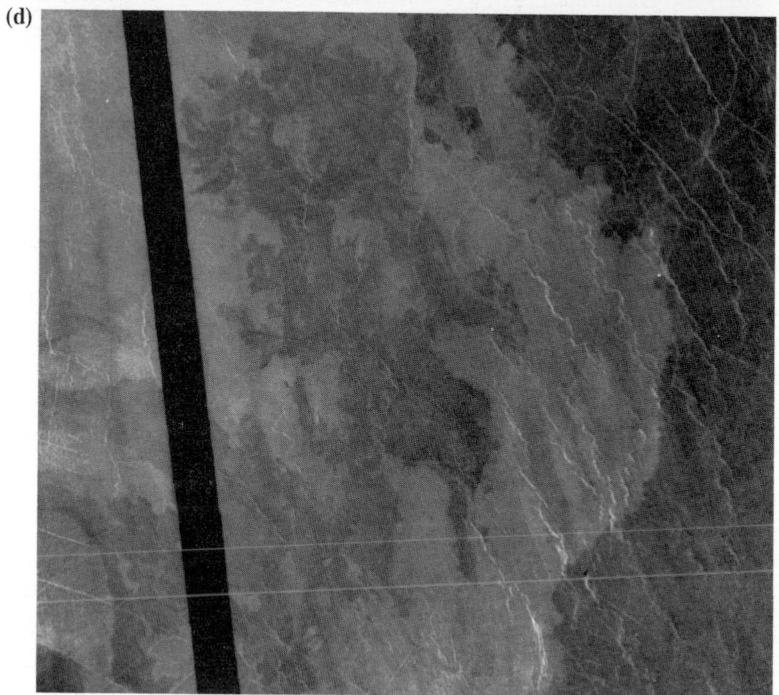

faces was unveiled for the first time, as was the highly complex volcanic and tectonic history of coronas and related features. In the initial Magellan Mission summary, Saunders et al. (1991) recognized four major plains types. *Smooth plains*, as the name implies, are relatively featureless, have no discernible volcanic flow features and also few linear structures or dome-shaped hills. They tend to be radar-dark but range up to moderately radar-bright. These are seen as having an origin in volcanic flooding, presumably by very fluid lavas or by the coalescence of a plethora of low volcanic shields. *Reticulate plains* are recognized by having one or more sets of somewhat sinuous, radar-bright lineaments, which often cannot (at the radar resolution of Magellan) be identified positively either as ridges or grooves; whichever these are, however, they are spaced on average less than 5 km apart. Their morphology suggests an origin in volcanic flows or low shields that either have embayed older units, or have been tectonically deformed, or both. *Gridded plains* are rather distinctive in having intersecting orthogonal sets of radar-bright lineaments, regularly spaced, which extend for hundreds of kilometres. The spacing of these features tends to be closer than those in reticulate plains and typically is less than 5 km. Complex deformation is a characteristic. *Lobate plains* comprise overlapping lobate flow features with variable radar signatures, which extend for tens to hundreds of kilometres. Such plains are traversed by few if any linear structures. Fractures, where developed, and local topography appear to have controlled the emplacement of the lobate plains materials, presumed to be complexes of volcanic flows. The morphological characteristics of each of the main plains types is illustrated in Figure 6.15.

Detailed study of plains in the region of Lavinia Planitia reveals that most of the different types of plain are traversed by *wrinkle ridges* (Squyres et al. 1992), which show up as radar-bright lines. They generally are less than 1 km wide, although a few are much wider (several kilometres); in length they range between a few tens of kilometres to in excess of 100 km. As a rule, individual ridges are located between a few kilometres and 20 km apart. As is the case on both the Moon and Mars, where similar structures abound on the plains units, the ridges show quite strong preferred orientations within different regions, as can be seen on Figure 6.16. In the region of Lavinia Planitia, Squyres et al. (1992) observe that the ridges decrease in concentration from older to younger units, while the trend of the ridges appears to be related to location rather than age. This implies that ridge formation – presumably by compressional forces – was, and may still be, an ongoing process connected to local stress fields that operated while plains units were being laid down.

Other common radar-bright lineaments that cross the plains as parallel families tend to be longer and less sinuous than wrinkle ridges. Where resolution allows, these can be identified as narrow grooves that may extend for anything between 25 and 200 km across the Venusian surface, with spacings of 30–100 km. In contrast to wrinkle ridges, these are extensional features and many (perhaps most) appear to be *graben*. As with the ridges, they tend to share a common strike over quite large areas, but always lie normal to the ridges, which gives the plains of such regions as Lavinia and Guinevere Planitiae a distinctive orthogonal imprint.

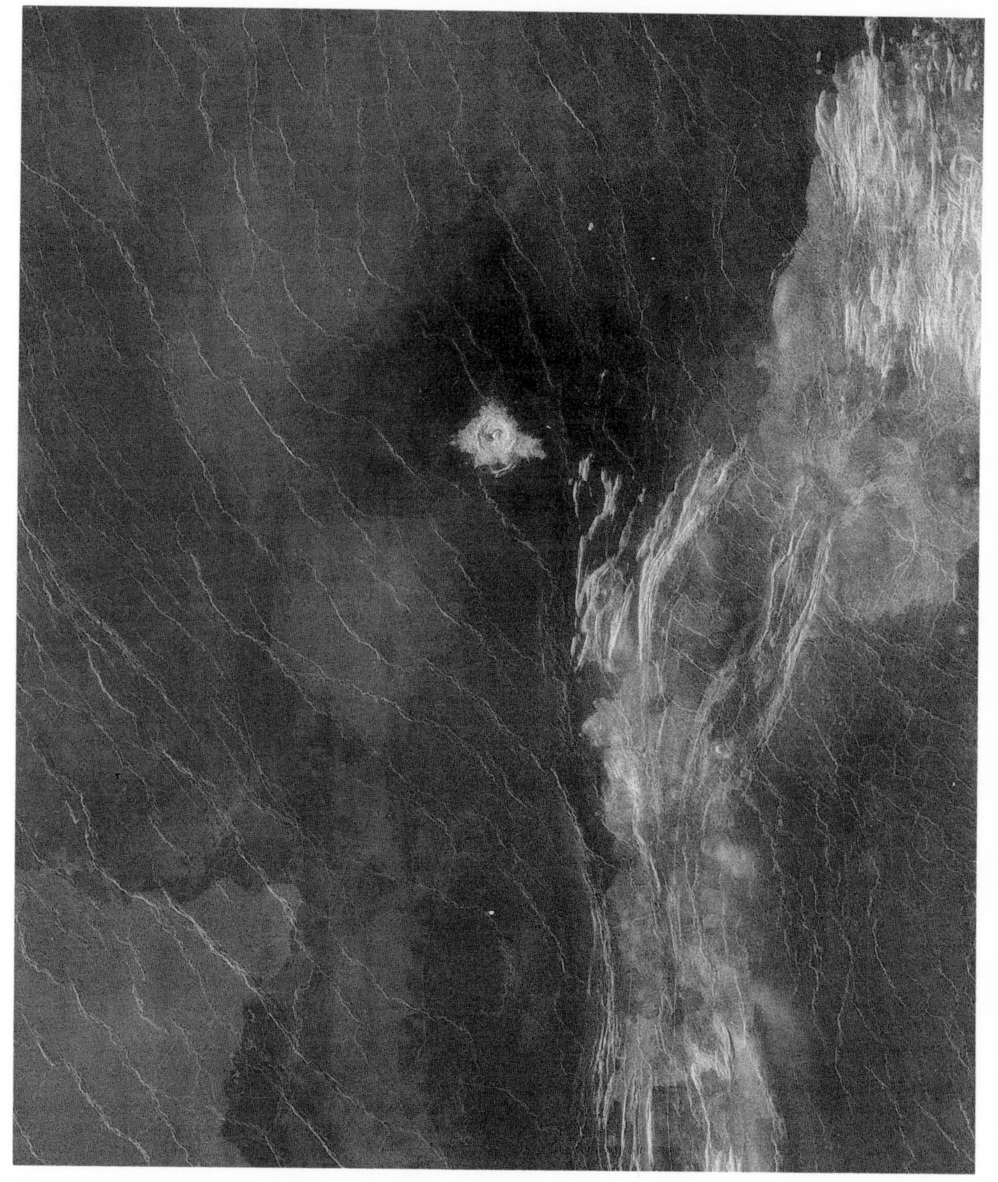

Figure 6.16 Wrinkle ridges on mottled plains
Magellan image of a part of Lavinia Planitia showing radar-dark plains with small domes and shields. Magellan image F-MIDRP 05N177;1.

With the benefit of all the available imagery, there can be little doubt that volcanism has been responsible for the generation of the Venusian plains; that is, over three-quarters of the planet's surface. The kinds of primary structures developed – long flow-like features, caldera depressions, domes and low shields – are typical of plains-forming basaltic volcanism

(Basilevsky & Head 1988, Barsukov et al. 1986), and a basaltic composition is also implied by the Venera and Vega lander geochemical data. In addition to the smaller-scale features, there are many larger volcanic centres, first revealed clearly by Venera 15/16, which imaged 800 volcanoes in the size range 200–1000 km diameter, distributed over just 25% of the globe. Magellan has refined interpretation of landform types and has extended this tally to 156 large volcanoes over 100 km across, 274 volcanoes in the range 20–100 km, 86 caldera-like landforms between 60 and 80 km diameter that are not associated with large shields, 550 clusters of small (< 20 km) volcanoes, called *shield fields*, 175 annular concentrations of fractures and ridges termed *coronas*, 259 *arachnoids,* and 50 *novae* (foci comprising radial fractures forming stellate patterns).

6.4 Ridge and fracture belts

Characteristically developed within the Venusian plains are linear deformational features termed *ridge belts*. These comprise narrow ridges and arches that rise a few hundred metres above the surrounding plains, are up to 300 km wide, and may extend for hundreds to thousands of kilometres as radar-bright swaths across the planet's surface. They show up dramatically in Arecibo mosaics (Fig. 6.17) and were described by Campbell et al. (1991). The zones of deformation that ridge belts represent are not distributed uniformly within the plains areas; rather, they tend to be concentrated in or near major lowlands (Solomon et al. 1992). In particular, there are two major concentrations, one in Lavinia Planitia, the other in the region of Atalanta and Vinmara Planitiae. In the former case, the belt is roughly coincident with the long-wavelength topographic low, but in the latter, it tends to sit well to the east. Other deformation belts are located southeast of Alpha Regio and south of Aino Planitia, in the southern hemisphere, and north of Bereghinya Planitia, in northern latitudes. The very broad spatial distribution of these features has important implications for the stress state within the planet, a point to which we shall return a little later on.

6.4.1 Types of belt

The belts have two regional manifestations: first, parallel or subparallel networks and fan-shaped patterns within lowlands such as Lavinia and Atalanta Planitiae and, secondly, as more nearly orthogonal networks in close proximity to raised tessera blocks. There is also a third style of development comprising long, rather narrow, belts that consist of a single sharply defined ridge, commonly with superposed smaller ridges and which are found around some large coronas.

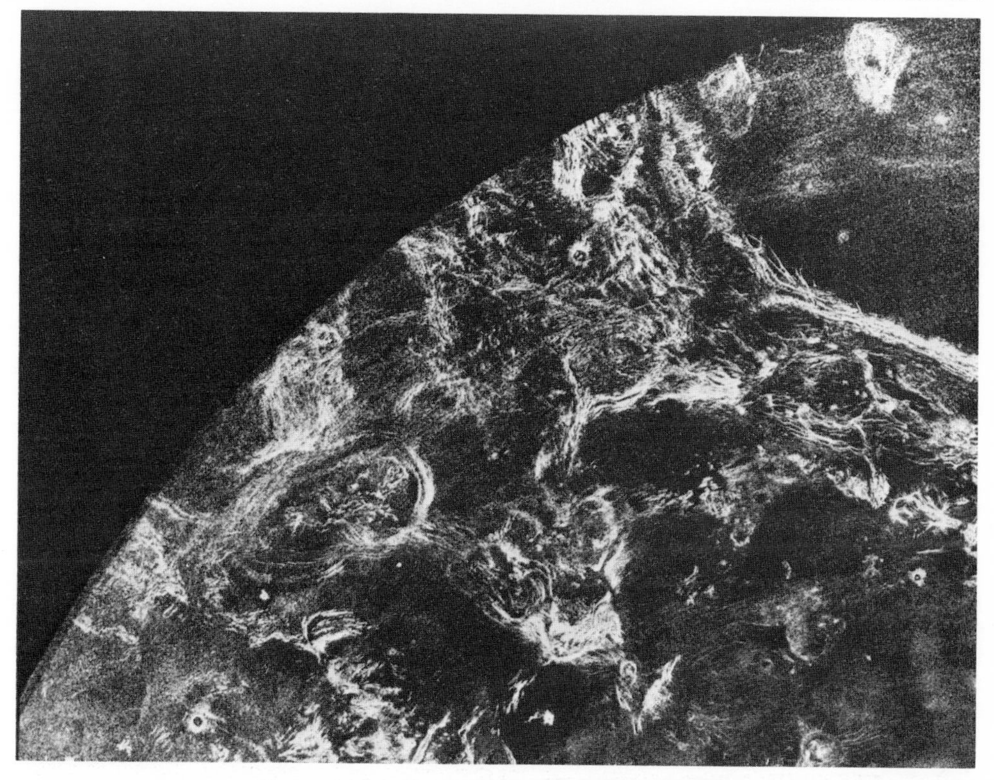

Figure 6.17 Ridge belts crossing Sedna Planitia
Arecibo mosaic showing the southwest border of Lakshmi Planum at top right, bounded by the radar-bright mountain belt of Akna and Danu Montes. The large corona, Demeter is seen at left just below centre.

The first type of ridge belt is the one that shows up on Venera 15/16 images of the northern hemisphere and which has been further subdivided by some workers (e.g. Basilevsky et al. 1986, Kryuchkov 1988, Frank & Head 1990). This type is generally sufficiently broad and has enough vertical relief to show up on most Magellan altimetric data, except those obtained with high radar incidence angles. The second type, however, typically shows little or no relief, yet the individual narrow ridges are evident whatever the incidence angle of the radar beam. To facilitate distinguishing between the two classes, it is customary to term the first type *fracture belts* and the second *ridge belts*, the latter being the "true" ridge belts of many earlier publications (Solomon et al. 1991). Frequently, the two classes of belt are found together, either in combination as a single complex belt, or as distinct belts with often different strikes. Within such composite belts, the ridges may share the overall strike of the fracture belt or may strike at some (often markedly different) angle to it.

6.4.2 The ridge belts of Lavinia Planitia

To date, the ridge belts of Lavinia Planitia are the most widely studied manifestation of these peculiarly Venusian structures. The belts that traverse the Lavinia plains are shown in Figure 6.18; they have widths of up to 200 km and may extend for distances of over 1000 km. Characteristically, they meander, bifurcate and meld along their length, and show complex cross-cutting relationships that imply a lengthy period of deformation. Furthermore, within the belts are volcanic plains units, some of which show deformation by the forces that produced the belts, while others are superimposed upon and partially bury the ridges within the belts. As Solomon et al. (1992) point out, there are also instances where a belt has been produced and then been partly buried by younger volcanics that, in turn, have been deformed by a more recent belt that cross-cuts the older one.

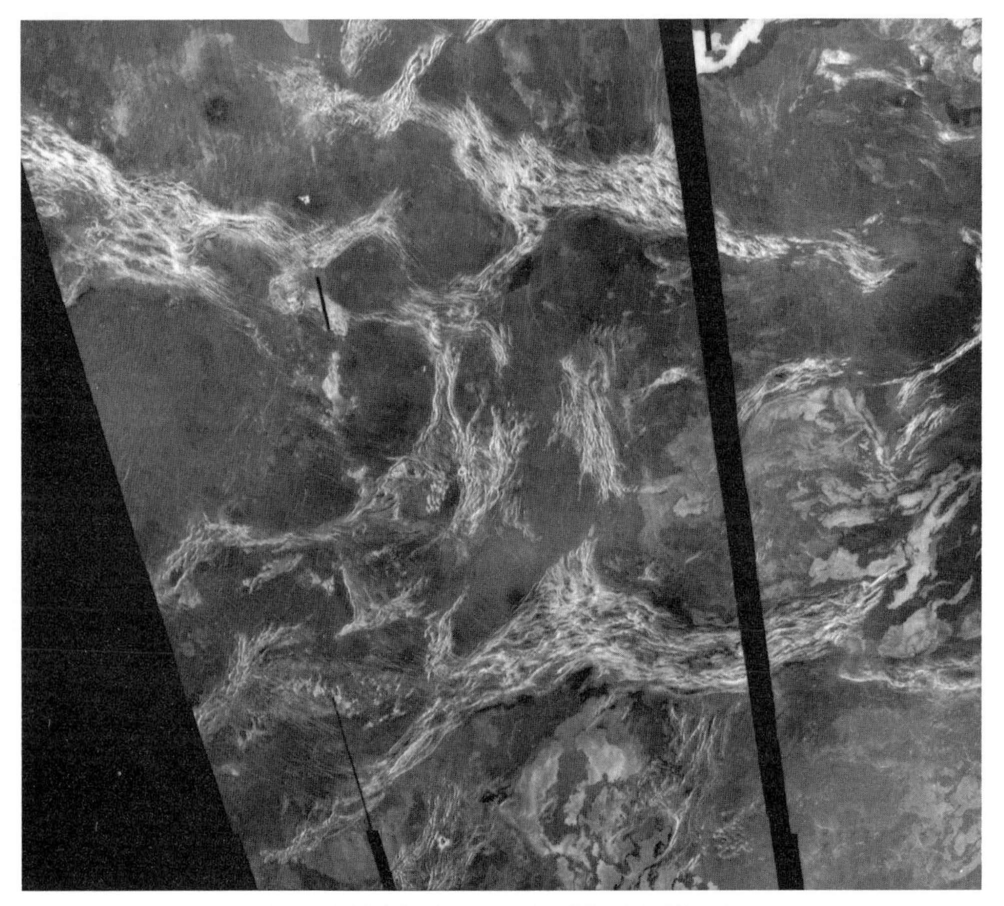

Figure 6.18 Magellan mosaic of Lavinia Planitia

Prominent radar-bright ridge and fracture belts can be seen crossing the plains. Also visible are intermediate-brightness lobate volcanic flows and blocks of tesserae. The image width is 1843 km. Magellan C1-MIDRP 45S350;1.

Although the individual ridges comprising the *ridge belts* are typically too narrow to be resolved altimetrically by Magellan's instrumentation, Squyres et al. (1992) have studied small displacements observed where ridges are cut by grooves on the plains units, enabling them to

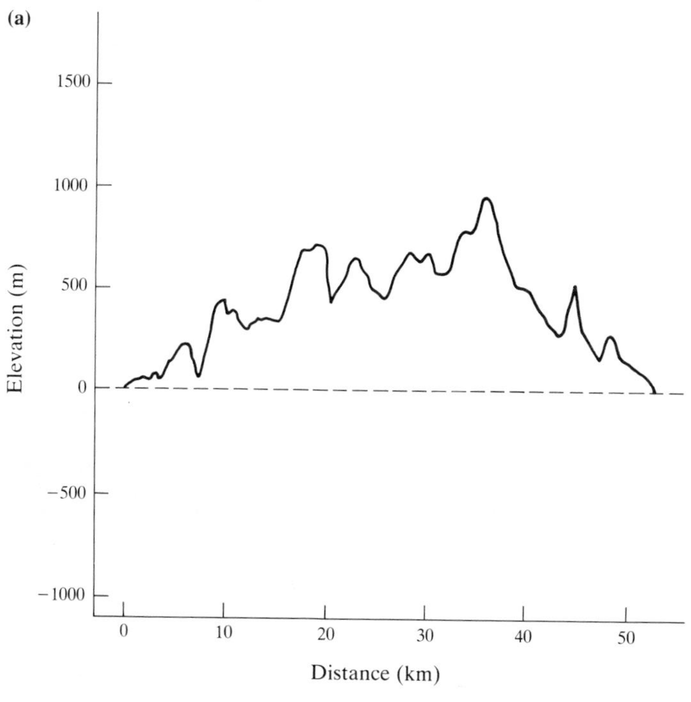

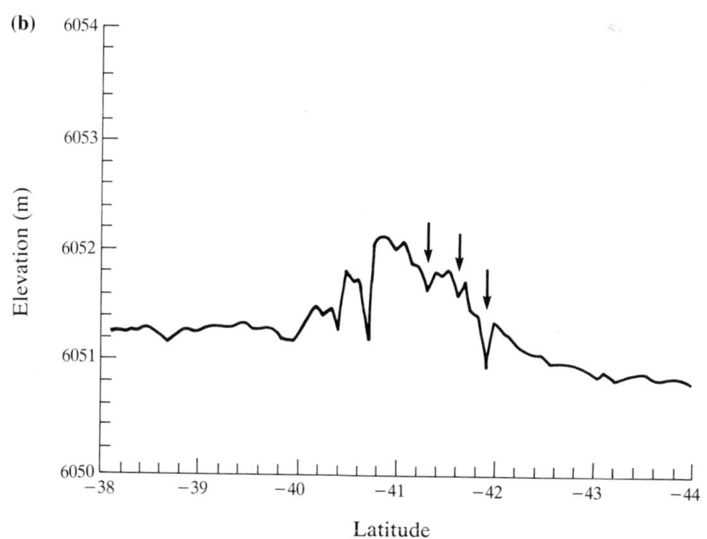

Figure 6.19 Ridge and fracture belts

Cross sections across (a) ridge belt, and (b) fracture belt. Inferred topographic profiles from Squyres et al. (1992).

construct profiles across some ridges. Their findings suggest that most are symmetrical, having arch-like profiles with heights of 200–300 m, and widths of several kilometres (Fig. 6.19).

The *fracture belts* found in Lavinia comprise families of grooves that range in width from a few kilometres down to the limit of Magellan's resolution. That they were formed either over a lengthy period of progressive deformation, or during repeated phases of deformation, is indicated by their typical winding courses and cross-cutting relationships. In some places the spacing of the fracturing is very close indeed, with the intensely fractured bands being separated by relatively little deformed rocks, e.g. as in Hippolyta Linea. The report by Solomon et al. (1991) that such intensely deformed zones corresponded to topographic highs (possibly anticlinal folds) now appears to be invalid, having been based on incorrectly processed data (see Squyres et al. 1992).

One other interesting observation in connection with the fracture belts is that the prominent faulting that is their characteristic is extremely similar to the less strongly developed fracturing in the adjacent plains. Indeed, there are instances where the faulting of a fracture belt penetrates into the plains, but with considerably decreased intensity, producing widely spaced grooves. There seems little doubt, therefore, that there must be a close genetic connection between the faulting typical of fracture belts and the production of grooves on the Venusian plains.

6.4.3 Regional pattern of ridge belts

It was clear right from the time that Venera 15/16 data became available that ridge belts were spatially extensive in the rolling plains of the northern hemisphere (Barsukov et al. 1986). Here they form a fan of north–south striking belts that extend from the north pole southwards to the limit of Venera's coverage (Fig. 6.20). The more global coverage produced by Magellan has revealed that similar belts cross the plains of more southerly latitudes, tending to occur within topographically depressed regions, such as Lavinia Planitia and Vinmara–Atalanta Planitiae. Most of the tectonic features found within the plains, including the ridge and fracture belts, demand only modest strains and horizontal displacements of the order of tens to hundreds of kilometres, assuming crustal thicknesses of between 20 and 30 km. We shall return to the tectonic history of Venus in the final three chapters.

6.5 Tesserae or complex ridged terrain (CRT)

Figure 6.21 shows that the tesserae or CRT areas are characterized by intersecting, closely packed ridges and grooves with typical ridge crest-to-crest spacings of between 5 and 20 km, and base-to-crest heights seldom exceeding a few hundred metres. They display patterns ranging

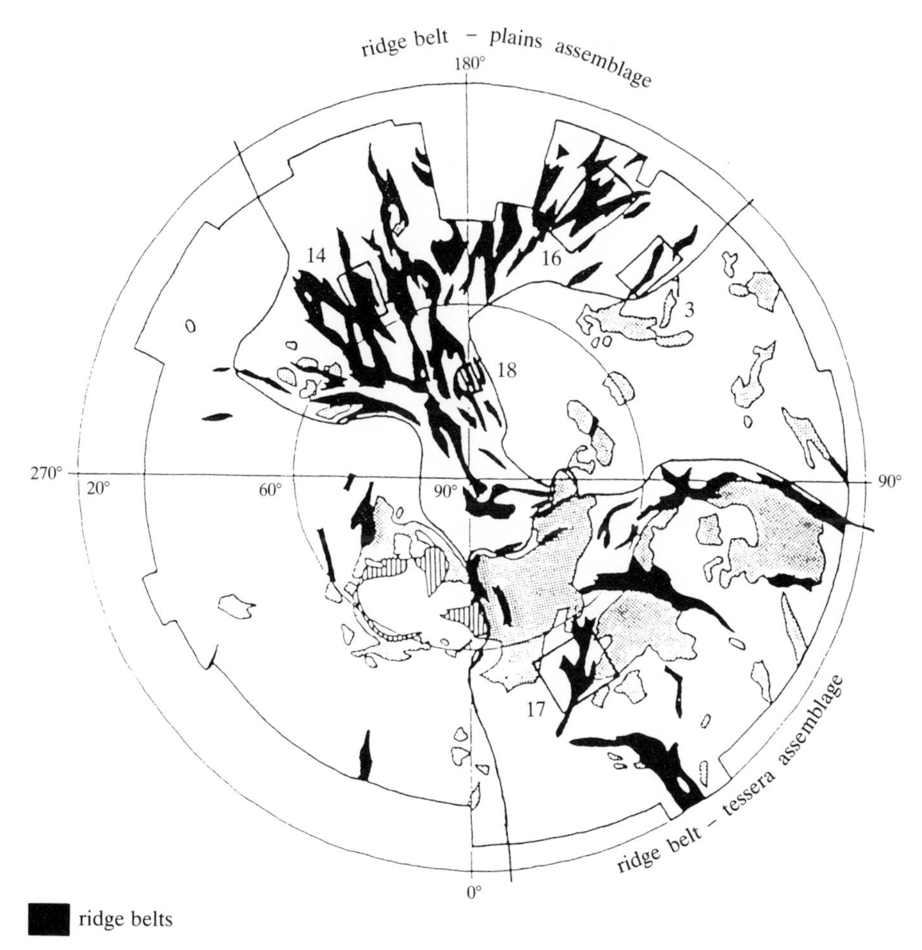

Figure 6.20 Distribution of ridge and fracture belts in the northern hemisphere of Venus
The map is drawn primarily from Venera 15/16 and Arecibo data. After Frank & Head (1990), Nikishin (1990).

from chevron-like, through diagonal or orthogonal, to chaotic; the ridges and grooves may reach hundreds of kilometres in length. They tend to be found in highland regions and, in general terms, it is found that the higher the unit, the larger it is (Nikishin 1990).

Initial mapping of tesserae was accomplished with Pioneer-Venus and Arecibo data (Bindschadler & Head 1988), using as a basis the knowledge that tessera regions typically have unusually low reflectivity (thought to be due to wavelength-scale roughness elements) and high RMS slopes. On the basis of Venera 15/16 imaging, tesserae were known to cover around 15% of the surface north of latitude 30°N (Sukhanov 1986). Analysis of the imagery led to subdivision into three main types: *linear ridged terrain*; *trough-and-ridge terrain*; and *disrupted terrain* (Bindschadler & Head 1988). Predictions also were made of the likelihood

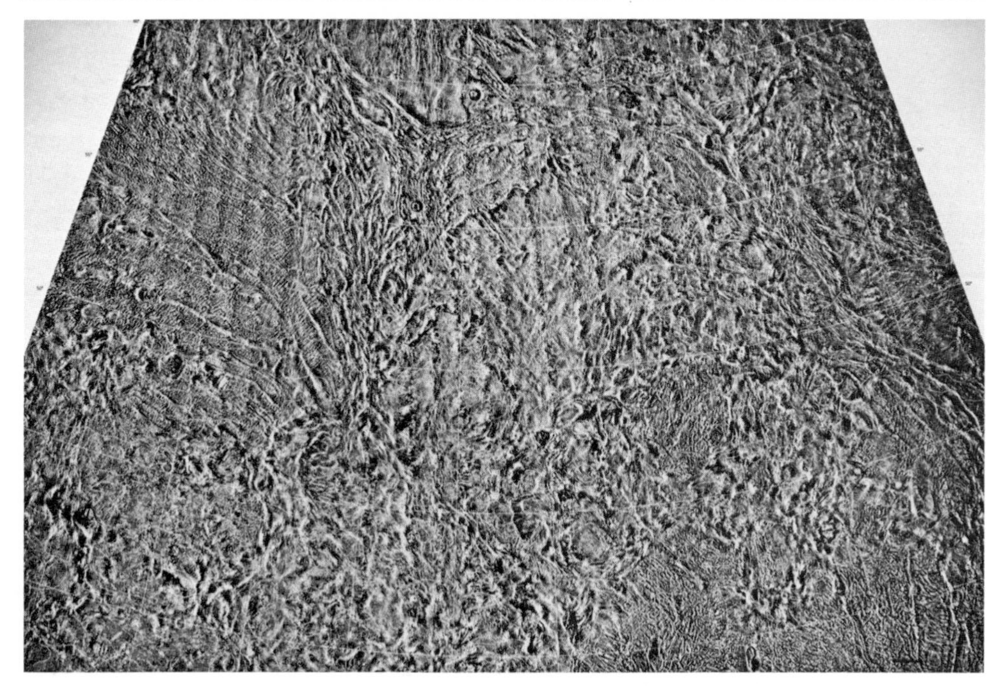

Figure 6.21 Northern tesserae (complex ridged terrain)
Venera 15/16 mosaic of the region between Laima Tessera (left side, above centre) and Tellus Regio (bottom right). Mosaic is 3600 km wide.

of finding tesserae in regions not imaged by Venera 15/16; Magellan largely has confirmed these earlier predictions.

Magellan subsequently has revealed that the global coverage is nearer three times that originally identified, and that the larger tesserae are typically steep-sided plateaux. The major regions are located in Ovda, Thetis, Phoebe, Beta and Asteria Regiones, north of Bell Regio, around Nokomis Montes (NW of Atla Regio), and in Lada Terra. The most extensive contiguous area is that which extends between the westward extension of Aphrodite and Ovda Regio; this has an area of 10^7 km^2. In contrast to rifted highlands such as Beta and Atla Regiones, their gravity anomalies are modest, and depths of compensation are often significantly less than 100 km (Smrekar & Phillips 1991).

The improved resolution of the new data has confirmed that the majority of the tesserae are characterized by compressional ridges and troughs upon which are superimposed (generally) younger extensional graben. The compressional ridges tend to be orientated more-or-less parallel to the boundaries of CRT plateaux (Fig. 6.22). The ridges are spaced 10 to 20 km apart, while the cross-cutting faults range down to the resolution limit of 200 m. While some structures may be traced with consistent strikes for up to 1000 km, most can be traced for only about one-tenth of this distance. It is likely that the tesserae represent blocks of thickened crust that have preferentially strained in response to regional deformational stresses.

Figure 6.22 Tessera structure in Laima Tessera
Magellan image of a part of northwestern Laima Tessera, showing NE-trending graben cross-cut by compressional structures. Magellan image F-MIDRP 50N054;1.

6.6 Rift and fault zones

Faulting is virtually ubiquitous over the surface of Venus. However, we can note several major zones characterized by large-scale faulting that are a strong feature of the global physiographic map. Some characterize broad zones of topographic uplift associated with strongly focused volcanism, which are aptly termed *volcanic rises*. Particularly strong rifting is associated with Beta Regio, whose scale of faulting and updoming is comparable with Earth's East African Rift Valley. The associated rift faults trend approximately north–south. Rifting is also associated with Atla Regio - witness Ganis, Dali and Parga Chasmata - and western Eistla Regio, where the NE–SW trending rift of Guor Linea joins Gula Mons to Sappho Patera. All of these regions typically have large positive geoid and gravity anomalies and large implied depths of compensation; they are also the sites of major volcanic constructs. They are believed to overlie major mantle upwellings, i.e. hotspots or plumes.

Major linear fracture zones also characterize parts of Aphrodite Terra, where the south-eastern corner of Thetis Regio has been rifted apart, and the southern flank of Ovda Regio was affected in like manner. Indeed, this zone of fracturing continues through Dali and Diana Chasmata, together with several large coronas and, in the other direction, strikes towards Atla Regio. These major tectonic zones have lengths measurable in several thousands of kilometres, and are further discussed in Chapters 10 and 11.

7 Surficial geology

7.1 Weathering and erosion on Venus

7.1.1 Introduction

Gravity operates on Venus, as elsewhere in the universe. In consequence, gravitational downslope movement will act on Venusian slopes, whether these be produced by volcanism, tectonism or impact cratering. The degradation of such slopes will be dependent upon the effectiveness of surficial mechanical and chemical processes on Venus, i.e. agents that operate at the interface between the atmosphere and crust. On the Earth, mechanical and chemical weathering strive to break down the surface rocks, generating a mixture of resistates and non-resistates. The former, although modified during weathering and transportation, survive largely unchanged; on Earth, the most abundant of these is quartz (SiO_2). Non-resistates are chemically reactive and may become hydrolysed to produce a mixture of secondary minerals and soluble or partially soluble cations that enter into solution.

To illustrate the path of such processes, we may cite the abundant constituent of granitic rocks, orthoclase feldspar ($KAlSi_3O_8$), which breaks down to a mixture of kaolinite ($Al_2Si_2O_5(OH)_4$), silica (SiO_2) and potassium ions (K^+). Of these varied weathering products, silica may remain in situ and form part of the regolith, but more probably will enter into solution and be removed; the clay, kaolinite, will enter the regolith, while potassium ions may be taken up by plants or enter other clay minerals. Whether or not a thick regolith develops at a particular locality will depend on a variety of factors, including local gradients, climate and the presence or absence of plants. As a rule of thumb, any particulate material gradually will be moved down slope under the influence of gravity aided and abetted by ice, wind or water, and soluble substances will be carried away in solution, either by surface fluids or by seeping into the substrate, to enter the hydrological cycle at some later stage. The ultimate fate of most of these components is entry into either the oceans or lakes, where they will be deposited by streams, rivers or glaciers, or blown by the wind. After a period of burial, they may be converted by a process of chemical and physical change, termed **diagenesis**, into coherent sedimentary rocks. The key to all this is the presence of water.

The effectiveness of rock weathering, transportation and deposition on water-rich Earth results in the production of extensive regions of interior plains on the Earth's continents, and of a veneer of marine sediment upon the oceanic crust. This invests in the Earth a characteristic distribution of 0.0° slope, with the interiors of the continental cratons appearing flatter, broader and more continuous than the floors of the oceans, largely because of the greater effectiveness of subaerial as opposed to submarine erosion (Sharpton & Head 1986).

7.1.2 Present conditions on Venus

The present lack of water on the surface of Venus naturally has important implications for the kinds of surficial processes that can operate now. There can be no integrated river systems, oceans, lakes or ice-sheets; in short, no Earth-style hydrological cycle. Furthermore, although interaction of Venusian atmospheric gases could generate iron oxides and sulphur-rich compounds from basaltic plains deposits, carbonate minerals are not stable on the incredibly hot arid surface of Venus and could not, therefore, lock up significant amounts of CO_2 in sedimentary rocks. However, there is a dense atmosphere, albeit largely of carbon dioxide and corrosive acids; therefore, there can be winds and there is obvious scope for chemical activity. A very important issue, however, and one that Magellan will address, is whether or not the greenhouse conditions that prevail today, and which preclude the existence of liquid water at the surface and in the atmosphere, have always prevailed. Should evidence of ancient fluvial landforms be forthcoming from Magellan's high-resolution radar-mapper, then this would imply that a greenhouse effect did not develop until late in Venus's history: what an extremely exciting discovery this would be!

That some form of surface modification has occurred on Venus was clear from Venera 15/16 and Earth-based images of impact craters. Fresh craters exhibit prominent radar-bright haloes, the brightness being a response to a degree of surface roughness (on the decimetre to metre scale) well above that of the adjacent plains. About 25% of all craters imaged by Venera 15/16 have these bright haloes. More mature craters, however, lack them, a characteristic that implies that some obliterative process has been at work, one possibility being aeolian deposition. On the assumption that there has been a steady state of weathering and erosion on the planet, a *surface* age of between 0.5×10^6 and 1×10^6 years (as derived by crater statistics), and an acceptance of the fact that only 25% of craters have such haloes, calculations disclose that it would take between 125×10^6 and 250×10^6 years for the roughness to be reduced to a point at which it would not be discernible on Venera 15/16 imagery. This suggests that obliteration has been at the rate of 0.04–0.08 cm/10^6 years (Ivanov et al. 1986).

Other theoretical considerations suggest that higher rates of weathering and erosion may have occurred. Calculations by Nozette & Lewis (1982) indicate that rock–atmosphere chemical interactions should be altitude-dependent. For instance, within high terrains such as Maxwell Montes, such interactions should be dominated by the generation of carbonates, sulphates and sulphides owing to reactions between atmospheric CO_2 and SO_2 and surface rocks. Pioneer-Venus radar observations confirm that rock surfaces at high elevations may well expose minerals with high dielectric constants, for instance, pyrite, ilmenite and magnetite (Pettengill et al. 1988). Should this eroded fragmental material subsequently be transported to lower altitudes by a combination of mass wasting and, more effectively, aeolian activity, then CO_2 and SO_2 should be liberated once more, as re-equilibration occurs. Fegley & Prinn (1989) conducted experiments to reproduce anhydrite ($CaSO_4$) by the interaction of SO_2 with calcite ($CaCO_3$). They estimated that a rate equivalent to 1 km per 10^9 years – roughly one-quarter to one-half the total estimated resurfacing rate for Venus – was entirely appropriate. If all this proves to be true, then chemical processes of this kind would provide a third potent alternative to the two currently accepted major surface-modifying processes, volcanism and tectonism.

What of the details of the surface at Venera and Vega landing sites? These, it will be recalled, set down in Beta Regio and Rusalka Planitia (northern flank of Beta Regio) respectively. The Venera 9 probe landed in a sloping region of rock detritus on the eastern slope of Rhea Mons, and returned panoramic images revealing abundant rock clasts with diameters from a few centimetres to tens of centimetres; between these lay pebbles and finer, somewhat darker fines (albedo 0.02–0.05), or regolith. The slabs of detritus, typically 50–70 cm across and not more than 15–20 cm in thickness (Florensky et al. 1983), showed evidence for two or three sets of intersecting "cleavage" planes (probably joints) and little or no effect attributable to any form of chemical weathering. The aspect of the Venera 10 site was similar, but with a greater preponderance of slabs. The greatest proportion of solid bedrock was encountered at the Venera 14 landing site. Also, at each of the Venera 10, 13 and 14 sites, bare soil-free surfaces were observed atop topographic highs, and regolith materials were seen in the lower-lying *wind-shadow* areas. This implies that at least small-scale aeolian processes are operating on Venus's surface, the assortment of rock slabs, smaller pebbles and clasts and crust-like regolith all providing suitable grist for the aeolian mill.

The Vega 1 and 2 sites have been characterized by studying Pioneer-Venus data reflectivity and RMS slope measurements. The sites give signatures most like the plains units and volcanoes unit mapped from Venera 15/16 radar imagery (Bindschadler et al. 1986). Little is known of the details of the surface structure, but the rocks themselves have been established as gabbroic in composition (Surkov et al. 1986b).

Pieters et al. (1986) analyzed the spectral reflectance properties of the Venusian surface at the Venera landing sites, noting that the surface appeared generally dark in the visible part of the spectrum, the finer material being darker than rock outcrops. However, at longer wavelengths ($> 0.7\ \mu m$), a brightening of the surface was noted, and was ascribed to the presence of ferric oxides. This implies that, assuming the surface is in thermodynamic equilibrium with the atmosphere, the lower atmosphere of Venus must be oxidizing.

Analysis of Pioneer-Venus reflectivity data has also helped with making predictions concerning the presence and distribution of regolith materials on a global scale. A study made by Bindschadler & Head (1988) suggests that not more than 5% of the surface is covered in regolith more than several tens of centimetres thick. Over the majority of Venus, it seems, bare rock surfaces tend to dominate. However, the strong correlation they found between low reflection coefficients, diffuse scattering and the tessera regions led them to suggest that, within these regions of strong tectonic deformation and steep slopes, there are abundant rock fragments that originate from a hitherto unknown but widespread mass-wasting process. Specific regions with these characteristics include Danu and Maxwell Montes, Gula Mons and Sif Mons. The regions of "radar-deep soils" pinpointed by Pettengill et al. (1988) also coincide with tessera units. It should be noted, however, that this does not unequivocally prove the existence of such material at these locations.

What of the wind velocities near the ground on Venus? Cup anemometers at both Venera 9 and 10 sites registered windspeeds of between 0.5 and 1.0 m s^{-1} (Avduevskii et al. 1977). At the Venera 13 landing site, slightly lower velocities were measured by an acoustic technique (0.3 to > 0.5 m s^{-1}). The acquisition of sequential TV pictures at the latter site elicited a bonus, when they demonstrated that a clod of regolith a few centimetres across that had been churned up during the landing of the probe was entirely removed during the hour-long interval during which observations were made (Selivanov et al. 1983).

7.1.3 Aeolian activity on Venus

Accepting that loose material exists on Venus and that winds blow across its surface, it remains to enquire how the two interact and what results ensue. It is known from terrestrial experience that there are three modes of aeolian transport: *surface creep*, which involves coarse grains (20–40 mm), *saltation*, which involves sand-sized grains (40 μm–20 mm), and *suspension*, which involves very fine particulate material (< 40 μm) (Bagnold 1941). The fundamental parameter that governs whether or not a particular aeolian process will operate is the *threshold wind velocity*. Saltation threshold wind velocity has been carefully analyzed for terrestrial conditions and for Mars, and some workers have made theoretical predictions for Venus. The saltation threshold for Venus is predicted to be a function of grain size, grain density, gravitational acceleration, and the temperature and density of the atmosphere (Sagan 1975, Hess 1975, Iverson et al. 1976).

Using basic assumptions, it is clear that the wind velocities recorded on the surface of Venus theoretically are capable of mobilizing at least fine sand and silt-grade particles. However, some more precise data are required. To this end, carefully controlled wind-tunnel experiments have provided insights into potential Venusian aeolian processes (Greeley et al. 1984a,b, Marshall et al. 1988). They have proved highly informative. First, they have revealed that within a pure CO_2 atmosphere, 0.6 km s^{-1} winds will entrain quartz grains 100 μm in diameter, while a wind of velocity 3 km s^{-1} will shift 1 cm particles. The most easily moved particles have a size of around 75 μm, under simulated Venusian conditions. Secondly, they have demonstrated that dune-like structures (*microdunes*) form in fine-grained quartz sands (50–200 μm diameter) under appropriate conditions; however, if wind velocities exceed about

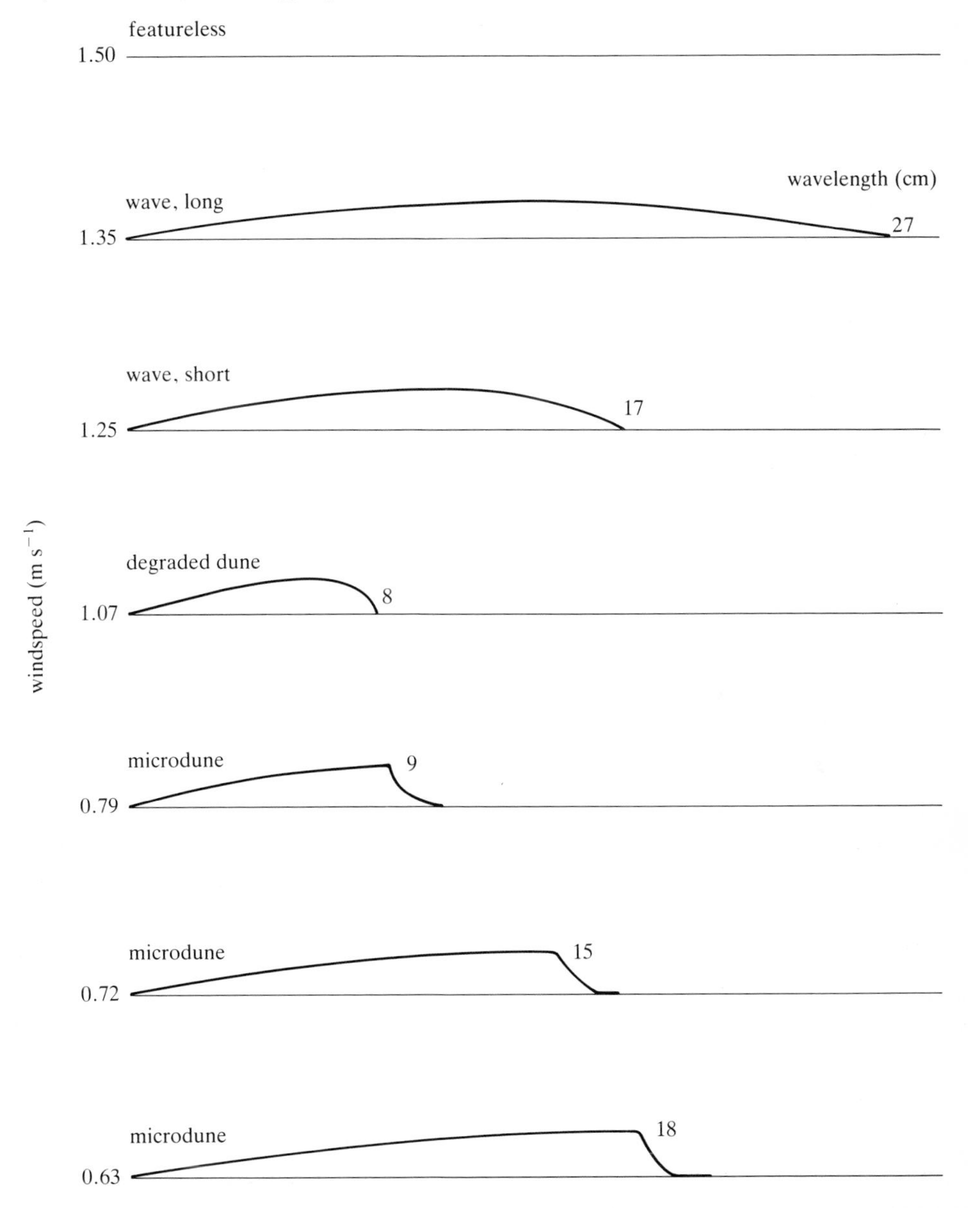

Figure 7.1 Bedforms produced under Venusian conditions

Laboratory simulations reveal a gradual change in bedform, wavelengths changing with free-stream windspeed. After Greeley et al. (1984b).

1.5 m s^{-1}, dunes apparently are suppressed, and resultant bedforms are flat and featureless (Fig. 7.1). The implication here is that, if relatively short periods of enhanced windspeed buffet existing dunes, these may be destroyed – a phenomenon that could account for the apparent paucity of such structures in pre-Magellan imagery. The generation of microdunes, as opposed to other bedform structures, i.e. surface ripples, waves and featureless surfaces (all produced during the experiments), is dependent primarily on windspeed.

The experiments also showed that Venusian aeolian transport differs significantly from that on Earth, particularly with respect to the rolling mode of sediment transport. This can be maintained for long periods under Venusian conditions without leading to the cascading effect, which, on Earth, will suddenly transform the surface of a bed into a saltation cloud. To some extent, therefore, aeolian environments on Venus share characteristics of both aeolian conditions on Venus and aqueous conditions on the Earth.

Of course, where particulate material is blown around by the wind, the moving particles have the potential to abrade mechanically any surface with which they come into contact, thereby redistributing surface debris and exposing fresh, chemically reactive material. Another series of controlled experiments conducted within a pure carbon dioxide atmosphere (Greeley et al. 1987, Marshall et al. 1988) involved observations on 3 mm angular basalt particles and the effects produced by these on striking basalt surfaces at velocities within the range 0.5–1.0 m s^{-1}. The results showed that not only are saltating particles abraded during transportation but that some of the abraded material becomes deposited on the surfaces of impacted rocks, forming an accretionary coating. On Venus, such an accretionary process seems to be a response to the high ambient temperatures involved. Accretionary "plastering" along these lines would provide a measure of protection for Venusian rock surfaces against in situ mechanical and chemical weathering. Furthermore, because the transported material leaves evidence of its motion in a trail of chemically modified terrain, complex surface chemical differences might be detectable by an orbiting probe, thereby allowing scientists to trace movement of aeolian material.

7.2 Magellan observations

7.2.1. Introduction

Even by the time Magellan had completed but half of its initial eight-month-long mapping cycle, it had revealed different kinds of surficial deposits not previously identified on Venus; several of these are associated with impact craters. Hummocky materials with high radar backscatter outcrop both within and surrounding many craters; these are presumed to have an impact origin. Also having bright radar signatures are lobate materials whose distribution appears to be topographically controlled. These have been interpreted as volcanically generated outflow deposits, the volcanism having its origin in impact (Saunders et al. 1991).

Sinuous channel features 1–3 km wide, with low radar back-scatter, have been identified in both highland and lowland regions, and are presumed to represent lava channels and sinuous rilles. Radar-bright areas associated with ridges and small domes, or occurring as linear units that may exceed 100 km in length, are interpreted as regions of increased surface roughness probably created by either erosive "roughing up" of surface units or deposition of coarse deposits by wind. The reverse of these are circular to elongate features with diffuse boundaries and characterized by low radar back-scatter; these represent relatively smooth areas most likely delineating patches of fine-grained sedimentary materials. Finally, linear outcrops of facetted materials form groups of lineations up to 100 km long; these may be sand dunes. Several kinds of aeolian feature are associated with the large impact craters, Carson and Mead.

7.2.2 Wind streaks and related features

As outlined above, the Magellan synthetic-aperture radar has already revealed a plethora of features attributable to aeolian processes; these include a few major dunefields, possible yardangs and various types of wind streak (Greeley et al. 1992a). Of the three, the most common are wind streaks – well known from both Mars and the Earth, and also recognized on Triton. Over 3000 of these have already been catalogued and, while the greatest incidence of aeolian features in general is on the smooth plains units of equatorial regions, wind streaks are found at most latitudes and elevations. This demonstrates that wind activity is widespread on Venus.

Wind streaks are well known from Mars (Fig. 7.2), where they range in size from a few centimetres long to 115 km (Thomas et al. 1981). Wherever they occur (Fig. 7.3), they represent the prevailing wind direction at the time of their inception, for which reason they are invaluable as local "wind vanes" and have been utilized by several teams of planetary geologists to establish details of local, regional and global atmospheric circulation (Sagan et al. 1972, Thomas & Veverka 1979, Greeley et al. 1992b). Their discovery on Venus is particularly timely since, prior to Magellan, there had been precious little hard information about the circulation of the lower atmosphere and its interaction with the surface. Furthermore, they allow identification of sediment sources capable of being mobilized by the Venusian winds.

Most streaks on Venus are located on smooth plains units situated between latitudes 23–30°N and 23–30°S. These were reported initially by Arvidson et al. (1991), who identified small-scale features as well as regional-scale, paraboloid-shaped, areas of low back-scatter focused on certain impact craters within the equatorial zone. Particularly prominent features of this type are associated with Sif Mons and with the impact crater Adivar (Fig. 7.4). Typically the vertices of large streaks are located on the eastern edge of each paraboloid, with the

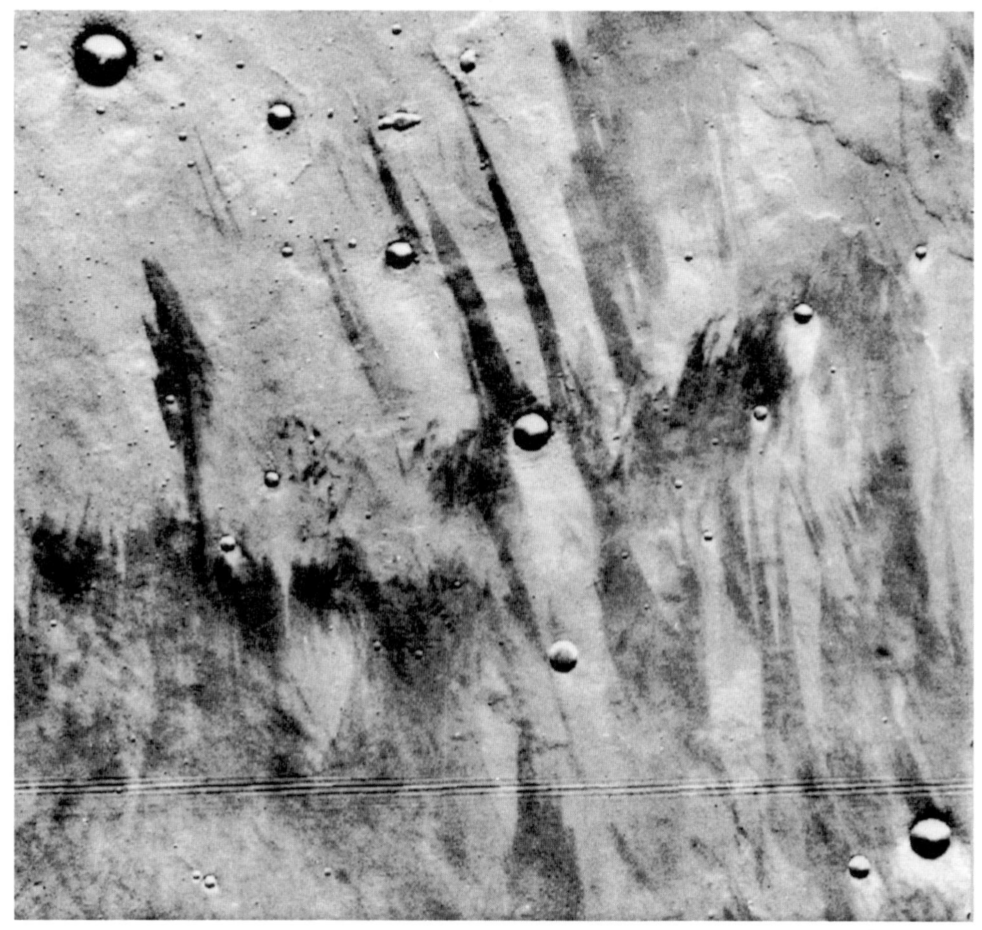

Figure 7.2 Martian wind streaks.
Viking orbiter mosaic 211-5601.

Figure 7.3 Typical wind streaks on Earth, Mars and Venus
(a) Cinder cone and dark streak, Amboy California; prevailing wind from west; general background consists of basalt with overlying windblown sand (white areas); dark areas are those swept clean. (b) Seasat radar image of same areas as (a). (c) SIR-A radar image of the Alriplano, Bolivia, showing 15 km long dark streaks associated with hills (bright features). (d) SIR-A radar image of linear streaks southeast of Laskar Gan, Afghanistan, formed predominantly in sedimentary deposits. (e) Dark wind streaks in Phoenicus Lacus, Mars; prevailing wind from southeast (left). (f) Bright wind streaks in Hesperia Planum, Mars; such streaks are thought to be dust deposited in the lee of topographic features. (g) Radar0bright wind streak on Venus at 23.9°S, 345.1°E; streak is about 10 km long and is associated with small hill, and occurs within the "parabolic halo" associated with crater Carson. (h) Wolf Creek impact crater, Western Australia; sand deposits (bright) and erosional areas (dark); prevailing winds are from the east (left). Composite photograph courtesy of Greeley et al. (1992).

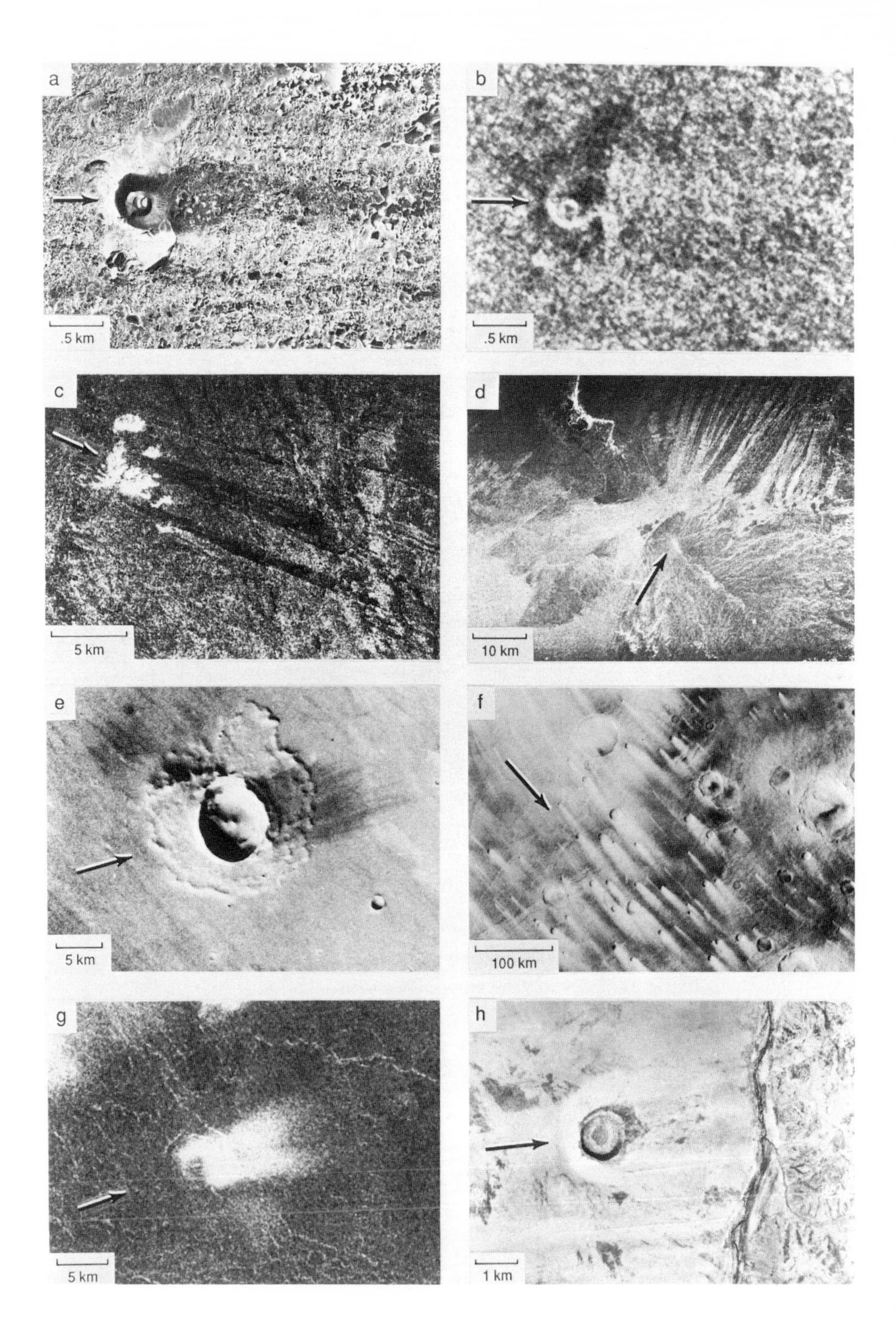
a
.5 km
b
.5 km
c
5 km
d
10 km
e
5 km
f
100 km
g
5 km
h
1 km

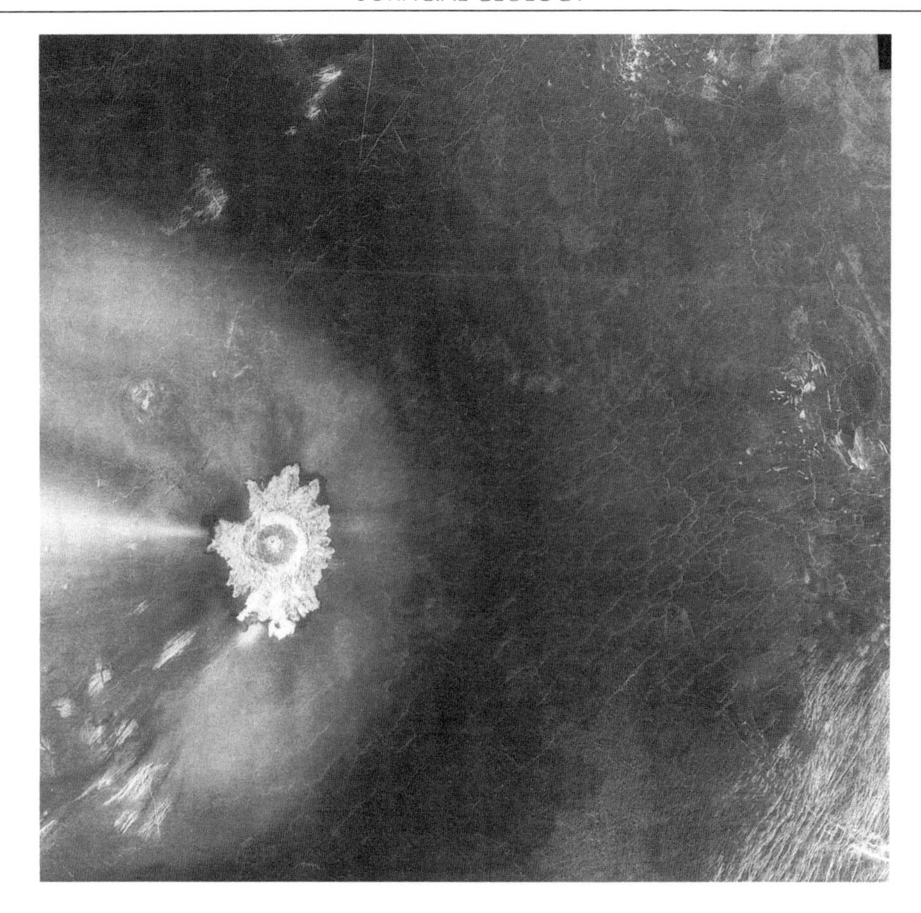

Figure 7.4 Wind streak associated with the crater, Adivar
Image centred at 8.95°N, 76.1°E. Magellan image F-MIDRP 10N076.

long axis striking north–south. As to scale, east–west dimensions of between 500 and 1000 km are typical. Recorded relatively low **emissivity** values of 0.77–0.80 (compared with 0.83–0.88 for immediate surroundings) and low back-scatter cross sections imply that these regions are smoother than their immediate hinterland. Additional evidence to support this is forthcoming from analysis of a combination of Magellan SAR and altimetric data, which demonstrates little difference in Fresnel reflectivity inside and outside each streak, implying that emissivity variation is attributable largely to roughness at the scale of the 12.6 cm radar wavelength.

Streaks occur in a variety of forms, which may show radar-dark, radar-bright or radar-mixed reflectivity against the background on which they have developed. Several basic outcrop patterns can be identified (Greeley et al. 1992a). These include fan-shaped, wispy, linear, transverse-ragged and transverse-smooth streaks (Fig. 7.5). Fan-shaped streaks charac-

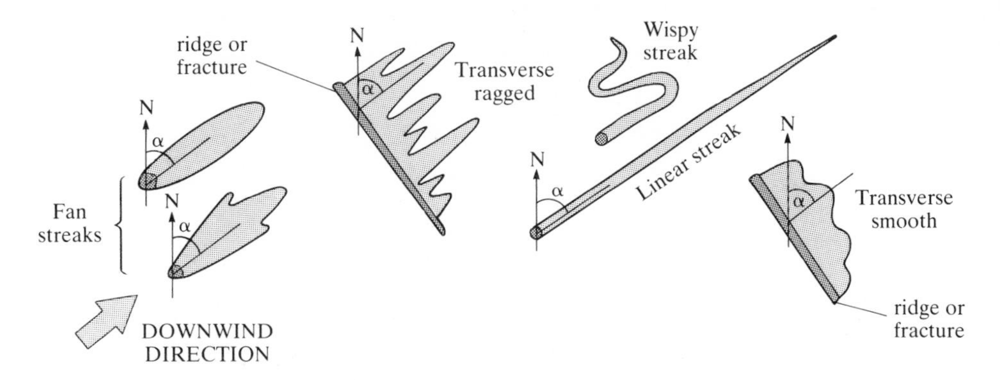

Figure 7.5 Planform shapes of Venusian wind-related features
In the diagrams α is the azimuth of the streak in the inferred downslope direction. After Greeley et al. (1992).

teristically are associated with positive landscape features such as small hills or domes; most have a radar-bright signature. Linear streaks, in contrast, are radar-dark and typically are at least ten times longer than they are wide, occurring in groups of at least six individuals. Wispy streaks meander across country, their width changing over quite small distances; they are often located adjacent to ridges with a parallel strike. Some groups of wispy streaks apparently are associated with impact craters, in which occurrence the streaks occur in groups that give the impression of being arranged approximately radially to the focal crater. However, because of their rather irregular nature, it is not usually possible to discern their point of origin with any real degree of certainty. Transverse streaks also occur in groups, often associated with fractures or ridges; these may be either radar-dark or radar-bright, and generally are orientated normal to the inferred prevailing wind direction.

The global analysis undertaken by Greeley et al. (1992a) shows that the streaks are found at all altitudes, and tend to be orientated down wind and towards the equator (Fig. 7.6). Most streaks also are randomly orientated with respect to slope, and located where the gradient is less than 2°. This is consistent with surface winds related to a Hadley circulation in the lower atmosphere. In such a pattern the circulation redistributes solar energy absorbed in the lower levels of the atmosphere and close to the ground near the equator. Such a meridional circulation is symmetrical about the planet's equator and it involves surface winds blowing towards the equator, upflow over the equator itself, poleward winds aloft, and air downflow at high latitudes (Schubert 1983).

Most wind streaks on the Earth and Mars are related to wind activity around obstacles, where local turbulence and consequent modified wind patterns may lead to a series of characteristic depositional and erosional patterns. On Mars, dark wind streaks are found where bedrock has been swept clear of loose sediment, or where lag deposits of coarse grains have

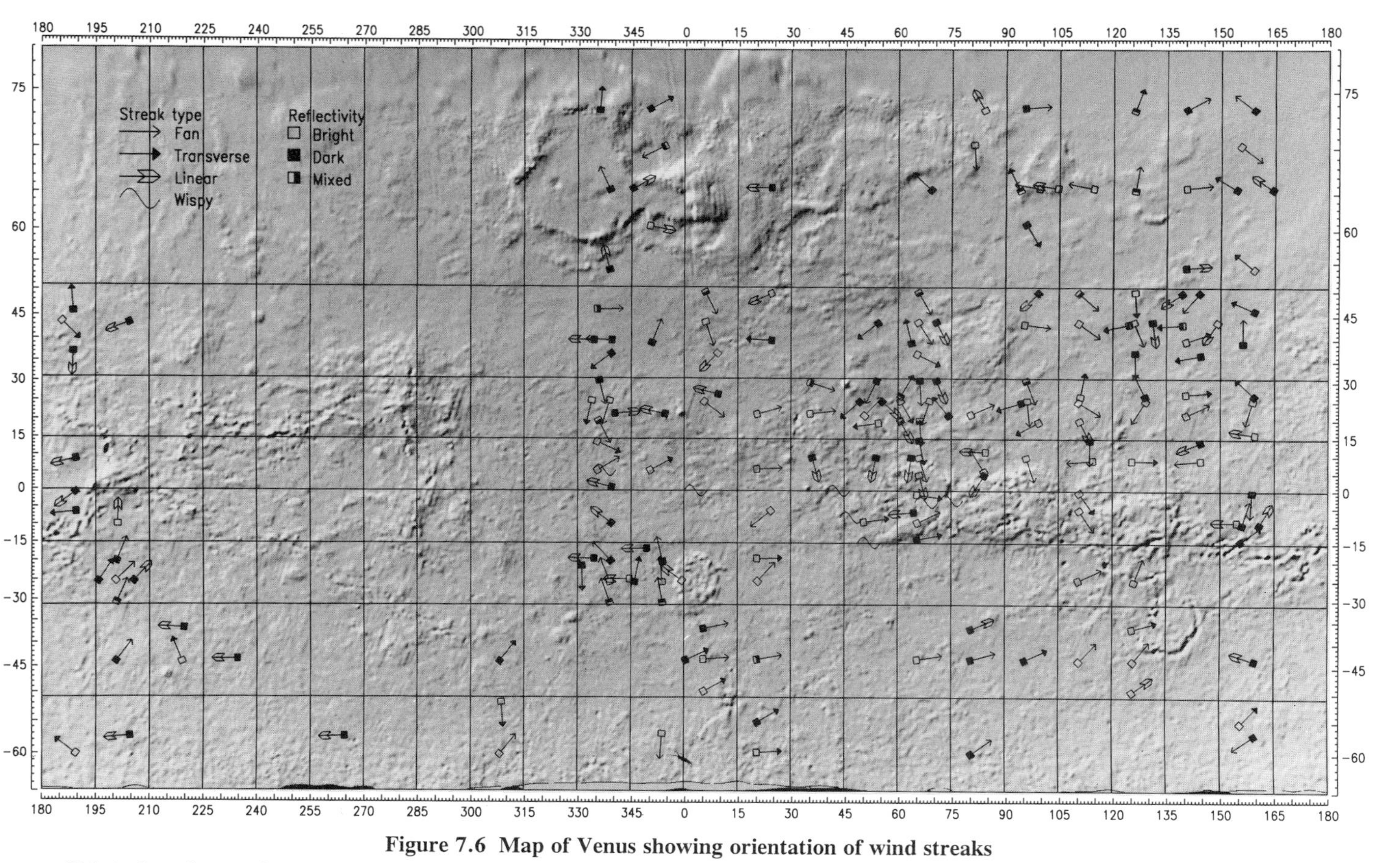

Figure 7.6 Map of Venus showing orientation of wind streaks

This is done for equal areas by latitude and longitude "bins". Downwind direction is indicated by arrows. After Greeley et al. (1992).

been modified by the removal of smaller, more reflective particles by deflation. In contrast, bright streaks are attributed to dust deposition in the lee of obstacles, probably under conditions of atmospheric stability (Veverka et al. 1981).

There are many terrestrial analogues, most of which are associated with the activity of turbulent winds around obstacles such as hills, craters with raised rims or domes. One of the finest examples is the remote Wolf Creek Crater, in northern Western Australia, where the deposition of bright, windblown sand has occurred in a belt up wind from the rim, and downwind as two trailing dunes; the horseshoe-shaped bright halo is particularly striking in satellite images, as is the dark, eroded area down wind of the crater's west rim (Fig. 7.7). Simulations of such features, using the Venus Wind Tunnel facility at Arizona State University (Greeley et al. 1984a), show how a horseshoe-shaped vortex forms around obstacles such as hills or crater rims, creating a zone of turbulence and high shear stress in the wake of an obstacle (Fig. 7.8).

Radar-dark streaks are more difficult to explain satisfactorily. Greeley et al. (1992a) consider these to be representative of loose sediments that have low radar back-scatter cross sections. Certainly the tendency for many radar-dark linear streaks to occur associated with ridges suggests that gaps in such topographic features may allow the wind to funnel loose

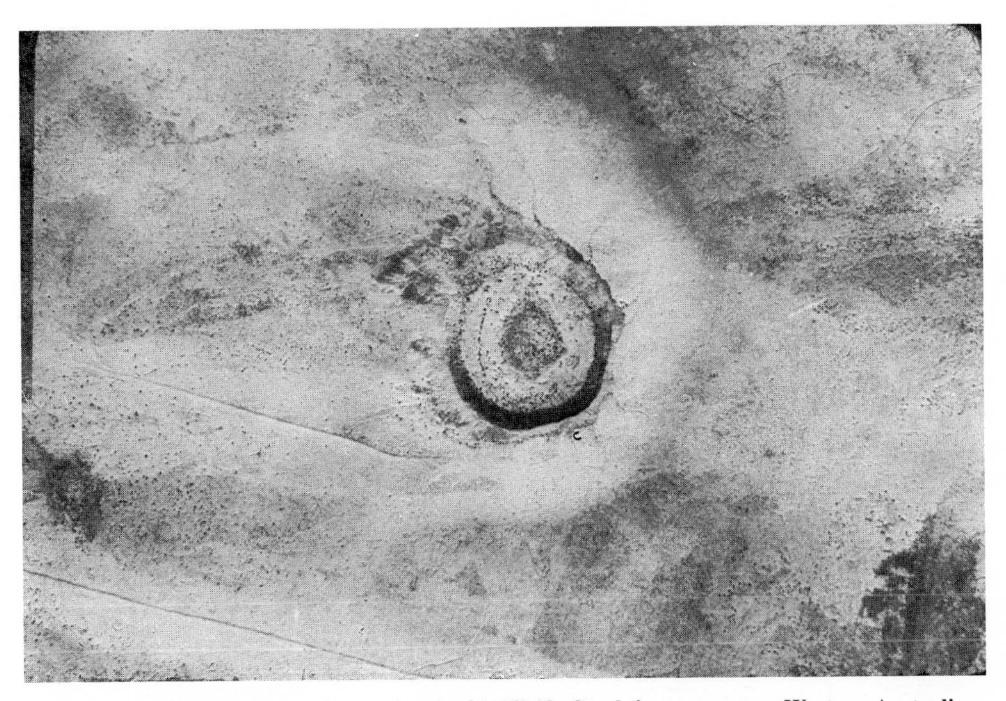

Figure 7.7 Paraboloid streak associated with Wolfe Creek impact crater, Western Australia
The bright halo is an area of sand deposition; dark areas are zones of erosion.

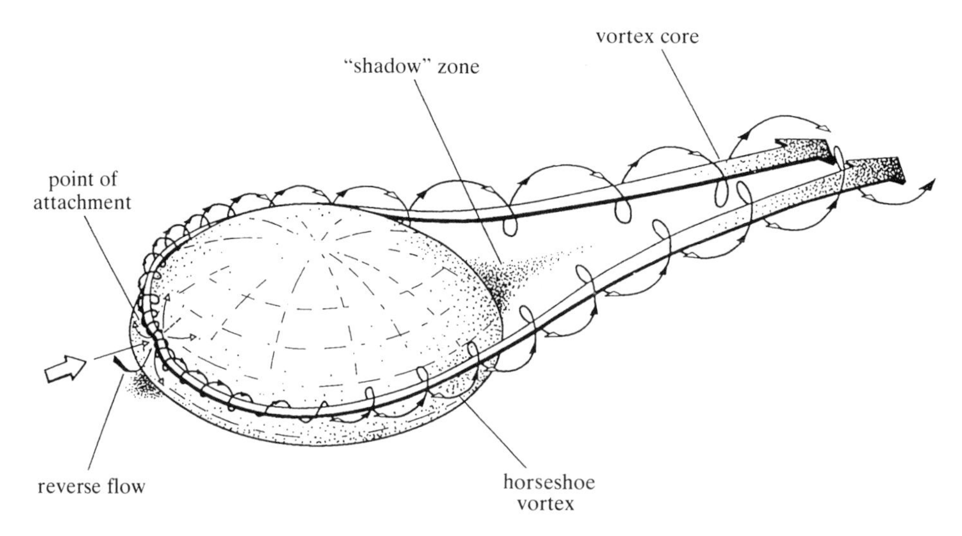

Figure 7.8 Wind flow around a topographic obstacle
A horseshoe-shaped vortex wraps around the obstacle, creating a zone of turbulence and high surface shear stress in the hill's wake. Material is eroded in this zone. Rising winds along the outer edges of the vortex may give rise to preferential deposition, creating a "halo" of loose material around the eroded zone. After Greeley (1986).

material along narrow corridors. A definitive answer to this may have to await further experimentation, but there is certainly the possibility that the dark materials, which seem to have a higher dielectric constant than the surface on which they rest, are mineralogically distinct from their surroundings.

7.2.3 Dunes

In the initial Magellan report, Arvidson et al. (1991) reported a possible dunefield associated with a 65 km diameter impact crater located west of Alpha Regio, near the crater Carson described in the previous section, and 100 km north of the crater Aglaonice. The diagnosis was largely based on the speckle-type radar return that this region gave and which terrestrial dunes – with their facets tilted towards the radar beam – are known to give. The area was subsequently studied in detail by Greeley et al. (1992a), who quote the centre co-ordinates of the field as 67°N, 340°E, and describe it as the "Aglaonice dunefield".

The Aglaonice dunefield is characterized by dunes 0.5 to 5 km in length whose spacing is difficult to assess owing to the specular nature of the radar returns. It covers an area of approximately 1290 km^2 and is associated with radar-bright *outflow deposits* associated with the impact crater from which it derives its informal name (Fig. 7.9). This crater is just one of an above-average concentration of such craters found in this region, and colloquially known as

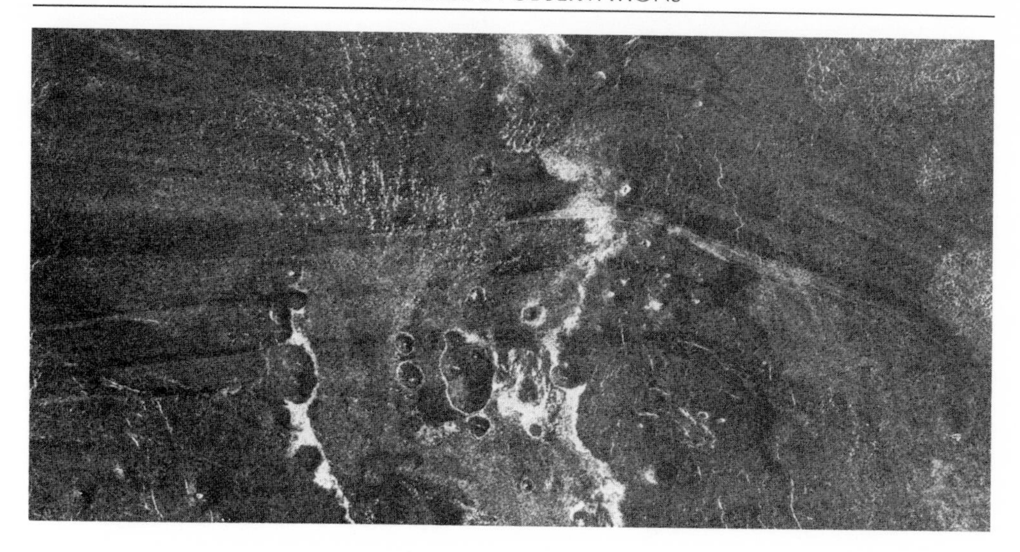

Figure 7.9 Aglaonice dunefield
The dunefield is seen as the radar-bright speckled region at top centre, and is associated with outflow deposits from the crater Aglaonice. Radar-dark streaks also cross the region. The frame measures 78 km × 180 km. Magellan image MRPS 34032.

the "crater-farm" by Magellan team members. Although there is no clear consensus as to how such outflow deposits are formed, it has been suggested that they represent deposition from either turbulent ejecta clouds or the outflow of lava (Phillips et al. 1992a,b). Either way, the outflow material evidently has been reworked by aeolian agents to produce both the transverse dunes and wind streaks. Since the source of the dune-building material, on account of the similar mode of transport by size distribution established by Iverson & White (1982) for Venus and Earth, would be anticipated to be of sand grade, such material must have been generated during the impact event that created the craters and made available for saltation. The orientation of the dunes, which are presumed to be transverse types, together with several bright wind streaks in the same region, suggest that the prevailing winds currently blow towards the west.

A second such field is located much farther north, in a valley between Ishtar Terra and Meshkenet Tessera; its central coordinates are 67.7°N, 90.5°E. This has been called the "Fortuna–Meshkenet" dunefield (Greeley et al. 1992a) and it covers an area of approximately 17 120 km^2. Individual dunes are between 0.5 and 10 km long, are around 200 m wide, and have an average separation of around 0.5 km. As with the Aglaonice field, it is associated with numerous prominent wind streaks. In the southern part of the field, the dunes and streaks indicate a flow of wind from the southeast, while towards the north the orientations of the aeolian landforms suggest a westward flow (Fig. 7.10).

The radar-bright wind streaks located within the field apparently have their origin in sev-

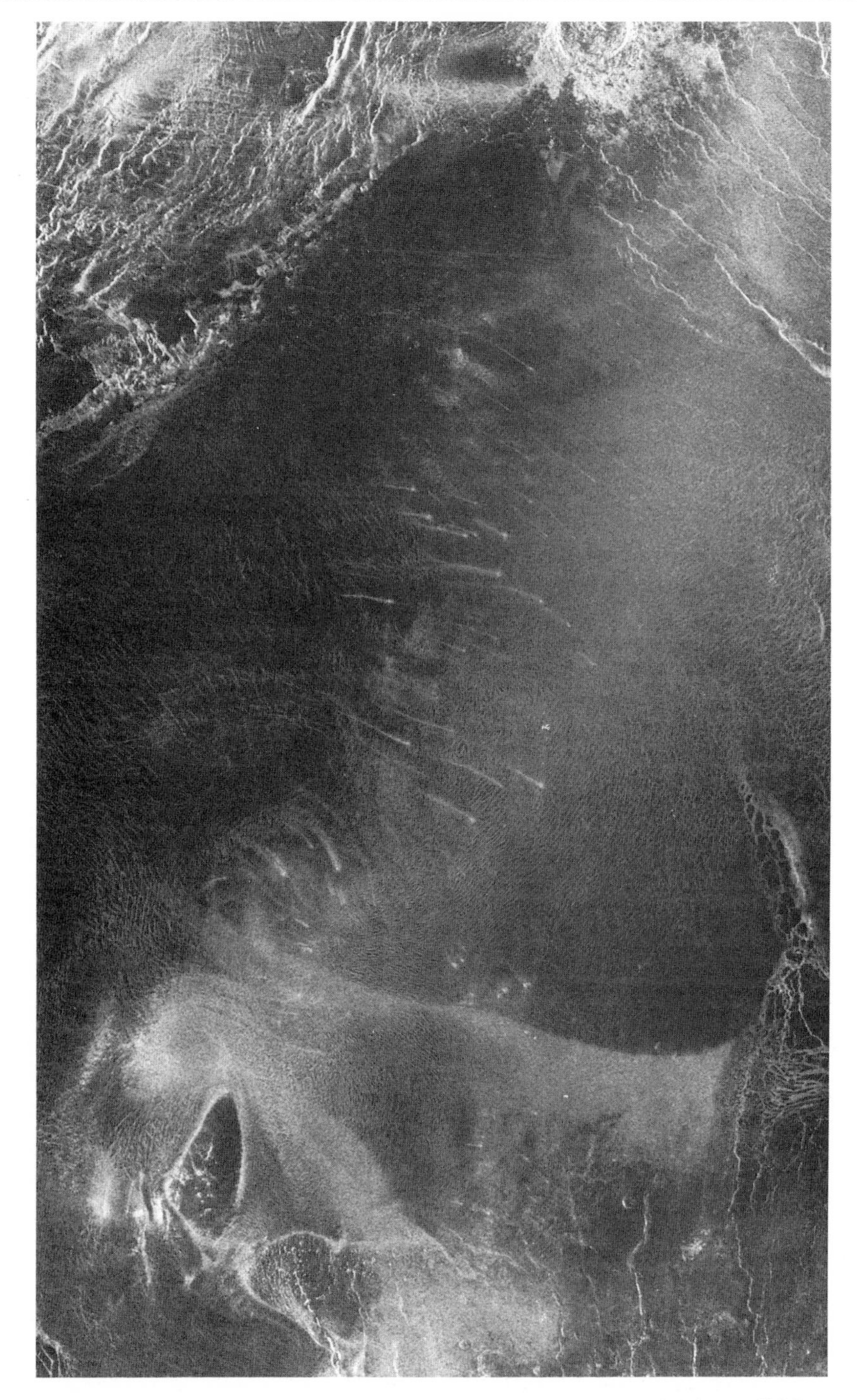

Figure 7.10 High-resolution view of dunes in the Fortuna-Meshkenet field
Both dunes and radar bright wind streaks are shown. Magellan image MRPS 39824.

eral small radar-bright cones; their similarity in radar return suggests that the cone material is their source. It will be noted that there is a sharp change in brightness across the centre of Figure 7.10; this does not have a topographic basis and is probably a boundary separating materials with different dielectric constants. Greeley et al. (1992a) suggest that the radar-bright materials are of different composition from the surface on which they outcrop. They also proffer the suggestion that the source material for the dunes is debris from the complexly deformed tessera regions that has become trapped within the valley separating them. That the tesserae blocks appear to provide sedimentary debris for aeolian transportation was earlier suggested in §7.1.2.

7.2.4 Yardangs

The area surrounding the impact crater Mead, situated on the northwestern flank of Aphrodite Terra (9°N, 65.5°E), hosts the greatest concentration of aeolian features on Venus. These include the only occurrence of potential Venusian *yardangs*. Terrestrial yardangs are narrow, often undercut, ridges of rock, separated by long corridors that have been eroded by wind-blown sand; they are roughly parallel and have their long axes parallel to the prevailing wind.

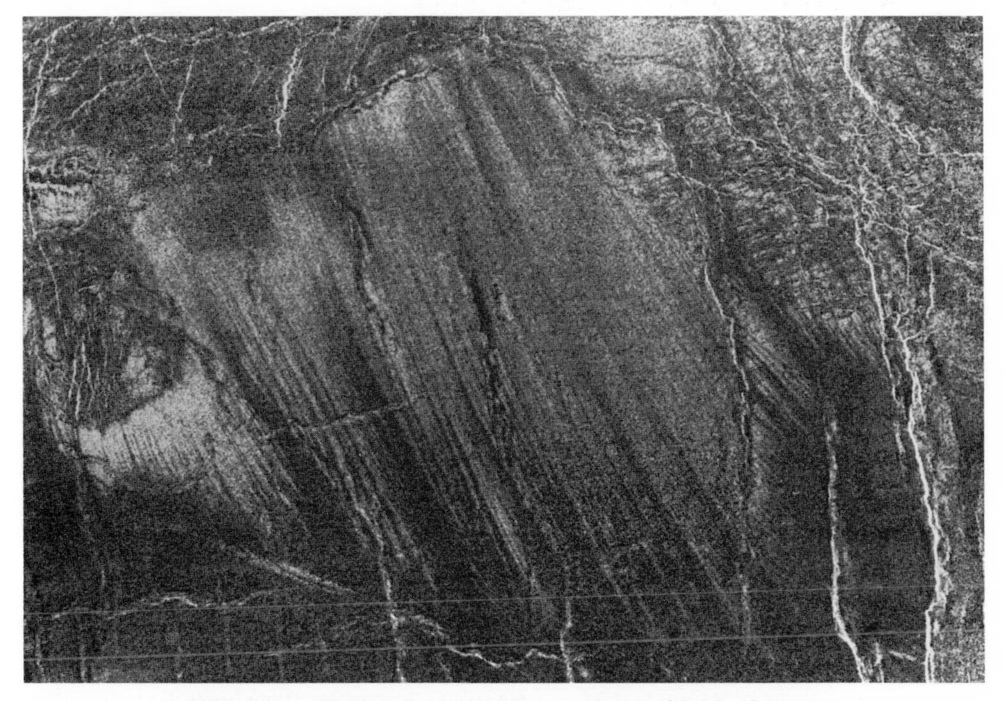

Figure 7.11 Multiple linear streaks associated with Mead crater
These are believed to represent yardangs. Magellan image MRPS 38160.

The Mead landforms take the form of somewhat sinuous, parallel ridges and grooves (Fig. 7.11). They have well defined boundaries and measure around 25 km long and 0.5 km wide; spacings of between 0.5 and 2.0 km are typical. The features believed to be Venusian analogues of this landform type comprise two sets, which together include 100 individuals; the orientation of both sets indicates a prevailing northeast to southwest wind flow. Since most terrestrial yardangs develop in relatively friable and often well jointed materials (e.g. lacustrine sediments, volcanic ashes), it seems likely that the deposits associated with the formation of Mead crater have provided suitable easily erodible materials for yardang formation and also suggests that winds in this area have remained constant in direction at least in the relatively recent past.

7.3 Summary

The studies that have been already completed show there to be a strong positive correlation between aeolian features and impact craters. The two large craters, Carson and Mead, with their parabolic haloes, are particularly noteworthy in this respect, but not unique. The suggestion has been made that such craters are relatively young and that their associated haloes of (presumed) ejecta deposits have not been homogenized by the Venusian environment (Greeley et al. 1992a). One possible model for the relationships that have been observed is as follows:

1. Prior to an impact, a "bow shock" is produced in the lower atmosphere by the infalling impactor.
2. Owing to the great atmospheric density, this shock would generate substantial turbulence where it interacted with the surface, dislodging weathered or loose debris, and injecting it into the atmosphere.
3. If the impactor was disrupted prior to reaching the ground, then a cloud consisting of a mixture of debris from the disrupted bolide and disturbed surface debris would be formed.

Such clouds could explain some of the "failed impacts" described by Phillips et al. (1992a,b) and discussed in Chapter 8. If, on the other hand, the impactor hit the ground, then a cloud of particulates, consisting of ejecta and bow-shock-generated material, would rise as an expanding hemisphere of fine debris. Under the influence of the prevailing easterly winds, this would be distributed asymmetrically to the west, giving rise to a radar-dark parabolic collar. The emissivity and reflectance data confirm that much of this material is of sand or silt grade.

Whether or not such a scenario proves to be the correct one, what does seem clear is that the impact process has been responsible for providing much if not most of the loose debris that has been moved by the wind. Where wind streaks are not associated with cratering, they tend to lie close to tesserae, as in Tellus and Ovda Regiones. The obvious inference here is that these highly deformed zones, having undergone intensive mechanical disruption, are another source of sedimentary debris. Furthermore, their steeper boundary slopes would facilitate mass wasting of any eroded material formed.

This now begs the question: Why are aeolian landforms not observed in association with deformed structures such as coronas? The answer probably lies in the lower degree of deformation and less-steep bounding slopes that are typical of these structures. The former would conspire to generate less particulate material in the first place; the latter would deter efficient mass wasting of any material that did form, and give rise to less-strong atmospheric turbulence, which might produce strong local winds.

Finally, it seems clear that low rates of erosion characterize Venus, largely owing to the absence of water at the surface. The distribution of wind-related landforms so far detected also suggests that a near-surface Hadley cell circulation currently is in operation, with prevailing winds blowing towards the Venusian equator. However, a more definite pronouncement on this matter must await detailed analysis of more imagery, preferably with different viewing geometries, so that spurious effects of any observational bias inherent in the spacecraft's trajectory can be removed.

8 Impact cratering on Venus

8.1 Introduction

Intense bombardment of planetary surfaces was an important phase in Solar System development. For the Moon, this period of surface modification began about 4.6 billion years ago, declined around 3.8 billion years ago, and has become known colloquially as the "Great Bombardment". This same activity would have affected Venus; however, Venus has a dense atmosphere whose presence not only filtered the population of incoming asteroids, meteoroids and comets, but also played a major rôle in determining that the sequence and kind of ejecta that surrounds Venusian impact craters are in several ways different from those ob-

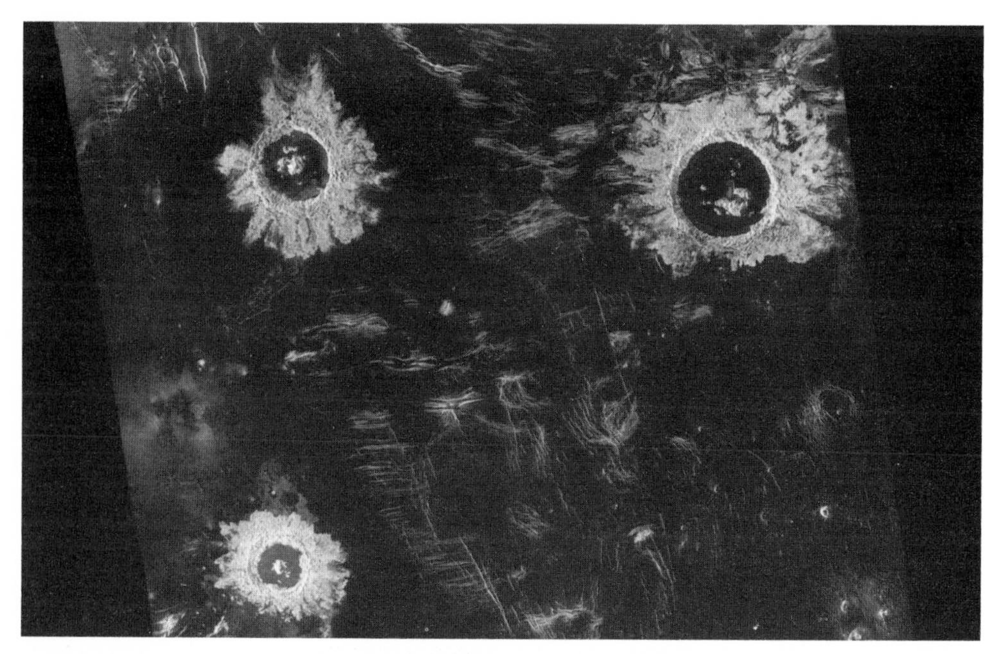

Figure 8.1 Three large impact craters
Three 37 to 50 km craters are seen in this Magellan image in the region of Lavinia Planitia. The typical radar-bright lobate ejecta patterns are distinctive and typical of Venusian craters of this size. Magellan image P-36711.

served elsewhere in the Solar System. Magellan images show that many craters and the pattern of ejecta that surrounds them are peculiarly Venusian (Fig. 8.1). Furthermore, modelling exercises and laboratory simulations reveal a complex interplay between impactors and the atmosphere of Venus that does not apply in vacuum or near-vacuum environments, such as those pertaining to either the Moon, Mercury or Mars. In the ensuing pages will be found a description of the main characteristics of the Venusian impact record and the phenomena that have produced it.

8.2 The Venusian impact record

Pioneer-Venus revealed large circular structures on Venus at roughly the same time as images showing crateriform structures were received at Arecibo (Campbell & Burns 1980). Some were interpreted to be impact craters (Masursky et al. 1980), although, owing to resolution constraints, it was not possible to be sure that such an interpretation was entirely appropriate. That a crater record should exist, despite the shielding effect of the atmosphere, had been predicted by Petrov & Stulov (1975).

8.2.1 Numbers and size range of craters

The arrival of Venera 15/16, with its greater resolution, aided by stereoscopic coverage, proved once and for all that such a record exists (Basilevsky et al. 1986, 1987, Ivanov et al. 1986). Although only 40% of the globe was imaged, at a resolution of 1–2 km, 146 circular structures larger than 8 km diameter were catalogued. By July 1992, Magellan had imaged 89% of the Venusian surface and extended the tally of craters to 798; their diameters range from 1.5 to 280 km (Schaber et al. 1992). Smaller craters have not evaded Magellan's scanner, they simply do not exist – a fact which, as we shall see, has important repercussions for the history of Venus.

8.2.2 Crater morphology

Most Venusian craters are virtually pristine and are characterized by sharp *rims* and well preserved *ejecta deposits*, rather like those on the Moon. Those craters larger than about 15 km in diameter are generally circular and similar to comparably sized craters on other planets; however, smaller ones tend to depart from circularity and have greater complexity. This is the reverse of the lunar situation, and is a response to the greater density of the Venusian

Figure 8.2 The impact crater Danilova
Situated at 337.2°E, 26.4°S, this crater is 48 km across. Note the radar-bright ejecta, terraced walls and central peak complex. The floor is radar-dark. Magellan image MRPS 35420.

atmosphere. Modifications that have affected some craters allow some general conclusions to be drawn regarding volcanic, tectonic and aeolian processes.

That larger Venusian craters share many of the features of large lunar ones can be appreciated by looking at the 48 km diameter crater, Danilova, which has a more-or-less circular outline, terraced walls and a central peak complex (Fig. 8.2). When all of the craters so far imaged are considered, there appears to be size-related a progression of morphological characteristics: thus the largest structures are multi-ringed, followed at smaller diameters successively by structures with double rings, craters with one ring, craters with a central peak, craters having structureless floors and, finally, irregular or multiple craters (Fig. 8.3). The final group represents a divergence from the familiar lunar pattern, where the smaller craters are simple, bowl-shaped and circular. As we shall see, the latter is a direct effect of the dense Venusian atmosphere.

The materials of crater rims characteristically have a strong radar back-scatter and therefore a bright signature; near-rim ejecta has a hummocky appearance and fans out into lobes or flaps. Farther out, while radar-bright patches still occur, ejecta facies tends to exhibit lower back-scatter and darker signatures (Fig. 8.4). Generally speaking, rim ejecta facies extends

(a)

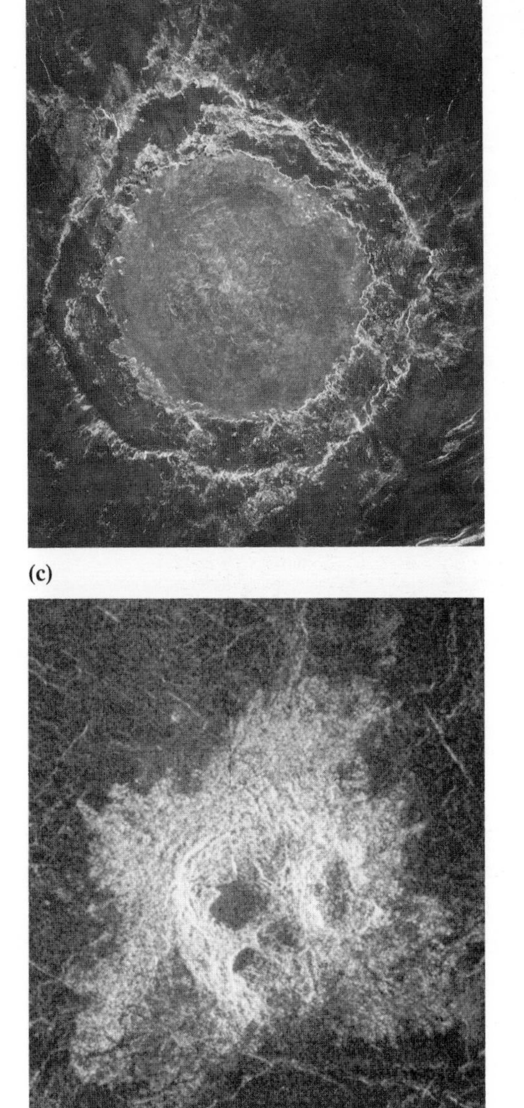

(b)

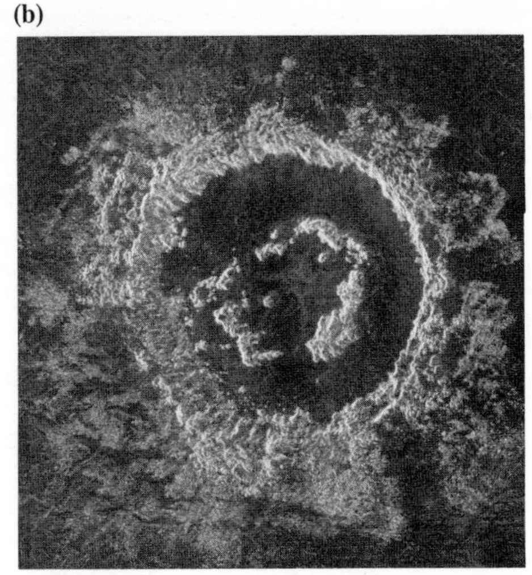

(c)

Figure 8.3 Morphological differences between craters of different sizes

(a) Crater Mead, 280 km in diameter, located at 57.4°E, 12.5°E. This is the largest crater on Venus and has a multi-ringed structure. Magellan image F-MIDRP 13N058. (b) Crater Barton, 50 km in diameter, located at 337.5°E, 27.4°N. This has an inner ring complex and modestly terraced walls. Magellan image F-MRPS 34799. (c) Crater cluster Lilian, located at 336.0°E, 25.6°N, diameter approximately 14 km. Magellan image MRPS 33958.

outwards for about three crater radii. The sharply defined, petal-like morphology of ejecta associated with smaller craters usually has a bilateral symmetry about a vertical line passing through the crater's centre. Strongly asymmetric ejecta distribution is taken to imply oblique atmospheric entry of the incoming bolide (Schaber et al. 1992).

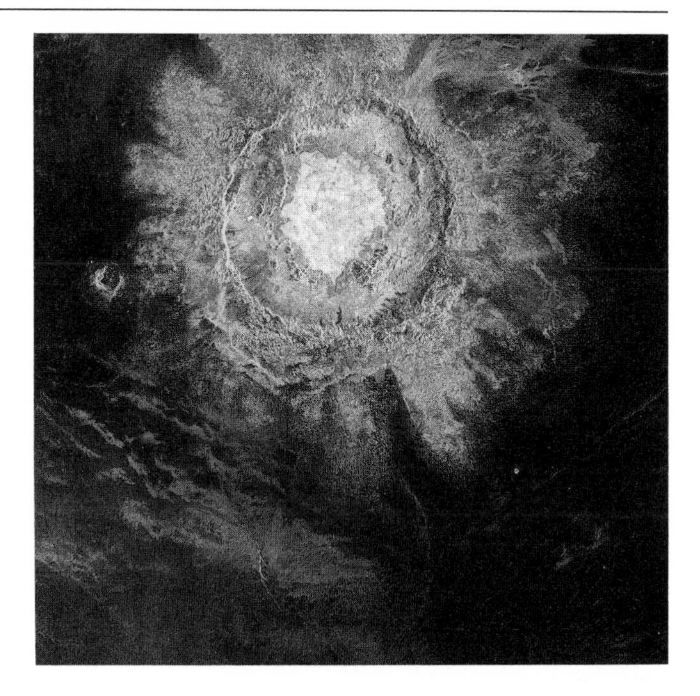

Figure 8.4 The multi-ringed crater Stanton
Located at 199.9°E, 23.4°S, this is a typical multi-ringed crater and is 110 km diameter. Magellan image F-MIDRP 25S198;201.

8.2.3 Parabolic haloes

Arguably, the most unusual and intriguing of the phenomena to accompany cratering are the extensive parabolic haloes of (usually) radar-dark material. These are particularly striking in the cases of the 38 km diameter crater Carson (Fig. 8.5) and 31 km diameter Aurelia, where they extend with an arc-like pattern for several hundred kilometres from the focus. They have been studied by Greeley et al. (1992a) during an appraisal of Venusian aeolian features, and by Phillips et al. (1992), during a cratering study. In the case of Carson, the radar-dark halo is superimposed on lobate volcanic flow plains, and has lower emissivity than the plains materials (0.80 vs 0.85). This could be interpreted as meaning that the material either has intrinsically higher reflectivity, or exhibits different surface roughness, or both. Associated with the surrounding area are large numbers of wind streaks whose orientation indicates winds blowing towards the Venusian equator, and a generally eastward flow. Greeley and his colleagues venture an explanation that the pre-impact bow shock, created by the incoming projectile, produced strong turbulence in the atmosphere, which lifted sand and dust high into the air. The finer particles, rising the highest, became entrained in the prevailing high-altitude easterly winds and were distributed progressively westwards, settling eventually to form the observed radar-dark parabolas. Such a suggestion does seem to be in accord with predicted behaviour of ejecta, based on both mathematical modelling and laboratory simulations (see §8.3.2 below). Another interesting point in connection with this is that the incidence of such features around fresh-looking craters, and the occurrence of large numbers of associ-

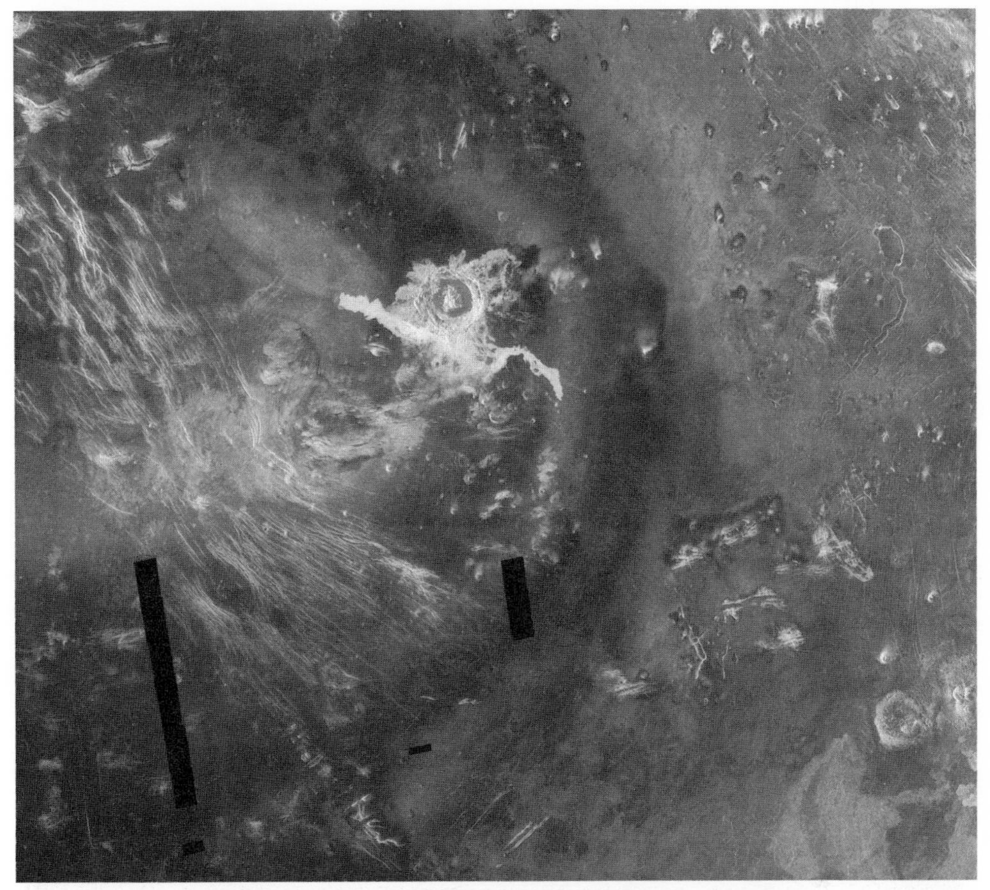

Figure 8.5 The crater Carson and associated parabolic halo
The arc-like pattern and radar-bright ejecta and outflow are both very striking. Carson is located at 433.1°E, 24.2°S. Magellan image F-MIDRP 25S345.

ated wind streaks, strongly suggests that the craters to which they are attached may be among the youngest on the planet, there having been too little time for time-dependent processes to eradicate by homogenization either the streaks or haloes.

8.2.4 Crater modification

In contrast to the Moon, only a very small percentage of craters appear to be undergoing removal, volcanic inundation or tectonism (Phillips et al. 1992). However, the total number of craters on the planet's surface is small compared with the lunar sample, and considerable caution must be exercised in interpreting the data. The limited crater sample predictably is a function of atmospheric filtering, but account must also be taken of the distinct possibility that resurfacing may have conspired to modify the population, before one can arrive at a credible interpretation of what the data mean in terms of surface ages.

When the locations, relative sizes and types of crater are plotted globally, first impressions are that the distribution is entirely random (Fig. 8.6). Visual inspection reveals that,

(a)

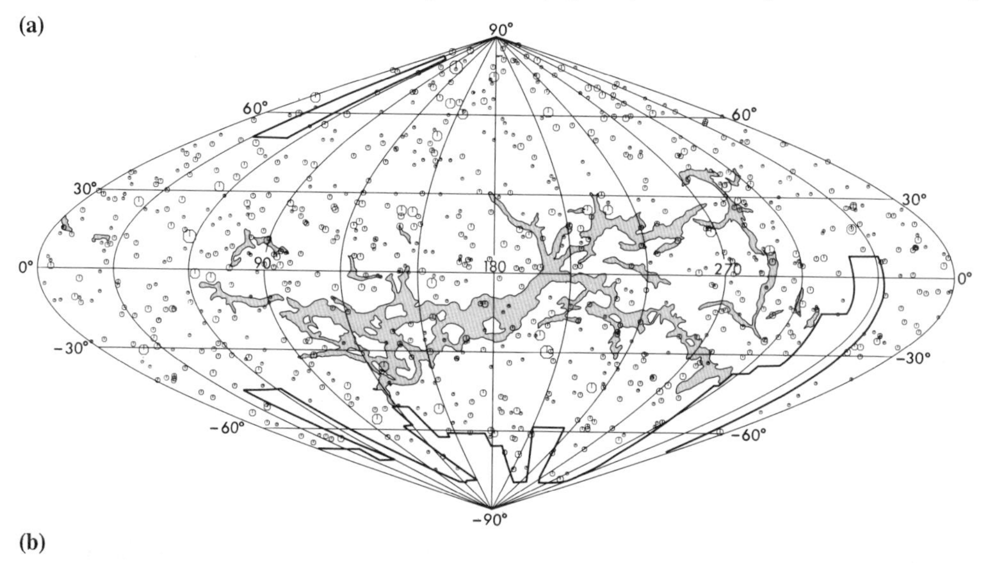

(b)

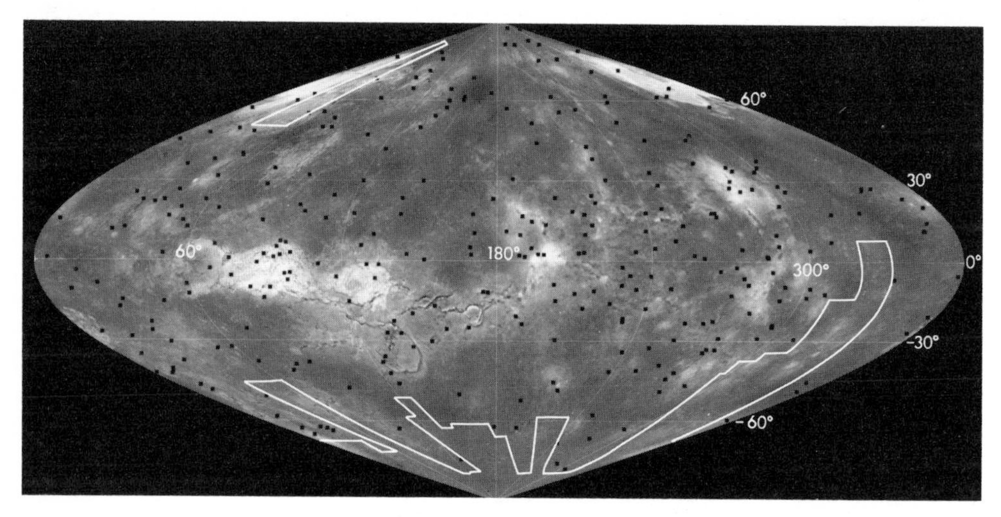

Figure 8.6 Distribution of impact craters on Venus
(a) Map in sinusoidal equal-area projection showing 842 impact structures on 89% of Venus's surface observed up to Magellan orbit 2578. Areas not mapped are shown within heavy dark line. Sizes of symbols are scaled to crater diameter: 1.4–11.3 km, 11.4–32.9 km, 33.0–64.9 km, 65.0–128.9 km, 129.0–362.0 km. Shaded areas represent belts of concentrated extension. (b) Map showing all craters that have been modified by fracturing, compression or embayment by volcanic flows. Maps after Schaber et al. (1992).

although pristine craters with radar-bright ejecta blankets typify large areas of the equatorial plains, approximately 17% of the population can be considered to have undergone modification, having either been partly filled by volcanic plains or had their ejecta deposits *modified*, i.e. embayed by lava flows (Fig. 8.7a), or *tectonized*, i.e. affected by fractures and faults that transect some or all of their walls, floors and ejecta blankets (Fig. 8.7b). The distribution of such modified craters is not spatially random. Specifically, the volcano-tectonic zones of Beta, Atla and Themis Regiones are not only deficient in pristine craters but actually have an excess of modified and tectonized craters (Fig. 8.8). This inevitably implies that there is a non-random distribution of such craters on the surface of Venus, and that a degree of *re-*

(a)

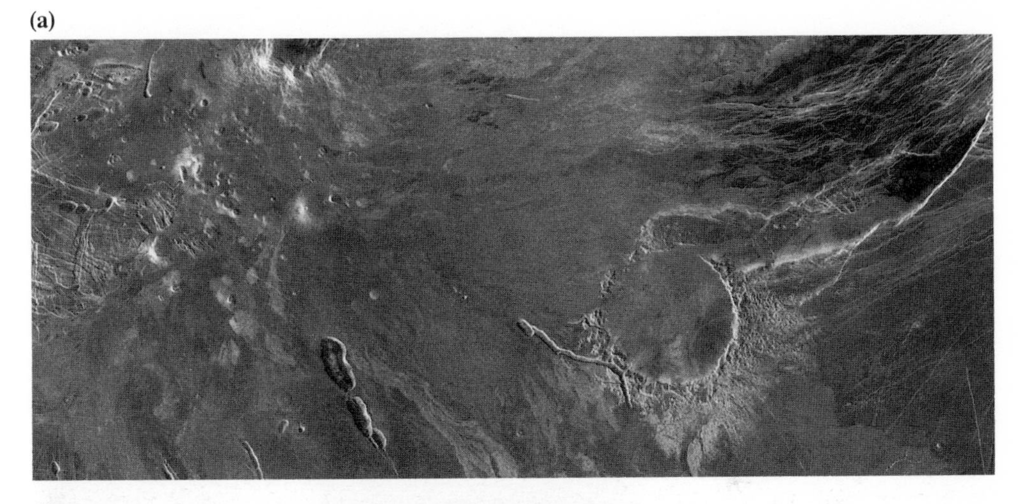

(b)

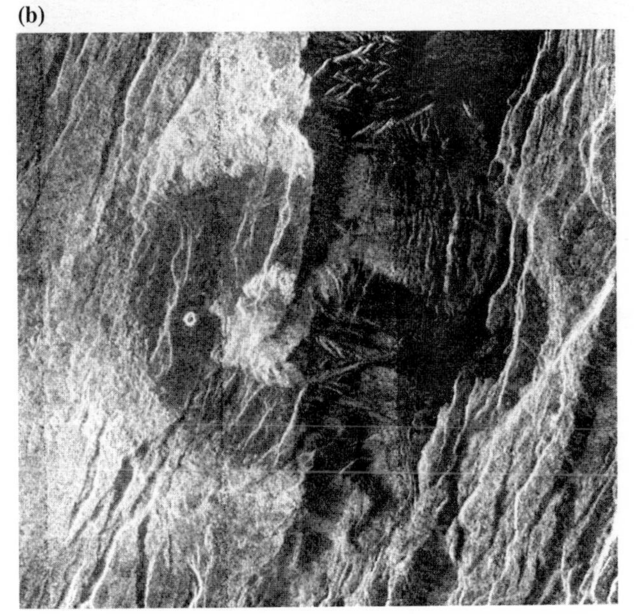

Figure 8.7 Modified impact craters (a) Crater Alcott, located at 354.5°E, 59.5°S, showing heavy embayment by volcanic plains deposits. Magellan image C1-MIDRP 60S347. (b) The 37 km diameter Crater Somerville, which has been pulled apart by faulting associated with the rise of Beta Regio. Location 155.5°E, 29.95°N. Magellan image P-38741.

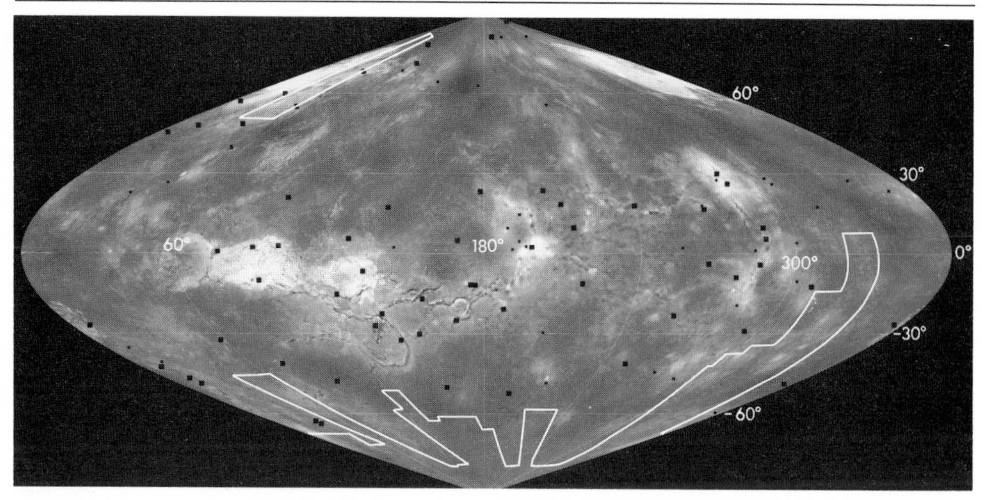

Figure 8.8 Severely modified craters on Venus
Those affected by fracturing are shown by solid squares, those by lava embayment by small squares. After Schaber et al. (1992).

surfacing has occurred. Statistical analysis of crater size, frequency and distribution confirms that the visual impression described above is real (Phillips et al. 1992), and that the western part of Aphrodite Terra – a region where volcanism is relatively lacking – shows evidence for craters having been removed by intense tectonism, while over the western portion of the Beta–Atla–Thetis Regiones zone, both tectonism and volcanism have worked together to eradicate a proportion of the record. This suggests a broad region of lesser age than the planetary average (Arvidson et al. 1992). An important question is when this may have occurred, and whether or not it was a catastrophic global event.

8.3 Atmospheric effects on the impact record and surface ages

Experimental and field data accrued rapidly during the years following the Apollo missions, as studies were made of nuclear explosions and terrestrial impact craters and from detailed mapping of the Moon. This led to a reasonable understanding of the cratering process in targets of varying type and, as planetary exploration gradually extended farther out into the Solar System, of the details of ejecta emplacement on planets and moons with varying conditions of gravity, target properties and atmospheric density (see Roddy et al. 1977). Now, some 20 years after the last Apollo lunar module lifted off from the Taurus- Littrow Valley, Magellan's exploration of Venus provides, for the first time, an opportunity to study the bombardment of a rocky surface protected by an extremely dense atmosphere and, as it turns out, one where relatively little modification has affected the most recent cratering record.

8.3.1 Atmospheric filtering

There are significant differences between Venus and those other planets and moons which have extremely tenuous or non-existent atmospheres. Using the well constrained data relating to both Venus and its atmosphere that now exist, it can be shown that the most probable entry velocity of meteoroids into the Venusian atmosphere is between 18 and 25 km s^{-1} (Hartmann 1977). Using this fact, Ivanov et al. (1986) have modelled the likely effects that the atmosphere would have upon incoming projectiles. Their research indicates that, at a height of around 80 km from the ground, a rapidly moving rocky object (meteoroid) begins to deform inelastically; as it approaches to within 30 km, deformation approaches hydrodynamic conditions. Projectiles therefore tend to become flattened rather than rounded like those objects that impacted the Moon and Mercury in vast numbers during the Great Bombardment. They were able to estimate crater diameter as a function of impactor dimensions, showing that the smallest craters that can be created, at least for low-strength meteoroids, are in the size range 7–13 km at zero altitude, and between 6 and 11 km at around 10 km altitude. The

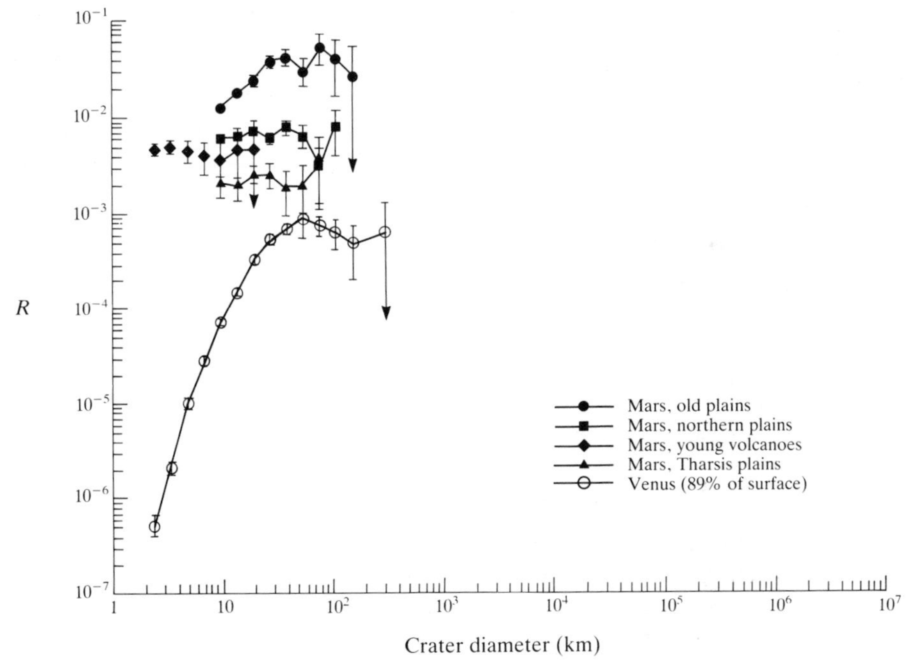

Figure 8.9 Impact crater size-density plot for Venus and selected regions of Mars
The size–density distribution for Venusian craters with diameters ≥35 km is nearly horizontal, as are those for unmodified populations on other planets. However, on Venus there is a distinct paucity of smaller craters. After Schaber et al. (1992).

$R = \frac{D^3N}{A(b_2-b_1)}$ where D is the geometric mean diameter, (b_2-b_1) is a bin ranging from b_1 to b_2, N is the number of craters in the size bin, and A is the area counted.

very different impact crater size–density distribution observed when comparing the Venusian data with that for, say, Mars, is indicative of this fact (Fig. 8.9). When the exercise is extended to non-rocky projectiles (Ivanov et al. 1992), using different modelling techniques, it can be shown that icy bodies with diameters of around 2 km would probably reach the bottom of Venus's atmosphere too, albeit in a partially disrupted state, and would produce craters about 20 km in diameter. In other words, comets also could penetrate the atmosphere and would be expected to have played a part in the cratering process.

More recent modelling of the atmospheric filtering process, using data of both the total population of craters and the population of crater clusters revealed by Magellan – which provides a cutoff limit for bolide break-up in the atmosphere – reveals that the smallest crater whose numbers are not affected by the filtering process is around 32 km across (Phillips et al. 1992). Using this as a basis, it is then possible to compare the size–frequency curve of craters 32 km diameter for Venus with that of the Moon to give an idea of the resurfacing history of the planet. On the basis of all of the tests so far conducted, an average surface age of 400–500 Ma was derived, implying that the latest resurfacing event must have occurred as recently as this. However, it is possible that this is an underestimate of the real age, since cratering efficiency on Venus may be less than on airless worlds (Schultz 1992). Phillips et al. (1992) accept that their derived resurfacing age may be only a lower bound, but even allowing for the fact that their modelling did not incorporate an input from cometary objects, it is not in error by more than a factor of 2.

It is clear from the relatively high incidence of modified craters within the volcano- tectonic zones stretching along the equator that the whole of Venus was not resurfaced in one gigantic event. Crater modification evidently occurred within quite small areas, not over the entire globe. Thus, we must reject the notion of some form of catastrophic resurfacing event akin to that which may have played a part in generating the Martian dichotomy. The geological evidence from the Aphrodite region of Venus favours piecemeal resurfacing, since here distinct ages of surface are preserved by the impact record. It remains to be discovered whether or not such largely volcanic resurfacing events take place regularly, on this kind of timescale.

8.3.2 Atmospheric effects upon the cratering process

A very fundamental difference between Venus and airless bodies such as the Moon concerns the way in which energy is partitioned between the incoming projectile and the atmosphere. In the vacuum conditions of the lunar environment, a bolide retains its original kinetic energy until it impacts, whereupon it is almost immediately transferred to the surface or to the ejectamenta that the impact creates. Certainly none is transferred to the atmosphere; it is absent. In contrast, under Venusian conditions, both kinetic and internal energy are transferred to the atmosphere at an early stage, the period during which this occurs being particularly lengthy

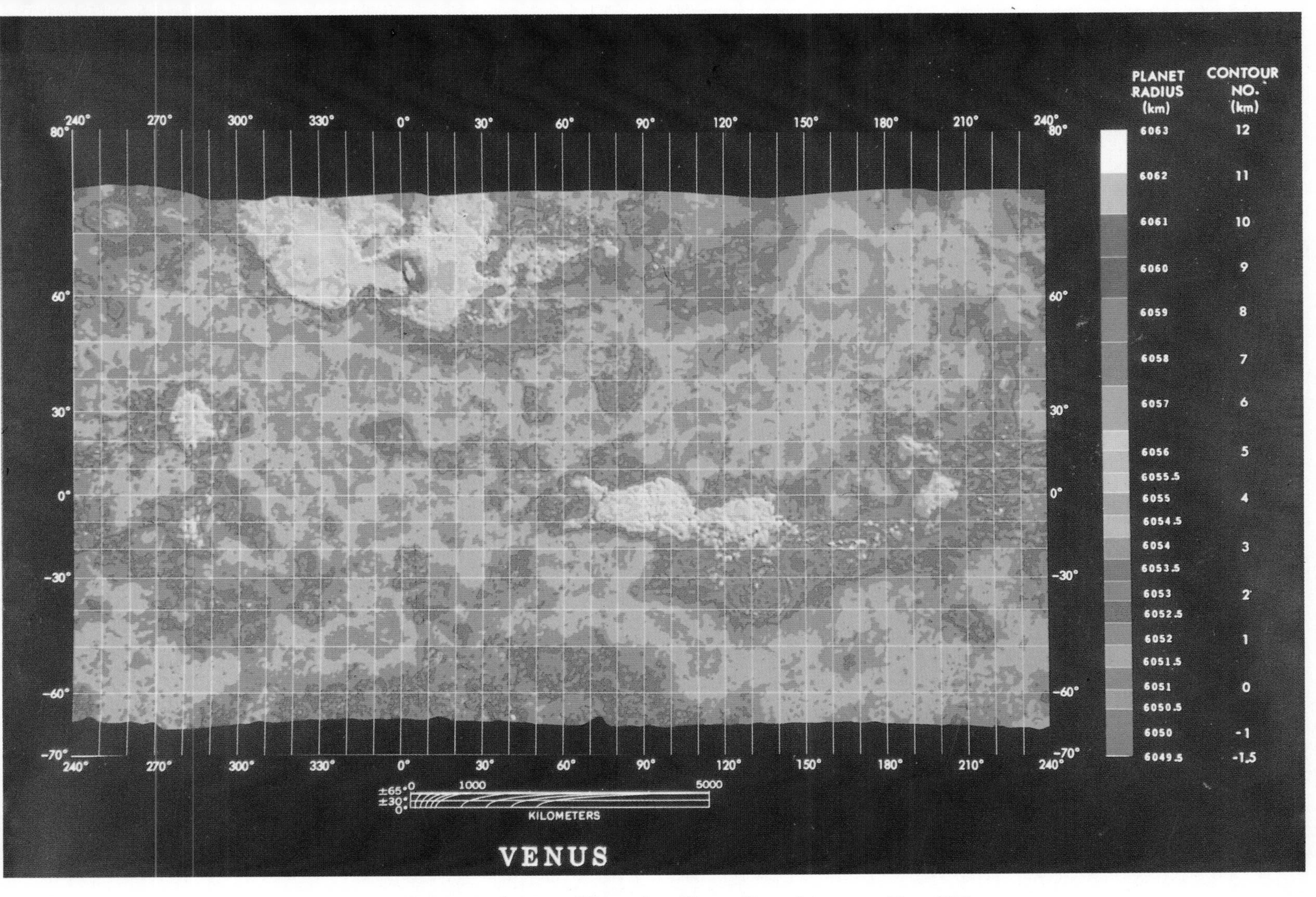

Plate 1 Contoured relief map of Venus, from Pioneer Venus data. USGS Map I-1324.

Plate 2 The Magellan spacecraft and the Space Shuttle launch vehicle. The large high-gain antenna is mounted at the top of the probe. The smaller horn antenna (used for altimetric mapping) is mounted on the side. Magellan P-33264BC.

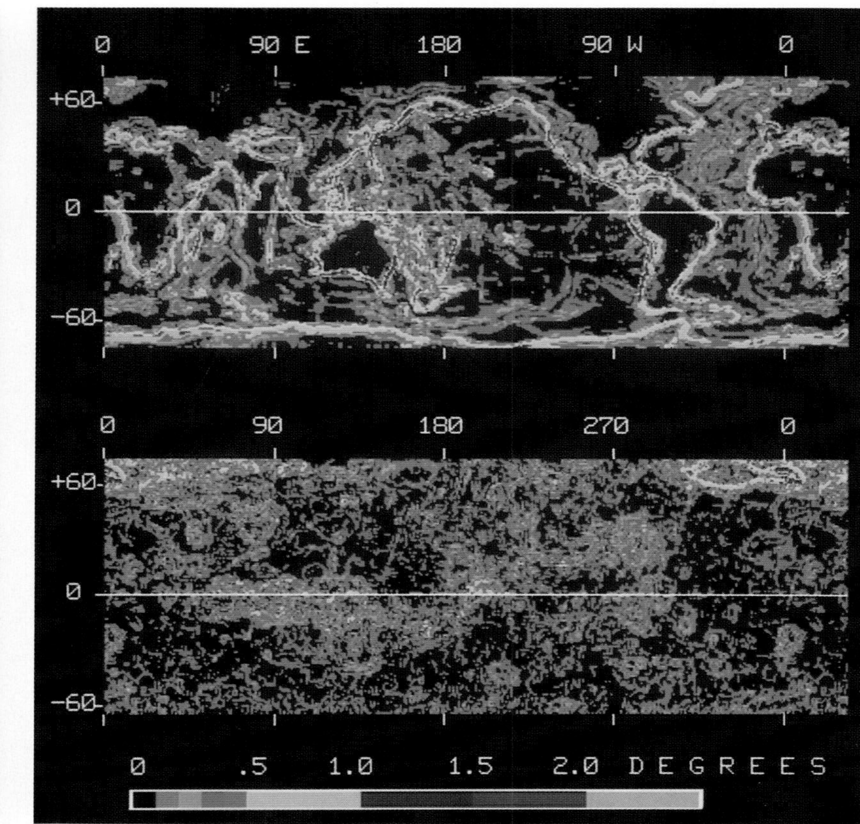

Plate 3 Regional slope maps for Earth and Venus showing the maximum slope measured over 3° by 3° regions. Range of regional slope values for Earth (top) and Venus (bottom) extends from 0° to 2.4°. However, the slope frequency distributions are quite different. After Sharpton & Head (1986).

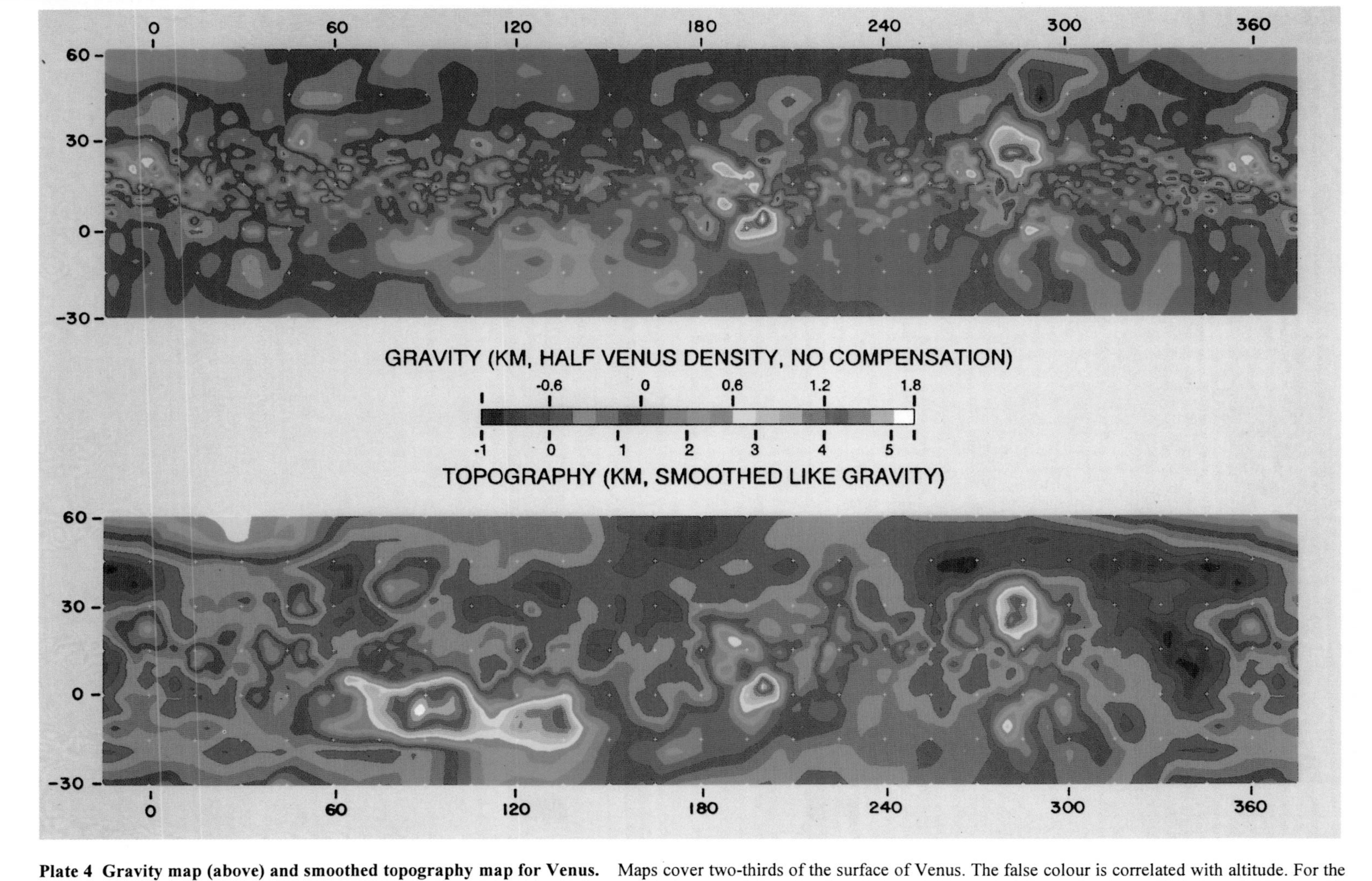

Plate 4 Gravity map (above) and smoothed topography map for Venus. Maps cover two-thirds of the surface of Venus. The false colour is correlated with altitude. For the gravity map, it is the altitude to which material of density half the mean density of Venus would need to be piled to yield the observed external potential. From Reasenberg & Goldberg (1992), with permission.

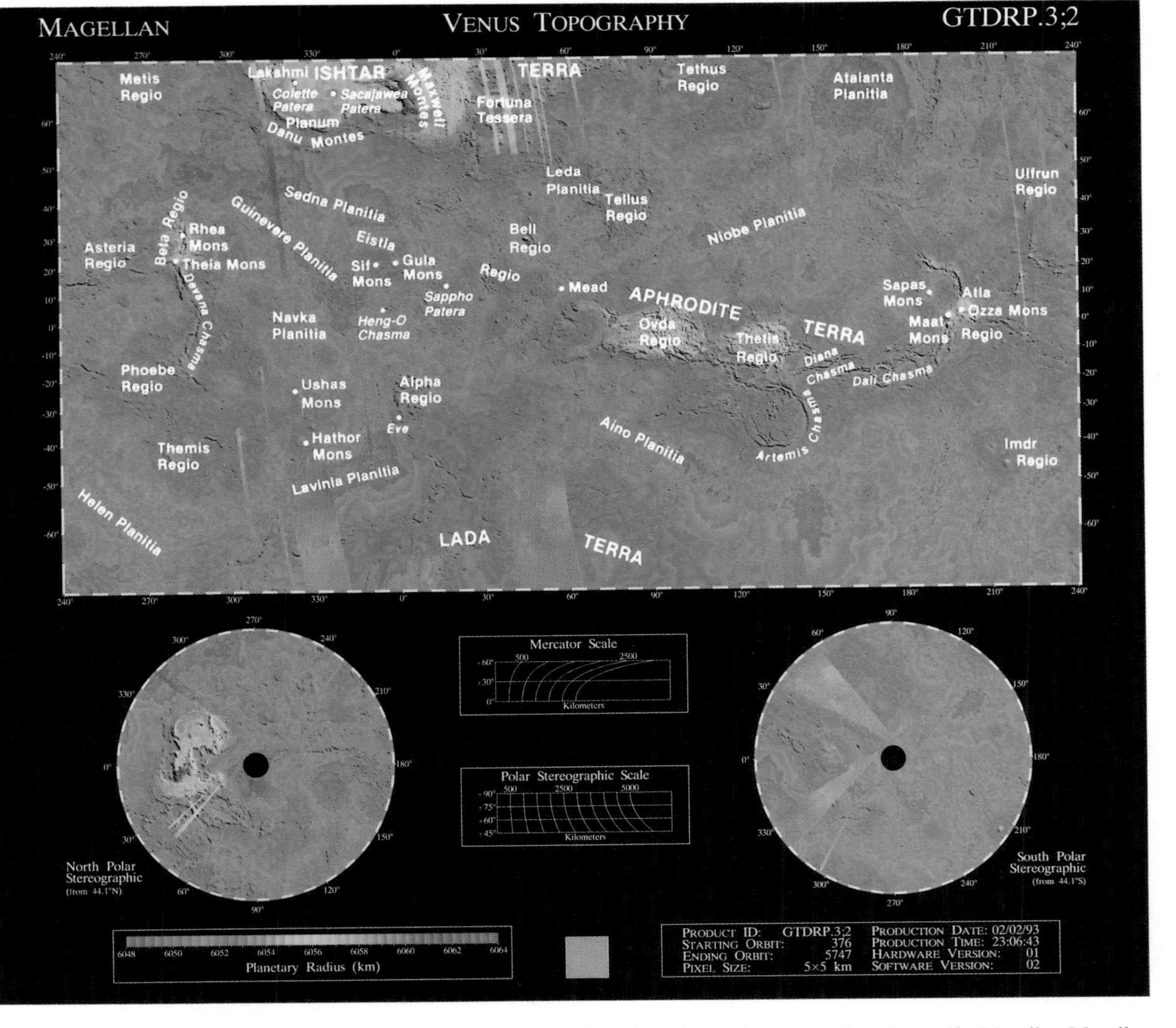

Plate 5 False colour representation of Venus topography. The regions shown in grey have not yet been imaged by Magellan. Magellan product P-41938.

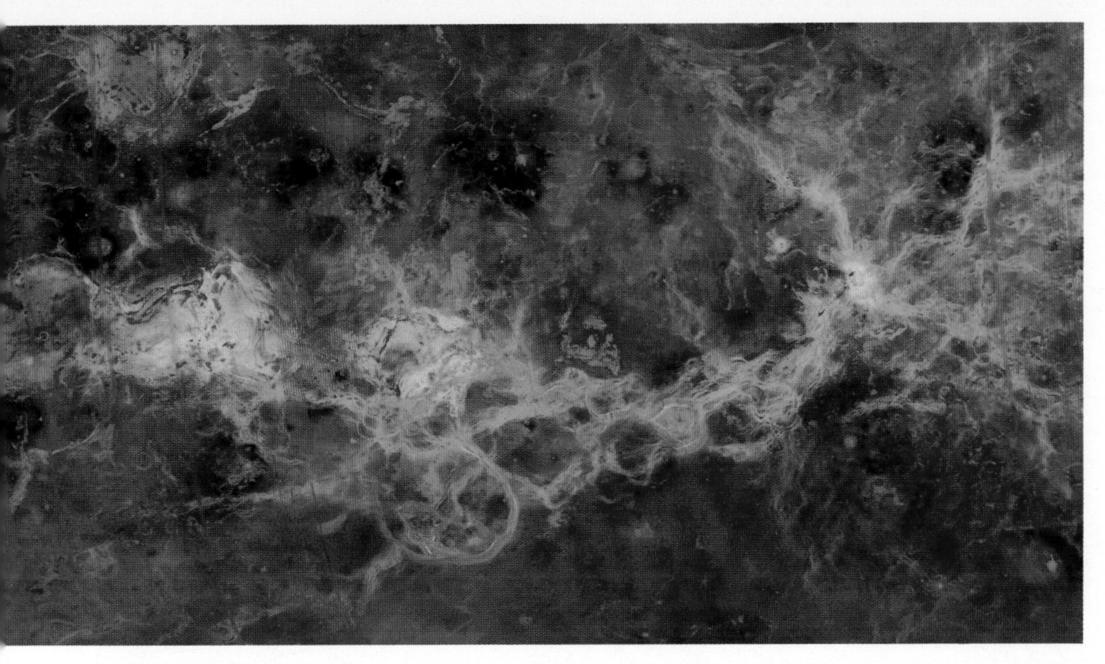

ate 6 False-colour radar mosaic of Aphrodite Terra. A Magellan mosaic (infilled with Pioneer Venus data) extending from 5°E to 240°E, and between 45°S and 30°N. Magellan MRPS 42310.

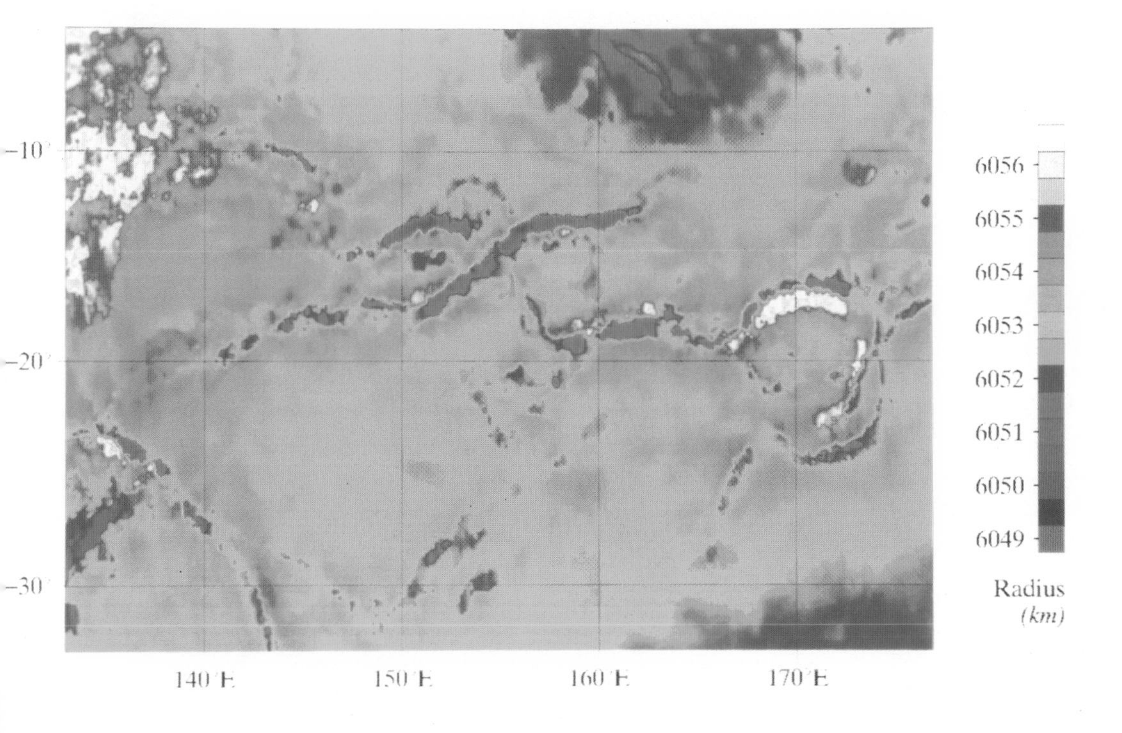

ate 7 Altimetric map of Aphrodite Terra. From Magellan altimetry on cylindical equidistant projection. Magellan product P-449.

Plate 8 Perspective view of Maat Mons. Maat Mons is a 5 km high volcano with a summit caldera complex. This view, looking south, shows radar-bright volcanic flows extending northwards and embaying ejecta from a 23 km diameter impact crater. Magellan image P-40175.

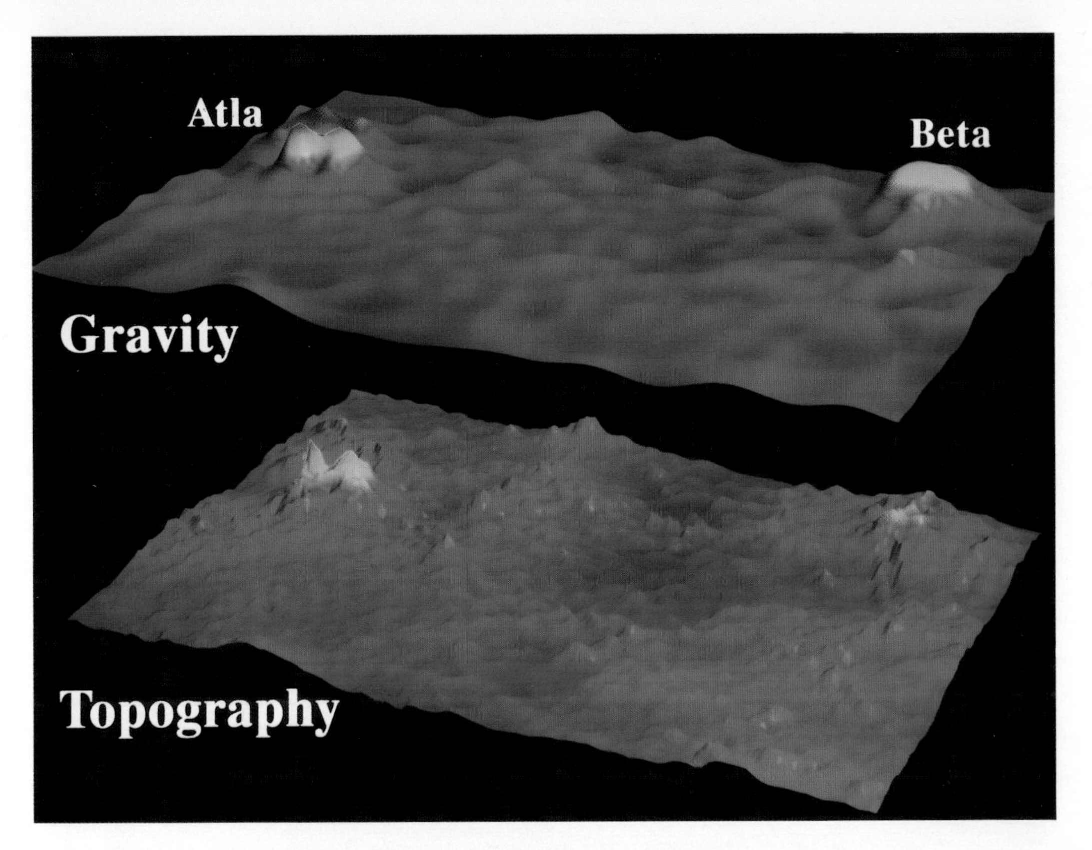

Plate 9 Topographic and gravity maps of the region between Atla and Beta Regiones. The high gravity and topography are typical of sites of rifting and volcanicity. Positive (light blue) and negative (dark blue) gravity anomalies correspond almost perfectly with topography. Magellan MGN-113, P-42356AC.

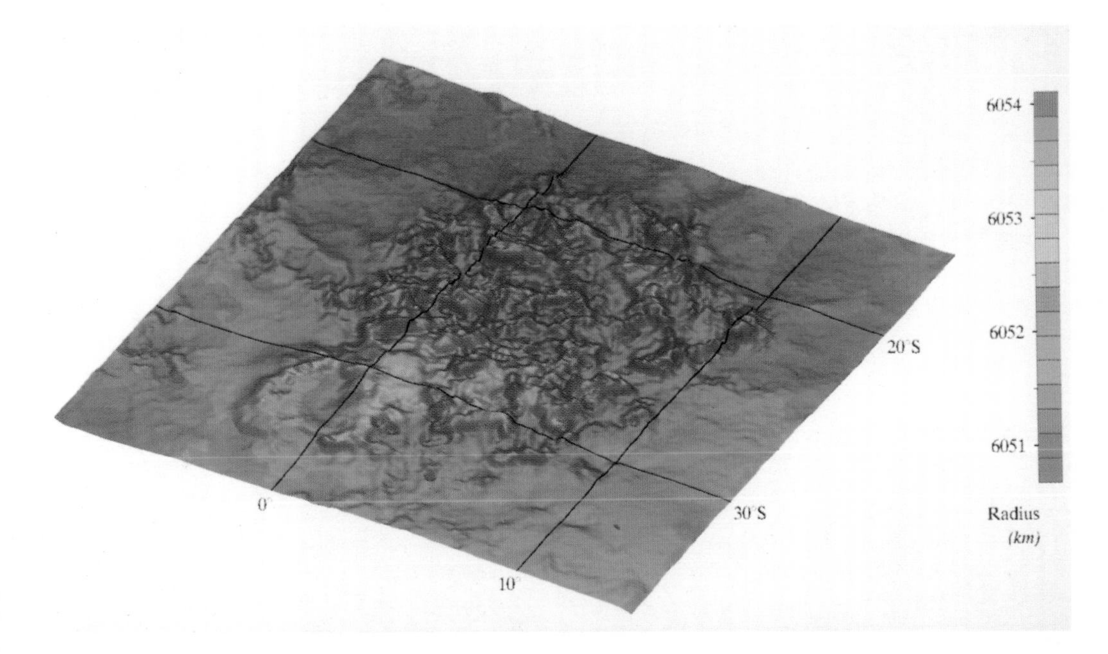

Plate 10 Alpha Regio. A colour-coded perspective image of Alpha Regio, one of the most topographically complex regions of Venus. Area is about 1000 km across. Magellan P-39374.

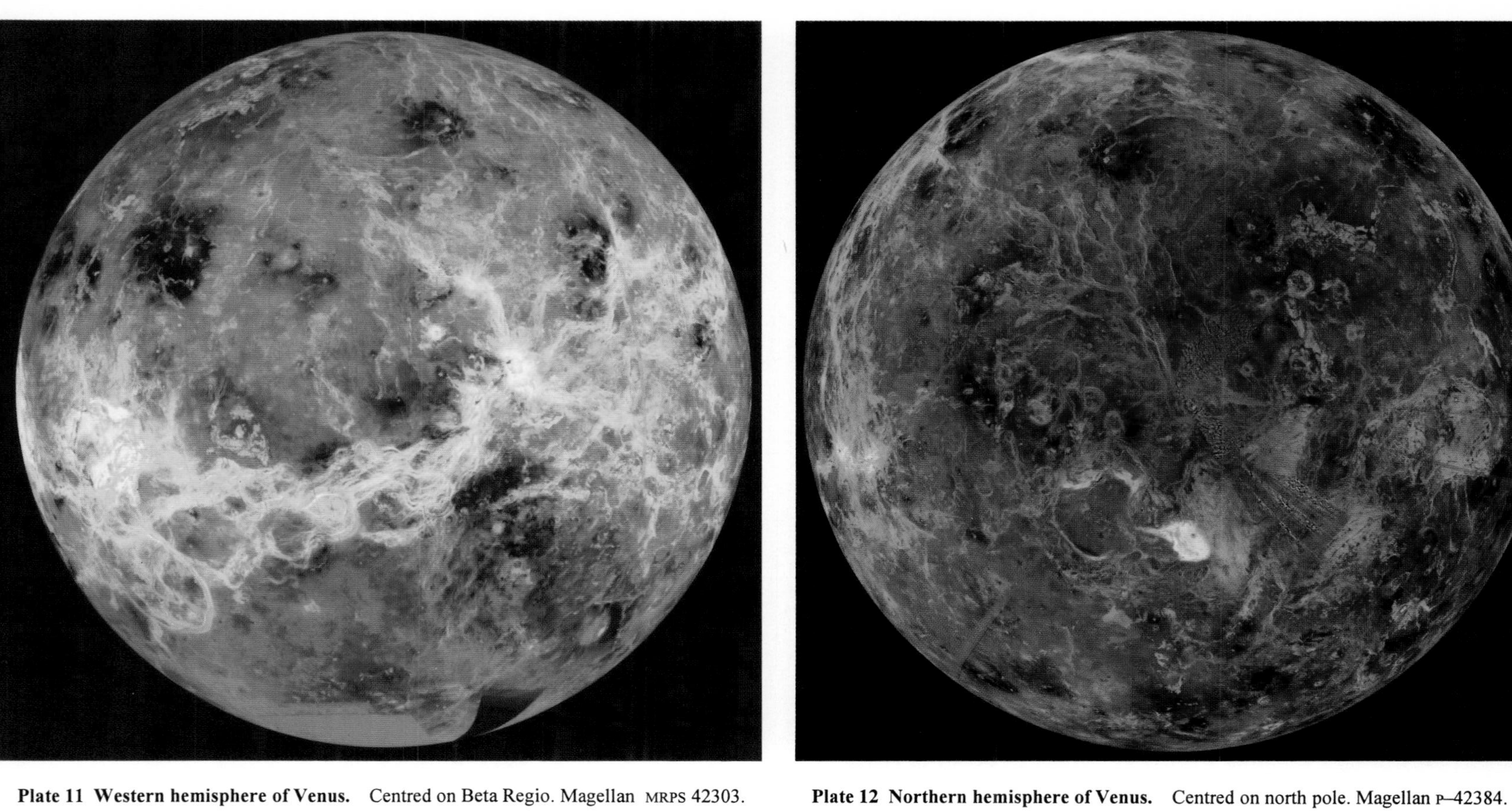

Plate 11 Western hemisphere of Venus. Centred on Beta Regio. Magellan MRPS 42303.

Plate 12 Northern hemisphere of Venus. Centred on north pole. Magellan P–42384.

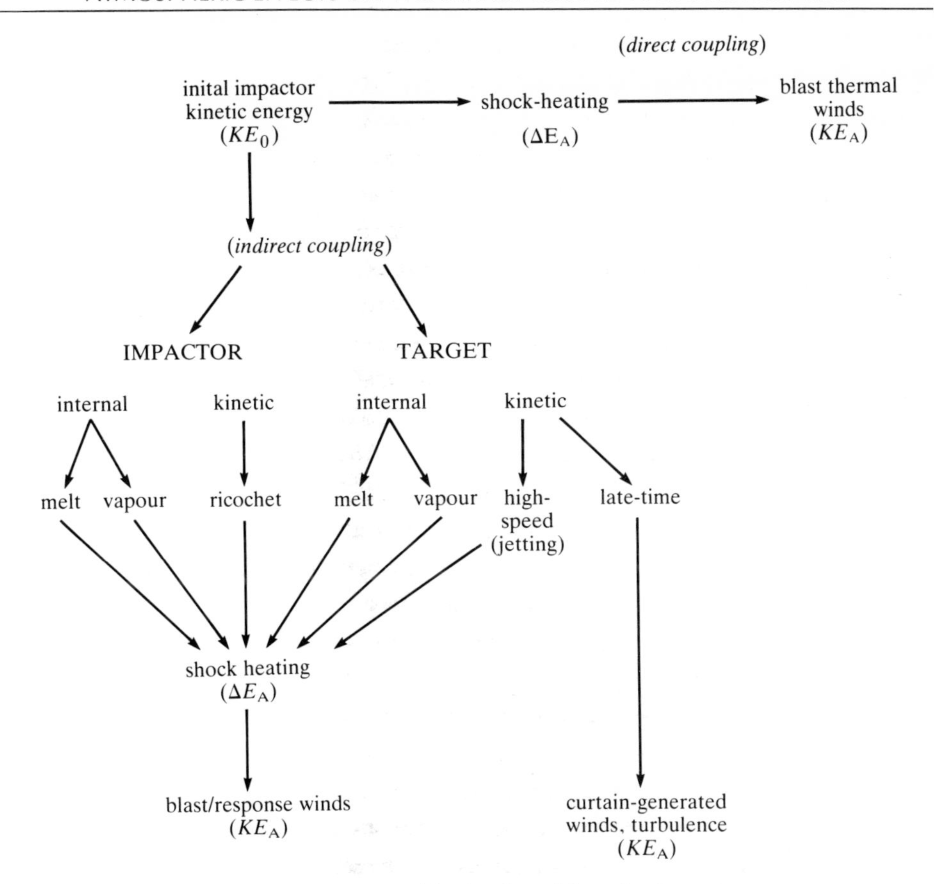

Figure 8.10 Energy partitioning for a Venusian impactor

Although most of the impactor energy eventually may be transferred to the atmosphere, signatures in the surface record will be diverse. After Schultz (1992).

in the case of an obliquely approaching bolide. A Venus-approaching projectile thus partitions a proportion of its kinetic energy directly to the atmosphere prior to hitting the ground (Fig. 8.10). In the much-quoted Tunguska event, where an incoming comet catastrophically disrupted above the Earth's surface, most if not all of the energy apparently was partitioned to the atmosphere, the shock wave flattening trees over a very large region (O'Keefe & Ahrens 1982). Should a projectile break up above the Venusian surface, and it seems that the maximum size of impactor that would undergo break-up is 3–4 km, then radar-bright zones of intense scouring will be associated with the craters produced, owing to catastrophic destruction of the bolide and collisions from the ensuing wake. This represents a kind of *atmospheric cratering* process, a phenomenon not found on either the Moon, Mercury or Mars. Evidence that this has occurred frequently above Venus is provided by both the abundant radar-dark splotches that have no focal crater (Fig. 8.11a), and by radar-bright haloes surrounding a radar-dark zone without craters (Fig. 8.11b). The radar-bright zones are thought to represent areas of

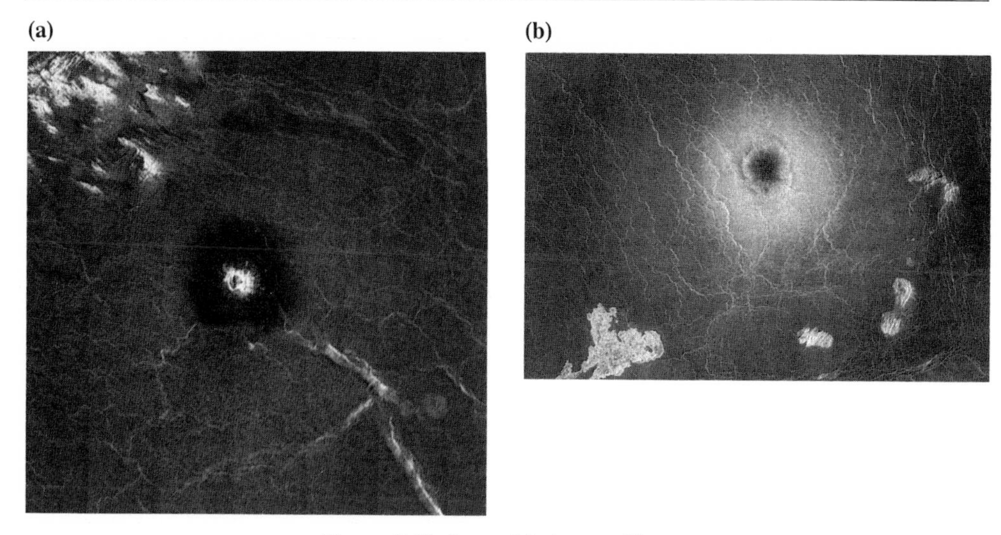

Figure 8.11 Impact haloes on Venus

(a) Radar-dark halo around bright feature, 35 km in diameter, located at 333.0°E, 47.2°N. Magellan image MRPS 33987. (b) Radar-bright halo surrounding radar-dark zone centred at 5°E, 37°N. Magellan image F-MIDRP 25N003.

removal of radar-bright surface debris by shock waves produced from bolides that either disrupted above the surface, or were slowed down so much during atmospheric transit that they were unable to effect excavation; while the radar-dark zones represent either a form of "atmospheric cratering", or transportation and deposition of fine debris by recovery winds (Phillips et al. 1991, Schultz 1992).

In the case of a non-disrupted body, the energy released upon impact is partitioned to the atmosphere by rapidly moving ejectamenta, a fast-expanding cloud of vapour, frictional drag, and from highly turbulent disturbances within the atmosphere induced by the outward-moving curtain of ejectamenta. This has far-reaching consequences for both the emplacement of ejecta and its effect upon the Venusian surface. Because a significant amount of energy is partitioned to the atmosphere during the passage of any projectile, the gas flow associated with passage of a bolide through the dense atmosphere will imitate the characteristics of a strong explosion. Furthermore, a *tube* of low residual gas density will form along the trajectory, whose radius is large compared with the diameter of the resultant crater (Fig. 8.12). The effects of this are significant, since ejecta entrained in the region 2–3 radii wide around the focus of a crater are not affected by the "normal" undisturbed atmosphere, a phenomenon that would be expected to lead to the generation of lunar-like ejecta blankets. Thus, the boundary of the flow of dense gas probably corresponds to the inner zone of hummocky ejecta (Fig. 8.13).

Upon impact, the uplifting of huge amounts of turbulent hot gas not only disturbs ejecta deposition but also transports large volumes of dust into the upper atmosphere. Numerical simulations conducted by Ivanov et al. (1992) show the possible form of an expanding ejecta

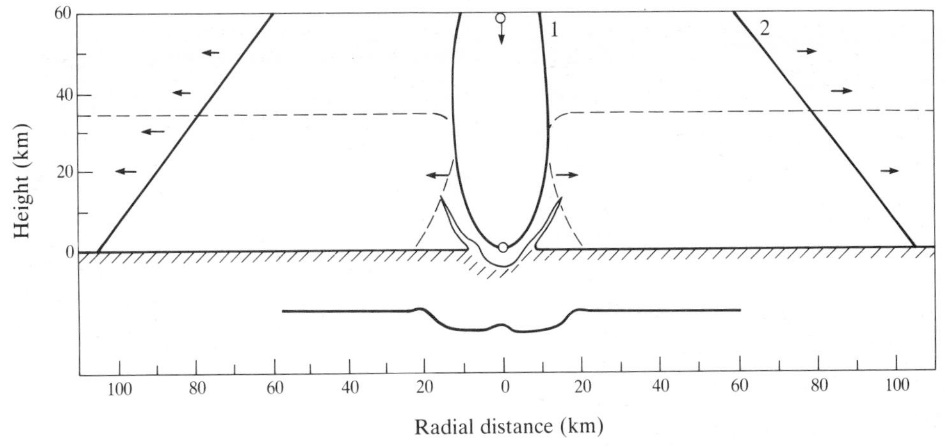

Figure 8.12 Form of low-density gas tube during Venus impact event
Position of atmospheric shock wave at moment of impact (1) and at the end of transient cavity formation (2), approximately 37 s after impact event. The dashed lines mark the position of air density 10 times smaller than the normal surface density (64.8/10 = 6.48 kg m^3). The crater profile is shown at the bottom of the diagram. The stony meteoroid has a radius of 1 km and an impact velocity of 20 km s^{-1}. After Ivanov et al. (1992).

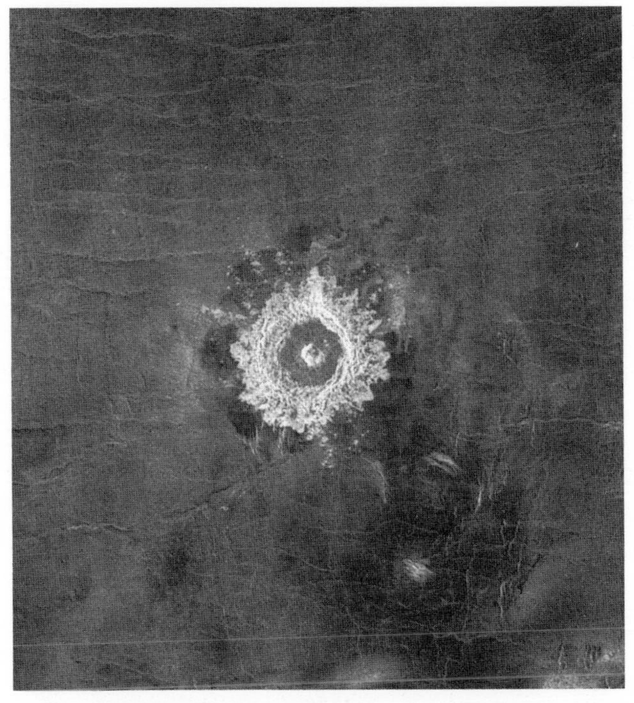

Figure 8.13 28 km diameter crater Barrera
The symmetrical form of the ejecta implies a high-angle impact trajectory. Magellan F-MIDR 15N111.

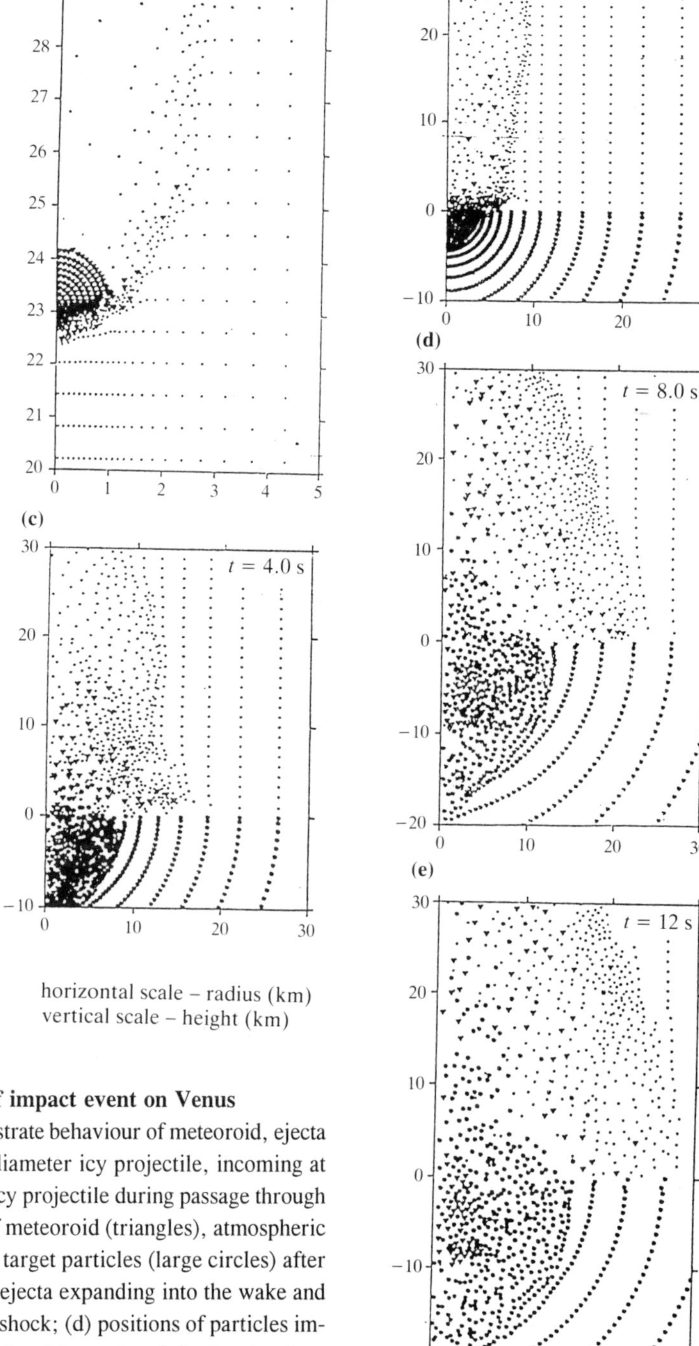

Figure 8.14 Simulation of impact event on Venus
Numerical simulation to illustrate behaviour of meteoroid, ejecta and gas flow for a 2.5 km diameter icy projectile, incoming at a velocity of 20 km s^{-1}. (a) icy projectile during passage through atmosphere; (b) positions of meteoroid (triangles), atmospheric particles (small circles) and target particles (large circles) after impact; (c) development of ejecta expanding into the wake and propagation of atmospheric shock; (d) positions of particles immediately after atmospheric breakthrough; (e) further development of ejecta curtain and movement of atmospheric gas. After Ivanov et al. (1992).

cloud above the Venusian surface immediately after impact of an icy meteoroid (Fig. 8.14). Because of the great density of the Venusian air, a very considerable pressure differential will be set up on either side of the advancing curtain of ejectamenta, giving rise to extreme turbulence. This has the ability to scour the region within 4–6 radii of the crater rim, creating a coarse lag deposit that is characterized by a radar-bright signature, a process that experiments conducted by Ivanov et al. (1992) have confirmed. One unusual surface-damaging mechanism they cite is *back-venting*, whereby gas forced into ground pore spaces by the shock wave as it moves outwards from the impact focus is subsequently vented back into the atmosphere as the overpressure diminishes with time, in response to declining shock-wave pressure.

For oblique impacts, the distribution of ejecta and scour zones associated with this stage will be strongly asymmetrical; thus, owing to the lengthy period of energy transference to the atmosphere prior to impact, downrange offset of ejectamenta is to be anticipated and it involves features such as radar-bright haloes, scour zones, radar-dark zones (parabolas) and turbulently deposited ejectamenta. Downrange offsets of this type are seen associated with the 31 km diameter crater Aurelia (Fig. 8.15).

Immediately behind the advancing curtain of ballistically ejected material, separation of coarse and fine debris is likely to occur, with the result that runout of fine-grained particles entrained in the outwardly flowing gas goes well beyond the inner hummocky zone. It is also probable that the shock-wave-induced flow of finer ejecta beyond the hummocky zone may resemble that in turbidity currents (Schultz 1992), in which case such flows would have the power to erode sand-sized clasts, pebbles and boulders within this zone, thereby giving rise to the distinctive radar-dark collars that have been widely imaged by both Venera 15/16 and Magellan. The eventual deposition of this finer-grained material would lead to an annulus of

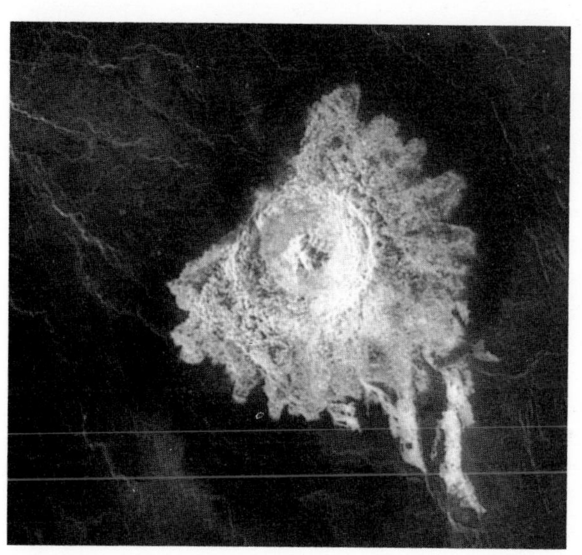

Figure 8.15 Venusian crater Aurelia An oblique impact generated this 31 km diameter crater, as is implied by the butterfly pattern of the ejectamenta and the zone of avoidance. Located at 332°E, 21.5°E. Magellan image F-MIDR-20N334.

smoother radar-dark terrain peripheral to the inner regions of blocky ejecta. That deposition of this fine fraction has subsequently been modified by further atmospheric activity is witnessed by the redirection of ejecta flows, these having been drawn downrange by recovery winds. The very latest fallout from the initial cloud would be recondensed vapour, which is believed to contribute at least in part to the distinctive radar-dark parabolas associated with some craters (an analogue of terrestrial tektite fields).

Another phenomenon is due to the deceleration of the projectile as it descends through the atmosphere and of any vapour and melt generated during descent. Because this occurs, highly mobile deposits may be emplaced well before impact and crater excavation. Such deposits are derived from the collapse of the column of low-density gas that follows in the wake of the bolide. Their form will be dictated, at least in the first instance, by the trajectory of the

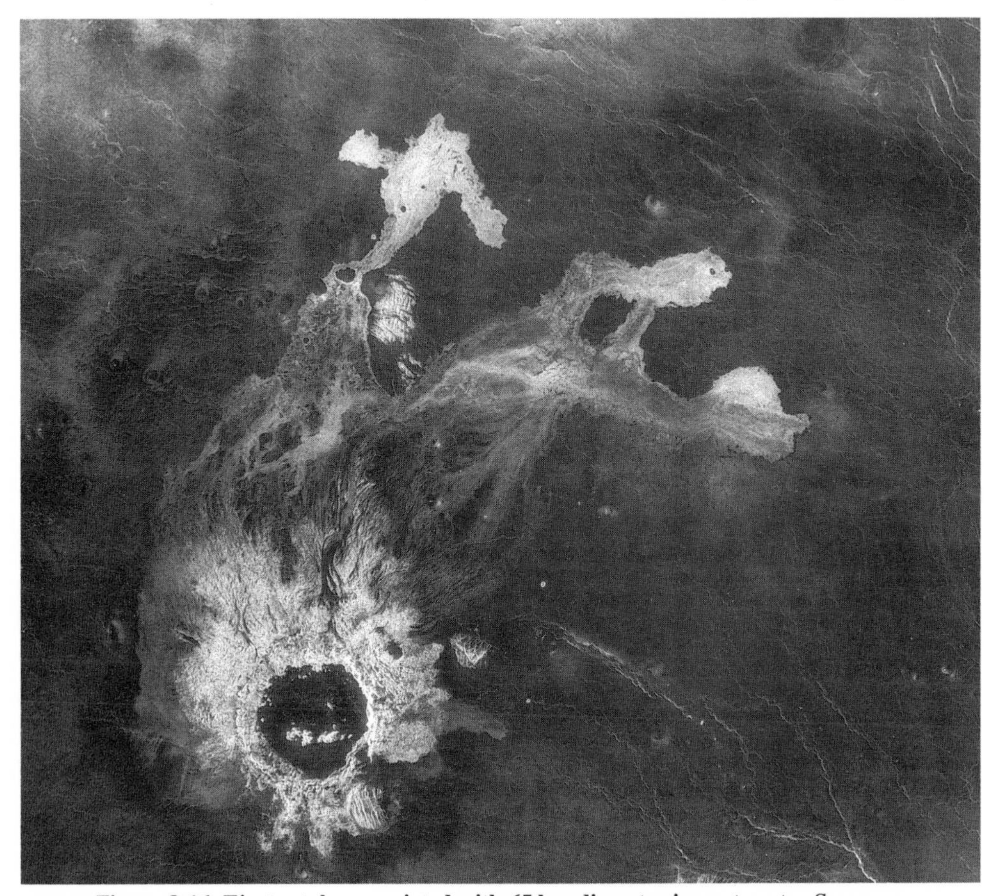

Figure 8.16 Ejecta styles associated with 65 km diameter impact crater Seymour
Up range and at right angles to impact trajectory (from the south), deposits are short stubby flows. These are surrounded by radar-dark flows which bury pre-existing fractures. Downrange, delicately textured intermediate-signature flows extend northwards, with radar-bright lobate terminations. Magellan image F-MIDRP-20N328;1.

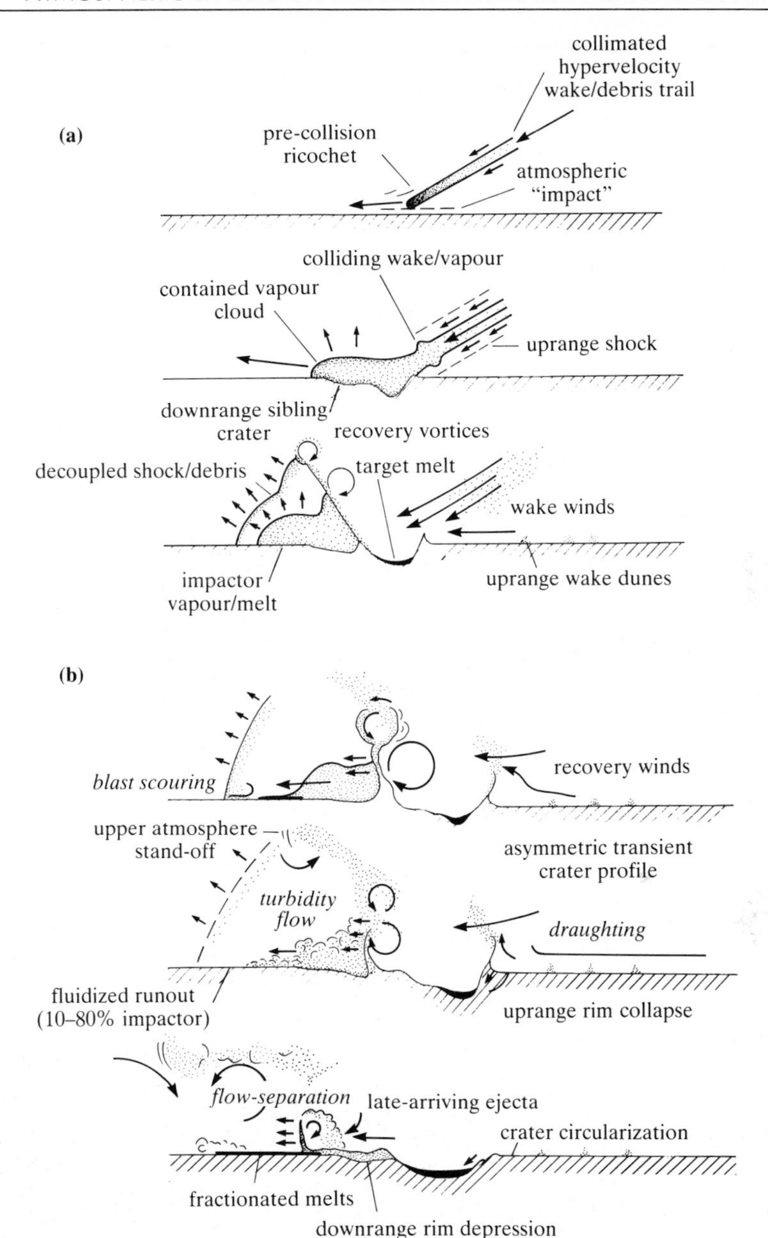

Figure 8.17 Idealized scenario for sequence and processes associated with Venusian impact

(a) Early period when initial transfer of impactor's kinetic energy to atmosphere and target occurs; (b) later period showing complex response of atmosphere to strong shocks created earlier. After Schultz (1992).

bolide, the direction controlling flow out to about one crater radius from the impact point, after which local gradients dictate this. The surface morphology of such flows, as revealed by Magellan imagery, suggests that, while many are comparable with terrestrial turbidity currents, others proceeded by laminar flow; both types are seen associated with the 67 km diameter crater Seymour, situated on the plains east of Beta Regio (Fig. 8.16). Schultz (1992) is of the opinion that turbidity-style flows are indicative of volatile-rich impactors (25% of the crater population), whereas laminar types indicate silicate- or iron-rich bolides (58% of the population). Composite flows represent a further 17% produced by either very high-velocity and/or volatile-rich silicate bolides. Late-stage turbidity-type flows theoretically may run out for very large distances, owing to turbulence generated by the latent heat released as the impact-induced vapour condenses to form a melt. A full diagrammatic representation of the processes of crater formation and ejecta deposition on Venus is to be found in Figure 8.17.

8.4 Epilogue

In the space available, this discussion on impact processes can only be rather general, and there are other important details that can be gleaned from the references cited. In closing, a brief mention of two other points must be made. First, another difference between Venus and the Earth is the amount of impact melt generated during cratering. Calculations of the differences between Earth and Venus with respect to the generation of impact melt indicate that melt generation on the latter should be between 20% and 40% greater than on the former, for craters of equivalent diameter. Furthermore, because of the high ambient temperatures, Venusian impact melt would stay molten for up to an order of magnitude longer than in the terrestrial environment, leading to extensive flows of such melt and the likelihood of a greater incidence of smooth crater floors.

Finally, the significant modification that the Venusian atmosphere effects upon the cratering process may also decrease cratering efficiency. Scaling exercises relating the morphology and dimensions of central peaks and peak rings as a function of impact angle and impactor dimensions enable first-order comparisons to be made between Venus and other planets for which suitable data sets exist (Schultz 1992). Schultz suggests that crater scaling relations for Venus are controlled primarily by atmospheric effects rather than by gravity, in which case the depths and diameters of large complex craters should resemble those of smaller simple craters. The degree of reduced cratering efficiency implied by this finding could have an effect on surface chronologies, since the reduced efficiency would reduce crater diameter for any given impactor size. Further work doubtless will shed more light on this particular aspect. In the meantime it is sufficient to note that parts of Venus have been resurfaced and that the ages of these surfaces, as indicated by the impact record, are of the order of 500 Ma.

9 Venusian volcanism

9.1 Some generalities about planetary volcanism

Terrestrial experience shows not only that volcanic activity takes on several guises, but that it is concentrated along relatively narrow but global-scale zones of enhanced seismic activity, high heat-flow and structural disturbance. In general these active zones are located along the boundaries between adjacent lithospheric plates (Fig. 9.1). However, some volcanic loci reside deep within plates (intra-plate settings); activity here is related to hotspots or mantle plumes – regions of hot upwelling mantle. The quiescent effusion of high-volume basaltic lavas from intra-plate centres such as Hawaii, or from spreading ridges astride which sit such volcanic foci as Iceland and the Galapagos, contrasts strongly with the violently explosive pyroclastic eruptions characteristic of island arc volcanoes, such as those of Indonesia or the Aegean, or of chains of volcanic mountains that parallel continental edges, such as Andean South America.

The contrast in eruptive behaviour is a response to the different tectonic settings of active zones. At spreading ridges and intra-plate hot spots, where mantle-derived magmas characterized by low silica and volatile contents rise to the surface without contamination by continental crust or suboceanic sediments, eruptions are typified by the quiet escape of highly fluid lavas. In island arc and active continental margin zones, however, mantle-derived magmas may become contaminated not only with subducted oceanic crust and ocean floor sediments (both containing volatiles), but also with silica-rich continental crust, leading to generation of modified and volatile-enhanced melts which tend to be of relatively high silication and viscosity. Such magmas tend to escape explosively at the Earth's surface.

In view of these differences, it is not surprising that the landforms attendant upon these different volcanic styles show similarly strong contrasts. Thus, low shields, evacuation/ subsidence calderas and extensive shield- and fissure-related flows, which often coalesce to form volcanic plains, are typical of basaltic provinces such as those represented by ocean floors and oceanic islands, and are produced during the early stages of continental rifting along continental margins. Active island arcs and divergent-plate-boundary mountain belts are typified by high-profile strato-volcanoes, stubby viscous flows rather than sheet floods, endogenous domes and landforms generated by pyroclastic flow deposits. Such is the diversity of vol-

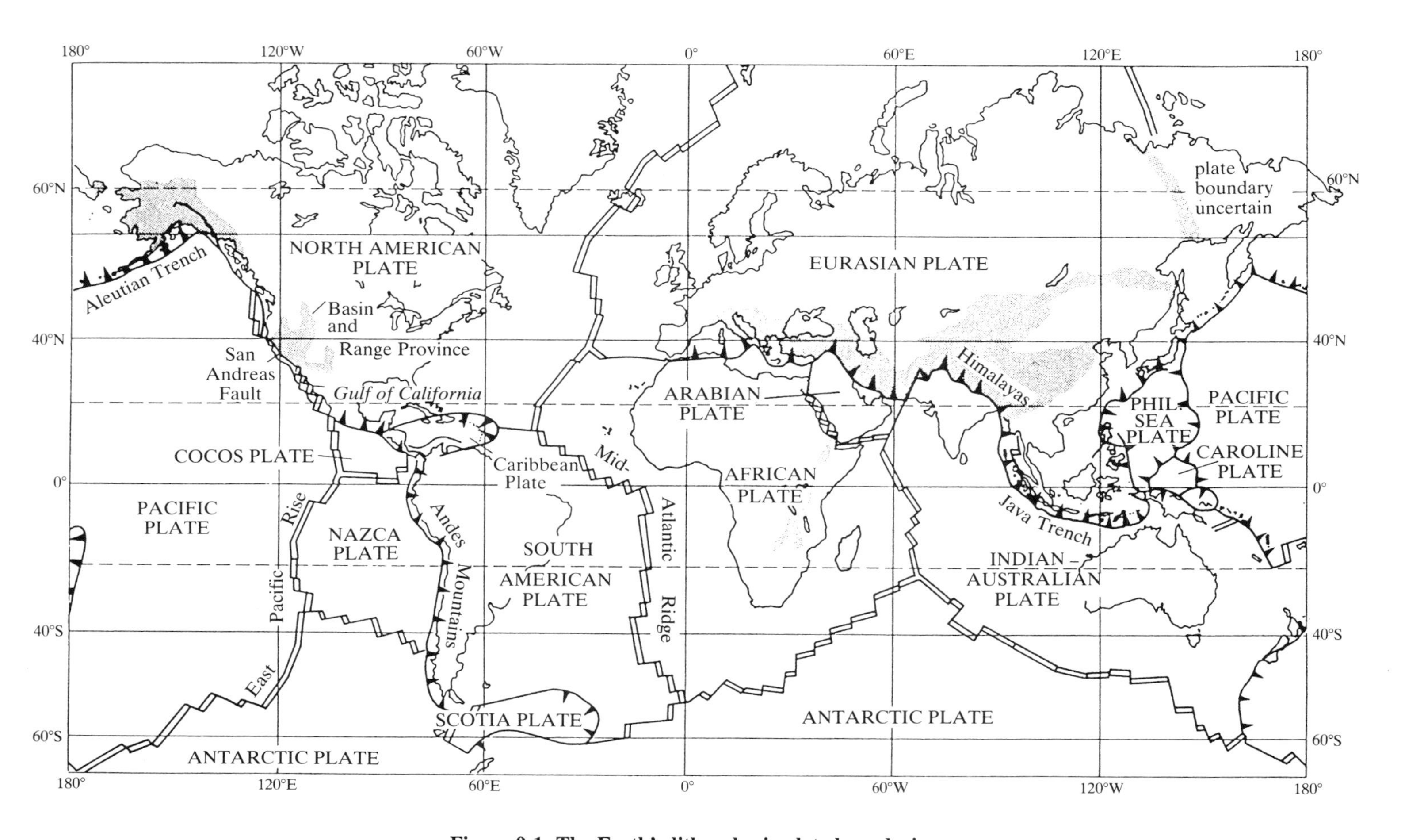

Figure 9.1 The Earth's lithospheric plate boundaries

These global-scale boundaries delineate zones of enhanced heat-flow, seismic activity, tectonism and volcanism.

canic style and landforms on a planet where crustal recycling is achieved by active *plate tectonics*.

While this is not the place to conduct a detailed survey of planetary volcanism – this has been done elsewhere (for instance, Cattermole 1989) – some general points can be made. On the Moon the predominant volcanic style was effusion of extensive floods of basaltic magma, which rose up along fractures and spread over the depressed floors of huge impact basins. Thus were formed the familiar maria. Landforms associated with such activity include relatively thin but extremely long, lobate and flat-topped flows, sinuous lava channels and tubes, low shield volcanoes and a very small number of steeper-sided domes. Certain spectrally red dark-mantling deposits, generally found at mare margins, are believed to be products of explosive eruptions.

On Mars too there are widespread volcanic plains, upon which are developed landforms similar to those described above. However, Martian volcanism took a different turn from the lunar sort, becoming centralized. Such activity saw the building of huge shield volcanoes and low-profile paterae whose dimensions dwarf terrestrial shields, even those of Hawaii. Furthermore, several volcanoes in the southern hemisphere of Mars show evidence for phreatomagmatic eruptions, indicating that, in the distant past, volatiles were more abundant on Mars than now. The immense size of the shields located in Tharsis and Elysium suggests that they grew over extremely long-lived mantle hotspots, above which the crust remained virtually static for many hundreds of millions of years.

On neither the Moon nor Mars is there any evidence that Earth-style plate tectonics ever occurred. Owing to the occurrence of global-scale linear belts of extensional and compressional features on Venus, and evidence that more than one kind of magma type may have evolved there, one of the more tantalizing questions concerning Venus, unanswered prior to Magellan, was whether or not plate recycling has played any part in that planet's geological evolution. The data now returned have gone a long way towards answering this query.

9.2 Eruption characteristics on Venus

Similar styles of volcanic landforms might be expected to occur on Venus as on other planets, and from their morphology it would be reasonable to infer the style of activity from which they derived. However, the very elevated surface temperature (470°C) and pressure (98 bar) experienced at mean planetary radius (MPR) have the potential to influence surface and near-surface volcanic processes on Venus. Theoretical consideration has been given to the rise and eruption of Venusian magma (Head & Wilson 1986), and to melt-generation and crustal melting under Venusian conditions (Hess & Head 1990). Suffice it to say that atmospheric pressure in the Venusian lowlands is so extreme (about 9×10^6 Pa) that at least 2 wt% H_2O (or 5 wt%

CO_2) would need to be dissolved in any crustal melt before pyroclasts could commence forming. In highland areas, where the pressure is lower, the appropriate figures would be 1 wt% and 3 wt% respectively. This contrasts sharply with the Earth, where fire fountaining would result from a mere 0.1–0.4 wt% dissolved H_2O. So, pyroclastic deposits would be less likely to form on Venus than on the Moon, Earth or Mars. In the event of pyroclastic activity occurring, any eruption clouds developed would not be expected to travel as far as those produced on Earth, owing to the enhanced atmospheric resistance they would experience. For the same mass eruption rate, it can be shown that an eruption plume would rise only one-third the height that it would on Earth, and one-fifteenth that attainable on Mars (Settle 1978) (Fig. 9.2).

The prevailing high temperatures would have different effects. Certainly they would inhibit the radiative cooling of Venusian lava flows; however, the high atmospheric density more than counteracts this, since it means that convective cooling initially will be more effective, with the result that, overall, flows will actually cool faster than they do on the Earth's surface (Head & Wilson 1986). The consequence of this is that, for the same mass eruption rate, a lava flow on Venus might be expected to be about one-fifth longer than its terrestrial counterpart, but to undergo a pahoehoe/aa transition more quickly.

The kinds of volatile contents typical of terrestrial melts mean that fire fountaining frequently occurs, owing to the explosive disruption of magma. Not surprisingly, this is accompanied by volatile loss, after which the bulk density of any melt is left relatively higher. On

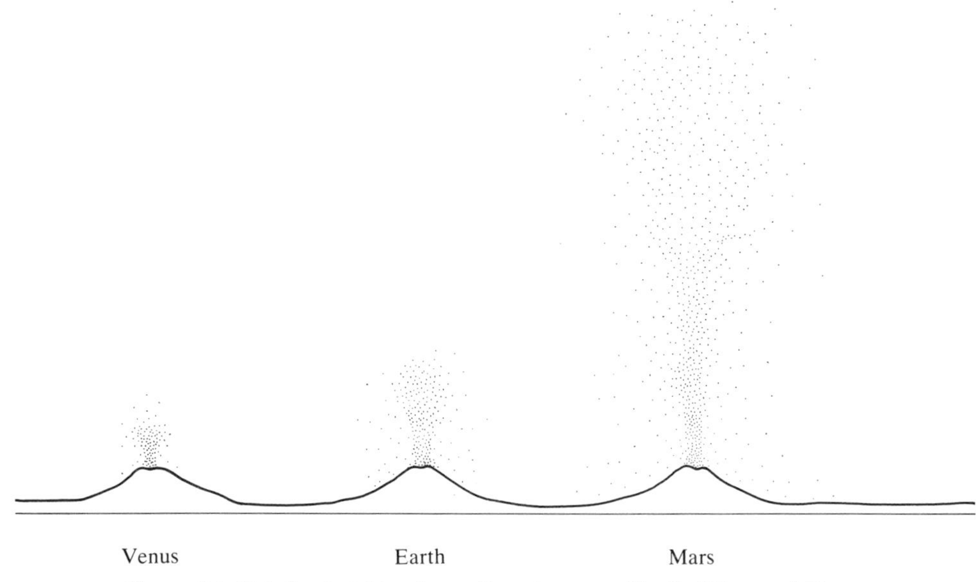

Figure 9.2 Relative heights of eruption plumes on Earth, Mars and Venus
For a given quantity of volatiles, a Venusian explosive eruption would rise only one-third the distance into the atmosphere as it would on Earth.

Venus, where atmospheric pressures are high, explosive disruption would be rare, with the result that lavas extruded onto the surface would retain most of their exsolved volatiles. For this reason their bulk density would remain lower than comparable terrestrial flows. On Venus also there is a marked pressure gradient with altitude, which has important repercussions, both for volatile exsolution and the density structure of the upper crust; furthermore, it also affects whether or not *neutral buoyancy zones* (magma stalling areas) can form and at what depth (Head & Wilson 1992). Near to MPR most magmas would rise directly to the surface, assuming volatile contents similar to those of Earth; however, at altitudes of 2 km or more above MPR, roughly half of any magma generated (assuming similar compositions to terrestrial ones) would be erupted directly, while the rest would stall at some depth, in neutral buoyancy zones. The relevance of this to the global distribution of volcanic rocks on Venus is that widespread extrusion of high-volume lava flows would be favoured in low-lying regions, while smaller-scale activity sourced by melts residing in shallow reservoirs/neutral buoyancy zones would characterize locales at greater elevation. Theory predicts that the depth of such zones would increase with increasing elevation, a factor of between 2 and 4 being appropriate for a height difference of between 1 km below MPR and 4.4 km above it (Head & Wilson 1992).

A recent theoretical modelling exercise conducted by Thornhill (1993) to investigate Plinian eruption cloud dynamics shows that development of high-convecting eruption clouds is unlikely to occur on Venus unless a combination of initially high temperatures, high volatile content and high-altitude eruptions concur. Such a situation appears very unlikely within the Venusian environment, and therefore it is to be expected that, should high-discharge eruptions take place on Venus, their resulting geological expression would be pyroclastic flows. These ought to be identifiable on Magellan images.

Now this kind of discussion may seem somewhat esoteric, but it is of the utmost importance to know as much as possible about the way in which potential magma reservoirs would grow on Venus, and also to understand what the relative proportions of extrusion to shallow intrusion might be. The latter, in particular, is of great interest, since only if the likely proportions of intrusion to extrusion are known can any idea of the implications of magma production rates be obtained. This information is vital if an understanding of crustal formation and evolution is to be reached.

Finally, it is important for the imagery coverage of volcanic features to be global, since only then can meaningful attention be paid to the volcanic associations developed on the surface. Particularly important is the relationship between volcanism and tectonism. On Mars, shield volcanism went hand-in-glove with extensional tectonism; on Earth, there generally is a close genetic relationship between faulting and volcanism, although volcanoes do not occur along every fault, as is common knowledge. Therefore, it is not surprising that Magellan imagery shows a strong correlation between volcanism and rifting on Venus; however, there are many major rifts that have no attendant volcanic deposits. There is a need to understand what is the true relationship between extensional faulting and volcanism, and how exactly the release of pressure within the crust affects generation and rise of magma towards the surface. Then again, an appreciation of the relationship between volcanism and structures such as novae and coronas – believed to be generated above mantle plumes and hotspots – can only serve to improve our understanding of mantle processes on Venus. In the ensuing pages will be found a description of volcanic landforms and their distribution, followed by a discussion of volcanic style on the planet Venus and the degree to which volcanic activity has resurfaced it.

9.3 Volcanic landforms of Venus

A wide variety of topographic features identified on Venus have been attributed to volcanism. The structure of some these strongly implies that they are the direct surface manifestation of subcrustal mantle activity. These range in size from less than one hundred to several thousand kilometres across. The smaller of such landforms come under the general heading of *coronas*, while the larger features are termed *volcanic rises*. The latter are represented by regions such as Bell and Beta Regiones.

Volcanic edifices on Venus share many of the characteristics of terrestrial structures, as do the extensive lava flows that have been identified, but peculiarly Venusian features also occur. The landforms that have been identified range in size from small shields a mere few hundred metres across, to massive centres whose diameters exceed 300 km and with which are associated extensive volcanic deposits. Coronas, novae and arachnoids – all fairly major volcanic features – appear to have no terrestrial counterparts.

Table 9.1 Size, distribution and area covered by volcanic features on Venus.

Feature	Number mapped	Approximate average diameter (km)	Approximate average area covered (km^2)
Shield fields	556	150	17 700
Intermediate volcanoes	274	25	490
Large volcanoes	156	400	125 600
Calderas	86	60	2 900
Coronas	176	250	49 000
Arachnoids	259	115	10 400
Novae	50	190	28 300
Flood lavas	53	350*	128 200

*Length

9.3.1 Global distribution of volcanic landforms

One of the tasks undertaken by Magellan team members was the compilation of a global catalogue of all the volcanic features so far discovered on Venus. Such a database was to take account of the location, dimensions and general characteristics of each type of feature listed. At the time of writing, the final catalogue had not been published, but Table 9.1 summarizes the size, distribution and areas of those volcanic features identified by them on Magellan imagery.

9.3.2 Flood lava flow fields

The radar back-scattering properties of Venusian plains vary from low (radar-dark) to high (radar-bright). While several factors contribute to this, surface roughness is believed to dominate the radar signature over most of the planet's surface. A recent quantitative analysis of both Arecibo and Magellan back-scatter data, and a comparison with terrestrial flows, support the notion that the very extensive radar-dark plains are floored by lava floods with relatively smooth surfaces, akin to terrestrial pahoehoe (Campbell & Campbell 1992). Confirmation of this comes from the images received from Venera landers that showed slabby surfaces very reminiscent of the Snake River slab pahoehoe shown in Figure 9.3. However, not all volcanic plains regions are radar-dark, although it does appear to be the case for the oldest flows. This latter characteristic, by the way, may be a function of either slow chemical changes affecting the flow surfaces that have remained longest in contact with the dense and chemically reactive Venusian air, or of mantling of the older surfaces by aeolian materials. More research needs to be done to assess the relative merits of these two ideas.

Figure 9.3 Slab pahoehoe surface in the Snake River Plains, Idaho.
Courtesy of J. W. Head.

Younger flows may be either radar-dark or radar-bright, the signature being a reflection of surface texture. Some flows show a change in back-scattering from proximal to distal locations, whereas others show local changes that may be attributable to the breaking of slab crust as it either encounters obstacles or flows down a steeper slope; some are homogeneous over very large areas. Strangely, most show the radar signature of terrestrial pahoehoe flows, despite the theoretical prediction that there should be a rapid pahoehoe to aa transition on Venus (Head & Wilson 1986).

Fifty lava channel systems also have been identified; some are as long as 250 km. However, only in relatively few locations has it been possible to detect the central flow channels bounded by raised levees so commonly associated with many Martian flows. It seems unlikely, bearing in mind the considerable width of Venusian flows, that this simply is a func-

(a)

(b)

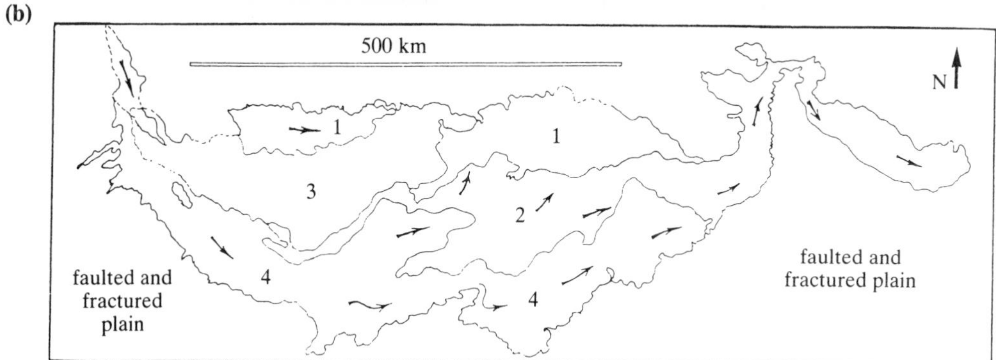

Figure 9.4 Lava flow fields in Atla Regio

(a) Flood-type lava flow fields in southern Atla Regio. Frame width is 900 km. Magellan C1-MIDRP 00N197;1. (b) Sketch map of above flow fields. The numbers indicate sequence of eruption. After Head et al. (1992).

tion of resolution, and it is more likely to be a manifestation of the rheological properties of Venusian lavas at the very high ambient temperatures found at the surface.

By terrestrial standards, many of the individual flow fields are large; some have volumes as great as 8000 km^3. One particularly prominent radar-bright flow field outcrops in southern Atla Regio, about 900 km to the south of the volcano, Ozza Mons (Fig. 9.4a). This family of lobate-margined flows has a total length of around 1000 km and covers an area of approximately 180 000 km^2. Superposition/intersection relations indicated in Figure 9.4b suggest that there were at least four discrete emplacement episodes, each of which saw extensive flows inundating fractured and faulted plains. This particular flow field appears to have been erupted from a series of graben associated with regional extensional faulting. In one location (near the centre of Fig. 9.4a) it is possible to discern a well defined flow channel with levees, and also to see at a bend in the channel that the flow crust has been broken into huge slabs, exposing radar-dark lava beneath. At another site, one of the flows floods a small 500 m high shield. By noting that nearby unflooded shields have a basal diameter of around 500 m, it has been estimated that the flow thickness is of the order of 100 m.

In some respects, even more spectacular are the flows associated with Mylitta Fluctus, in southern Lavinia Planitia, which inundated an area of at least 300 000 km^2. Like the Atla flow field, Mylitta Fluctus flows are radar-bright and fairly uniform in texture (Fig. 9.5). The complex actually is composed of six smaller flow fields, each of which appears to have emanated from a large shield volcano 400 km across that has a single summit caldera 40 km in diameter situated at the southern end of the complex (Fig. 9.6) (Roberts et al. 1992). The individual flow-fields have very large areas, the largest covering about 120 000 km^2. While there is little evidence that eruptions were fissure-fed, Roberts et al. (1992) admit that rift-related effusions may have preceded centralization, as is believed to have happened at certain Martian eruptive foci. The source volcano itself is located along a proposed rift zone that runs along the northern flank of Lada Terra.

Using studies made by Walker (1973) relating flow length to effusion rate, Roberts and her colleagues estimate effusion rates to be of the order of $460–4600 \times 10\ m^3\ s^{-1}$. This is of the same order as was reported by Cattermole (1987) for areally extensive sheet and tube-fed lava flows developed on the flanks of the Martian volcano, Alba Patera ($17–2120 \times 10^3\ m^3\ s^{-1}$), which also are thought to have issued from a centralized source. Using these derived values, Roberts et al. then calculated preliminary eruption durations, estimating that the first major outpouring took place over a period of between 10 and 70 years and generated the largest volume of lava ($1.7 \times 10^4\ km^3$). This early event is believed to have flooded the original Mylitta Fluctus rift zone and constructed the asymmetric shield on its northern flank. Subsequent individual eruptions are thought to have lasted only a matter of days and to have gone on over periods of a few months, rather than years.

Since over 50 major flow fields now have been recognized on Venus, it is clear that the rapid effusion of large volumes of fluid, presumably basaltic, lavas, over quite short periods,

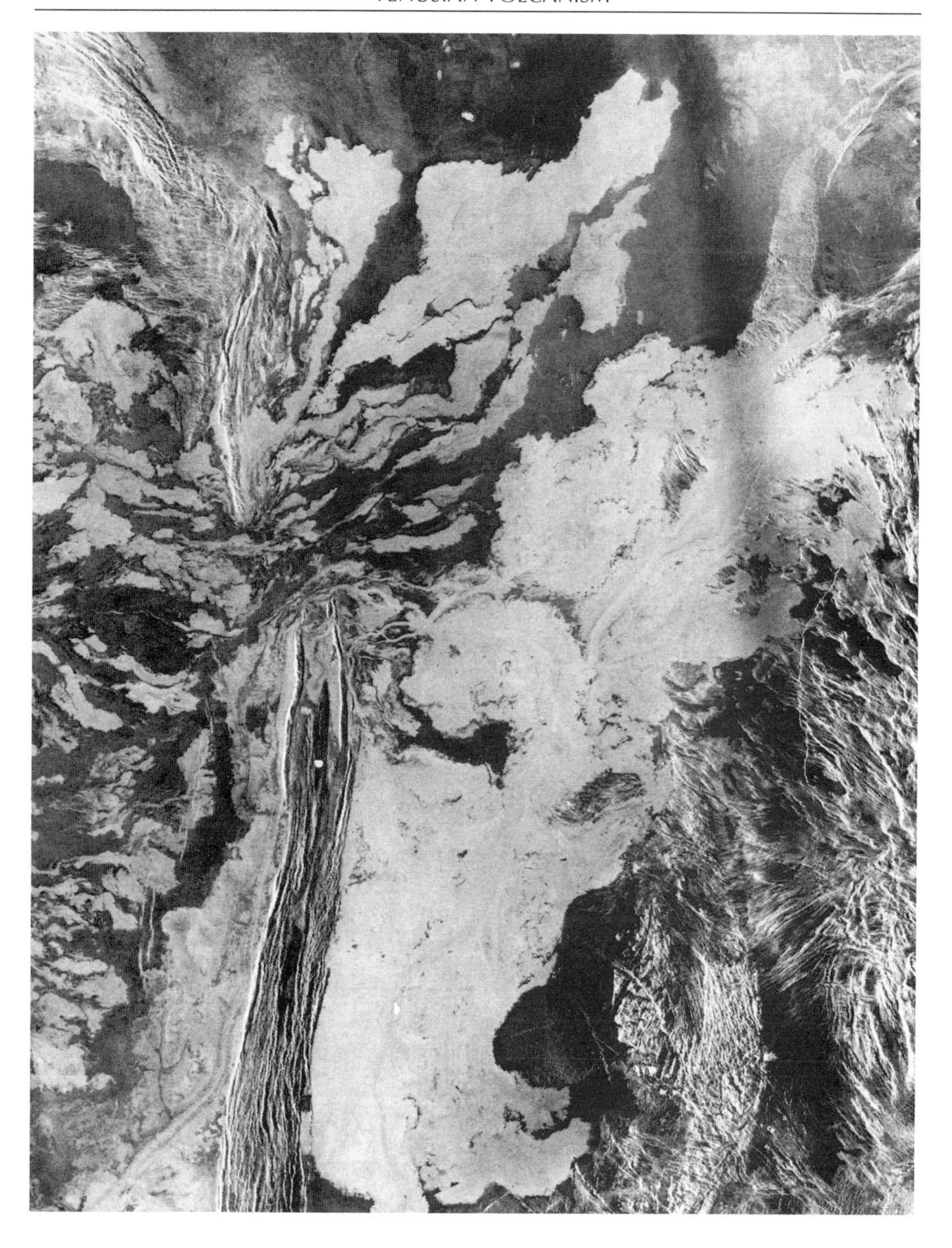

Figure 9.5 Mylitta Fluctus region

This part of the region north of Lada Terra is largely covered by a complex of lobate flows that have their source in or near a caldera located at 351°E, 58.3°S. Magellan image P-38088.

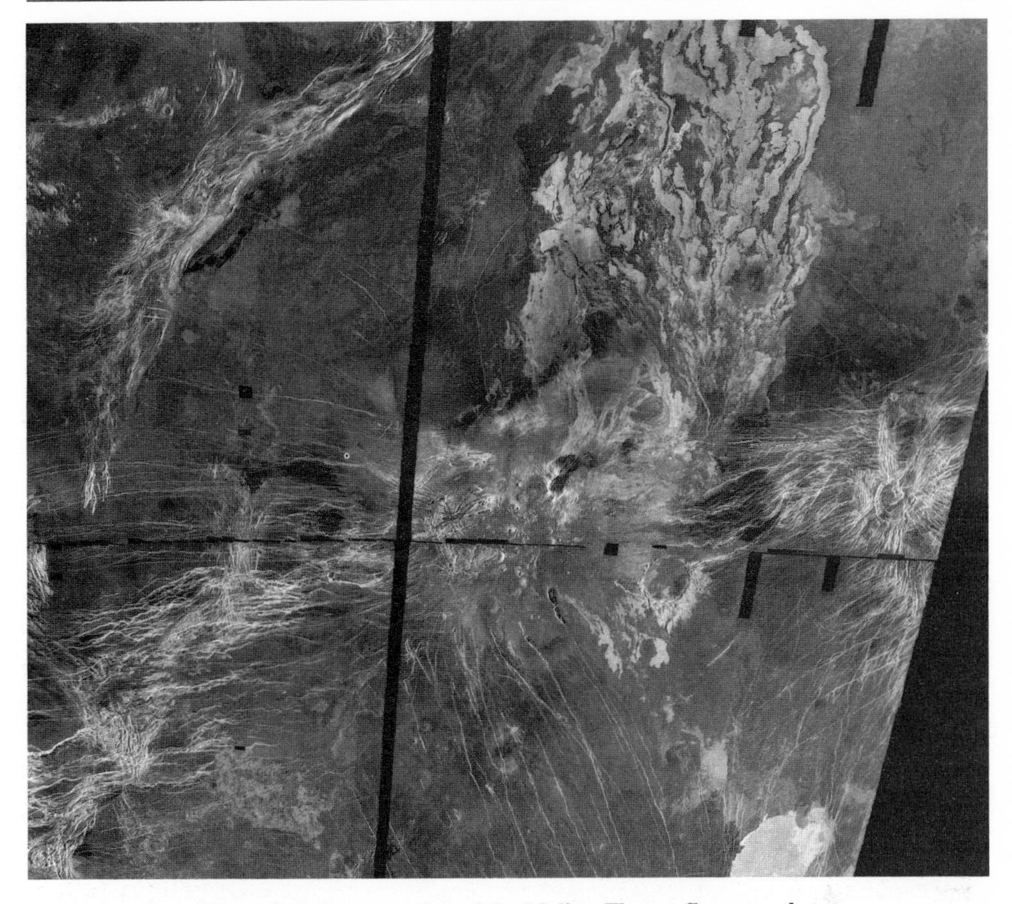

Figure 9.6 Source region of the Mylitta Fluctus flow complex.
This takes the form of a low-profile caldera structure which is located towards the centre of the image. Magellan C1-MIDRP 60S347;201.

was characteristic of Venus, at least during certain phases of its geological evolution. Continuing research is attempting to relate emplacement of such flows in terms of lava rheological and chemical properties, mass eruption rate and duration.

Comparable flows are located on Mars where they have been equated (Cattermole, 1987) with fluid terrestrial lavas erupted at high effusion rates, such as those of the Columbia River Plateau, at least in terms of rheology and composition (Shaw & Swanson 1970). The more voluminous of terrestrial flood lavas can be shown to have erupted from long fissures or from many vents sited along them; thus some of the Columbia River flows emerged from a fissure system 175 km in length. The same is probably true for many volcanic plains on Venus, although the source fissures are not discernible, probably since later flows have covered up the evidence, as they have on both the Moon and Mars. However, there is on Venus evidence for subsurface dykes that can be traced for upwards of hundreds of kilometres, and these support the notion of at least some measure of fissure-related activity. When the overall volume of these Venusian flow-field complexes is compared with terrestrial flood basalts such as those of the Deccan, Karroo or Columbia River, they are between one and two orders of magnitude smaller. The terrestrial flood basalt provinces cited are known to have been produced during the early stages of continental rifting, when huge volumes of mantle-derived basalt are erupted. Therefore, the more modest volumes of flood lava described here from Magellan imagery may more reasonably be interpreted as having originated in large-scale mantle upwelling, probably along pre-existing rift zones, rather than in any form of diverging-plate-margin situation.

As to the composition of the lavas: their morphology and extent is consistent with basalt-like chemistry and rheology, and such a composition is confirmed by analyses collected at the Venera and Vega landing sites. There is also the possibility – and this is true too for Mars – that a **komatiite** composition is appropriate for some, or that some may be of even more extreme carbonate-rich type (Hess & Head 1990). Whether or not such alternative compositions are actually found on Venus is unlikely to be established until further sampling is undertaken.

Finally, the distribution of major flow fields is not entirely random. There is a distinct concentration in the region of eastern Lavinia Planitia and Alpha Regio, on the flanks and sometimes on the surface of Aphrodite Terra and Atla Regio, in both Beta and Phoebe Regiones and in Sedna Planitia. Where it has been possible to determine the direction in which flow proceeded, it supports the notion that major outpourings of fluid lava have contributed significantly to the resurfacing of topographic lows such as the lowlands. In several instances it has been possible to trace lava channels for several hundred kilometres, indicating that these – in much the same way as lunar sinuous rilles – provided major pathways for the downslope emplacement of basaltic flows.

9.3.3 Small volcanic structures

A huge number of small volcanic structures are located within the plains regions. This applies both to the areally extensive low-lying plains and to smaller plains areas within more elevated regions. These can be classified into three groups: *shields*, *cones* and *domes*. Random sampling on a global basis serves to give an impression of the numbers of such structures and their sizes, the distribution of shield fields being shown in Figure 9.7. There is a clear inverse relationship between numbers and size, down to a diameter of 2 km, beyond which the numbers fall off dramatically. The latter is almost certainly a function of resolution rather than a real reversal of the otherwise inverse trend. The most abundant small structures are found to be those whose basal diameters fall within the range 2 to 8 km.

A more specific study of the type and distribution of the three landform groups within the

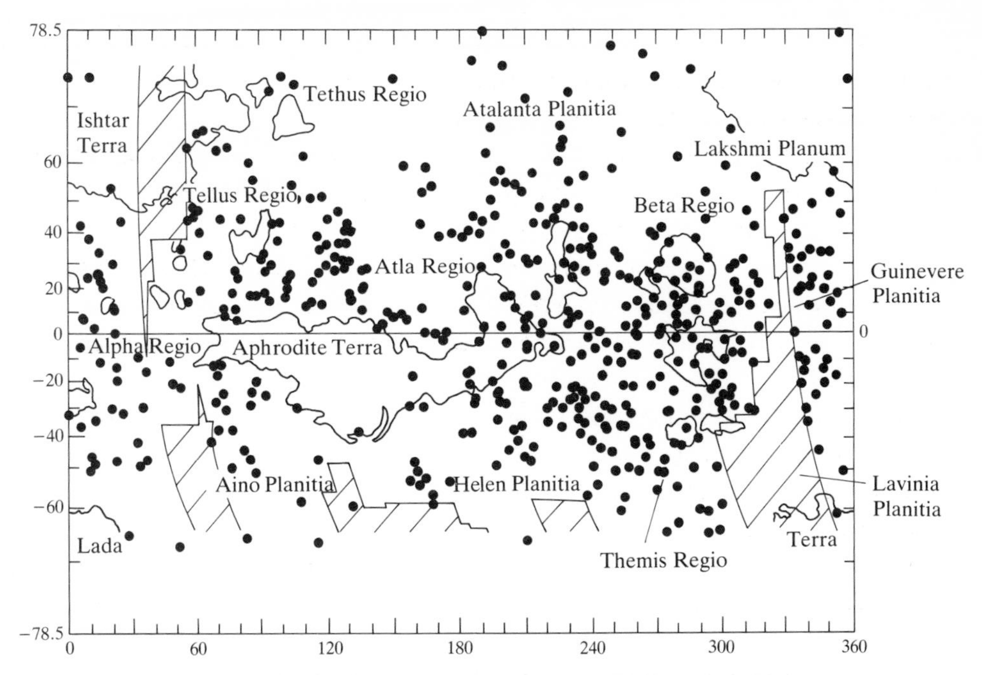

Figure 9.7 Global distribution of shield fields over the surface of Venus

Areas for which data have not been processed are shown by hatching. From Head et al. (1992).

Figure 9.8 Volcanic features in Niobe Planitia

The association of shield fields and fracture belts is clearly seen. Magellan F-MIDR 45N119e;1.

region of Niobe Planitia showed there to be 234 shields, 219 cones and 163 domes out of a total of 616 small structures present (Guest et al. 1992). In the same area, 393 summit pits were counted. Most of the volcanoes clustered together in large groups associated with fracture belts crossing the region (Fig. 9.8). This seems to be the characteristic of such small structures Venus-wide.

The most common of the three types are *small shields* – generally less than 20 km in diameter – of which over 22 000 were identified on the quarter of the globe imaged by Venera 15/16 (Slyuta et al. 1988); the Magellan tally is considerably greater than this, and on the basis of calculations by the above and by Aubele & Slyuta (1990), Head et al. (1992) derive concentrations of between 0.2 and 0.3 shields per 10^3 km^2. The majority are circular in plan and have gently inclined convex slopes; most have a single summit pit whose average size is around 700 m, but many of the smaller features appear either to lack a pit or to host one that is too small to be resolved by Magellan radar. While shields may be as large as 20 km across, most fall within the range 2–8 km and are less than 200 m high (Fig. 9.9). Other shield-like landforms are of similar dimensions but have flat tops, while a further expression sees much broader plateau-like summits with less uniform flank slopes. Some shields appear to have a central plateau-like region with radar-dark signature surrounded by a roughly circular region of higher radar back-scatter; these are believed to be shields with a summit lava lake surrounded by a spatter rampart. Guest et al. (1992) describe six different styles of small shield-like structure, and note that if most are analogous to terrestrial shield volcanoes then the individual lava flows from which they are built must be $<$ 75 m across since they apparently are irresolvable in Magellan images. However, they also note that some such landforms may be built from a single flow surrounding a central source vent.

While most such landforms are located on the Venusian plains, others are associated with coronas, arachnoids and other major volcanic edifices. Small shields frequently occur in groups, forming *shield fields*. This is true of large areas within Guinevere Planitia, for instance, where a group of 55 shields, ranging in diameter from 1.3 to 6.5 km, is superimposed on fractured dark plains.

On Earth, similar shields typically have summit depressions that are linked to subsurface magma conduits. Frequently the depression fills with lava, forming a lava lake, which may overflow the rim and spread down slope as a flow; when the reverse occurs, drain-back into the conduit forms the pit. Such is the pattern, for instance, at Mauna Ulu in Hawaii. Thus far, unequivocal shield-related flows have only been identified around some of the larger shields, perhaps due to lack of resolution or contrast. Those in question vary from the rest in being somewhat larger (generally tens of kilometres across), often rather elongated, and related to subsurface structures believed to be dykes (Fig. 9.10). They are also distinctive in having associated well defined volcanic flows that spread away from the summit region. Such flows typically are radar-dark towards the point of origin, but have a higher radar back-scatter distally, as if becoming slabby or rougher towards their terminations. This type of landform

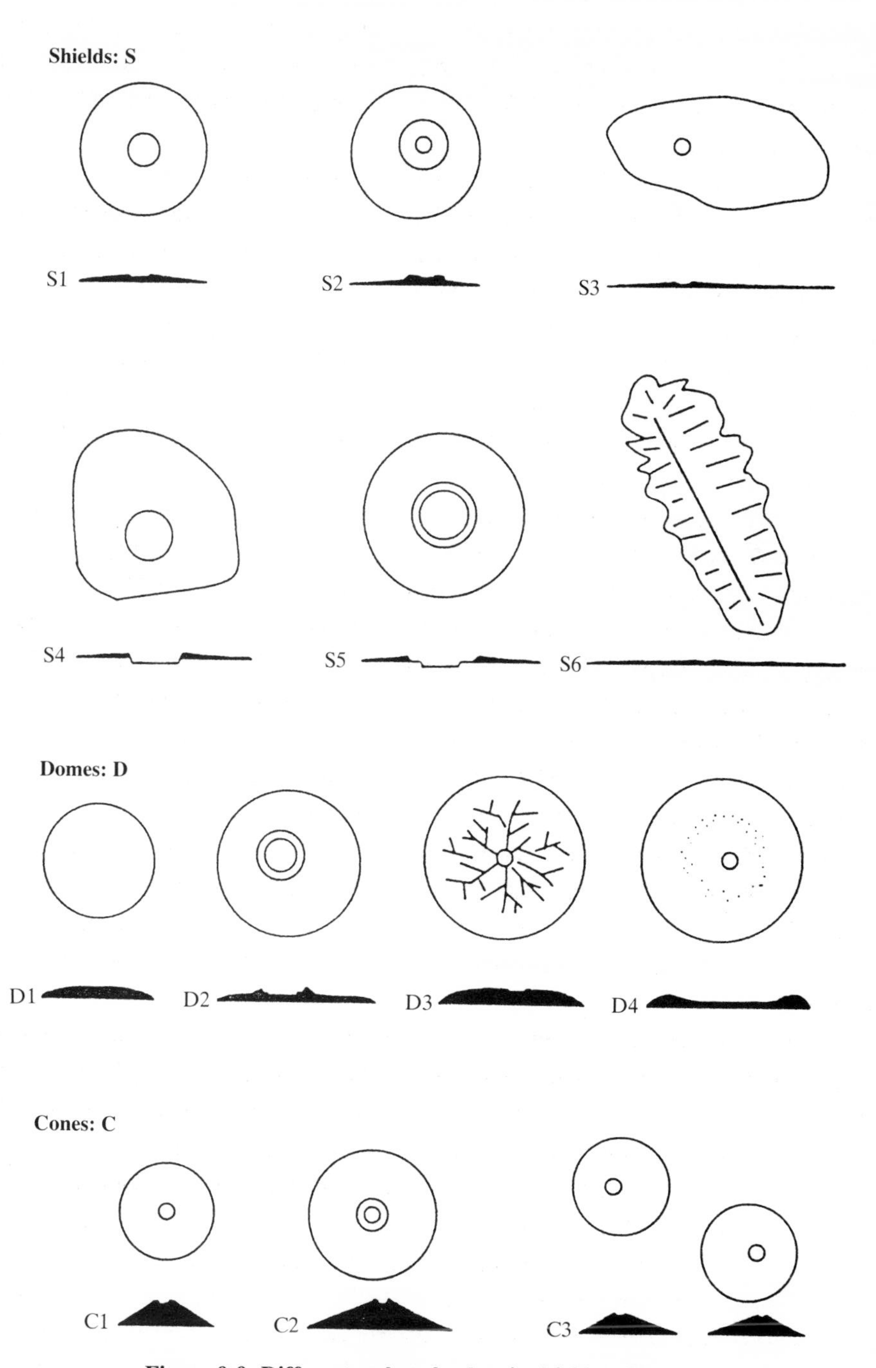

Figure 9.9 Different styles of volcanic shield on Venus
After Guest et al. (1992).

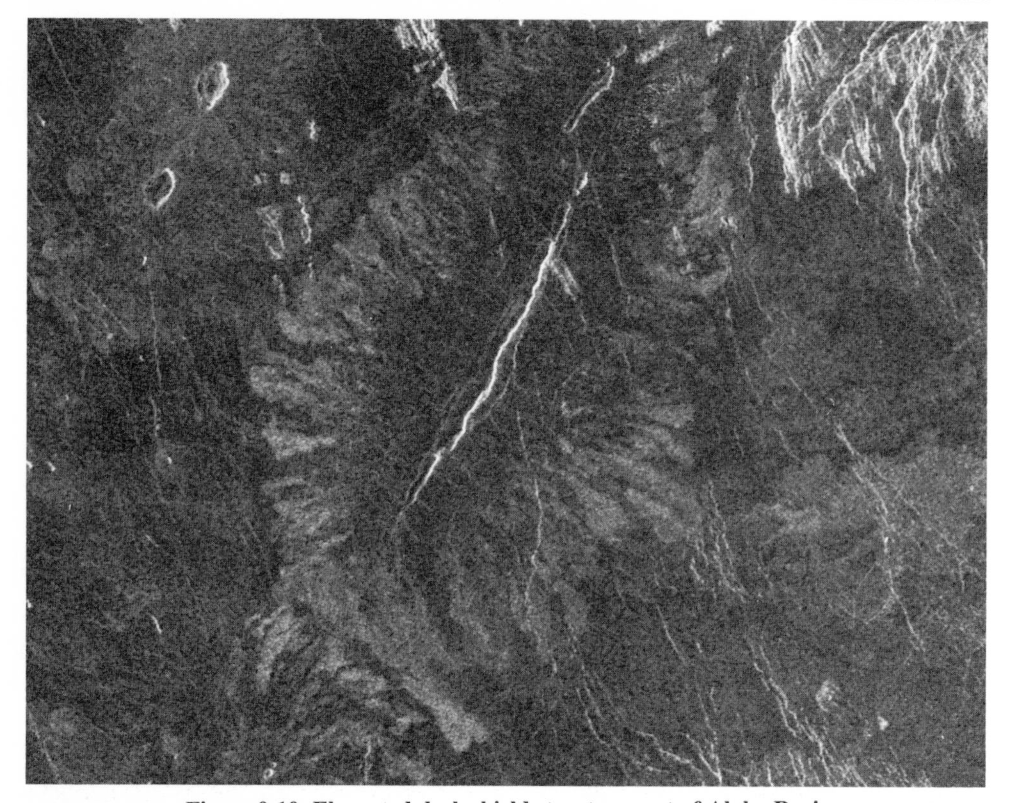

Figure 9.10 Elongated dark shield structure east of Alpha Regio
The radar-dark shield has associated lobate flows (radar bright) and an axial fractured zone 50 km in length. Magellan F-MIDRP 30S009;1.

bears the closest resemblance to terrestrial lava shields, for instance some of those located in Iceland.

There is also a strong association of shields with regional structural patterns. Thus, many summit pits are located on graben, which may also transect the summit area or have elongated lines of depressions running along their axes. These relations suggest that some faults may be the surface manifestation of dykes that supplied magma to the summit region. The most likely source for shield clusters is thermal anomalies in the Venusian mantle, whereas other, solitary, edifices could form anywhere along regionally propagating dykes. Either way, the shields and the surrounding radar-dark plains appear to have their origins in the same parent magma.

Most Venusian small volcanic *cones* are less than 15 km across; they tend to have a higher radar back-scatter than shield surfaces and higher flank slopes (Fig. 9.9). They are believed to have built up from cindery lava in much the same way as terrestrial cinder cones. Small *domes* are larger, typically 20–30 km in diameter and up to 1 km high, that is, larger than comparable terrestrial examples. Many have either a central summit pit, or one placed asymmetrically. They generally are quite circular and have steep margins, characteristics consistent with their having been produced by eruption of magmas more viscous than basalt, i.e. with the effective viscosities of dacite and rhyolite. The characteristically radar-bright returns from fractures and rough surfaces along their margins are consistent with furrowing and/or a degree of collapse of brecciated material at flow fronts. Guest et al. (1992) note that, if such domes represented extrusion from a single eruptive episode on Earth, it would have given rise to extensive ash-flow deposits such as ignimbrites. However, such activity may have been prevented by the dense Venusian atmosphere, with the result that massive domes were formed instead.

Some domes have deeply scalloped margins and an overall aspect that sets them apart from the rest, and from any known terrestrial counterparts. Such landforms – 80 have been identified – may have flat, convex or concave tops, and show all stages from steep margins with a single steeply backed alcove, to domes whose margins are entirely scalloped (Fig. 9.11). Their morphology is consistent with their having experienced peripheral collapse by slope failure, a view that finds support in the frequent presence of debris aprons beyond their col-

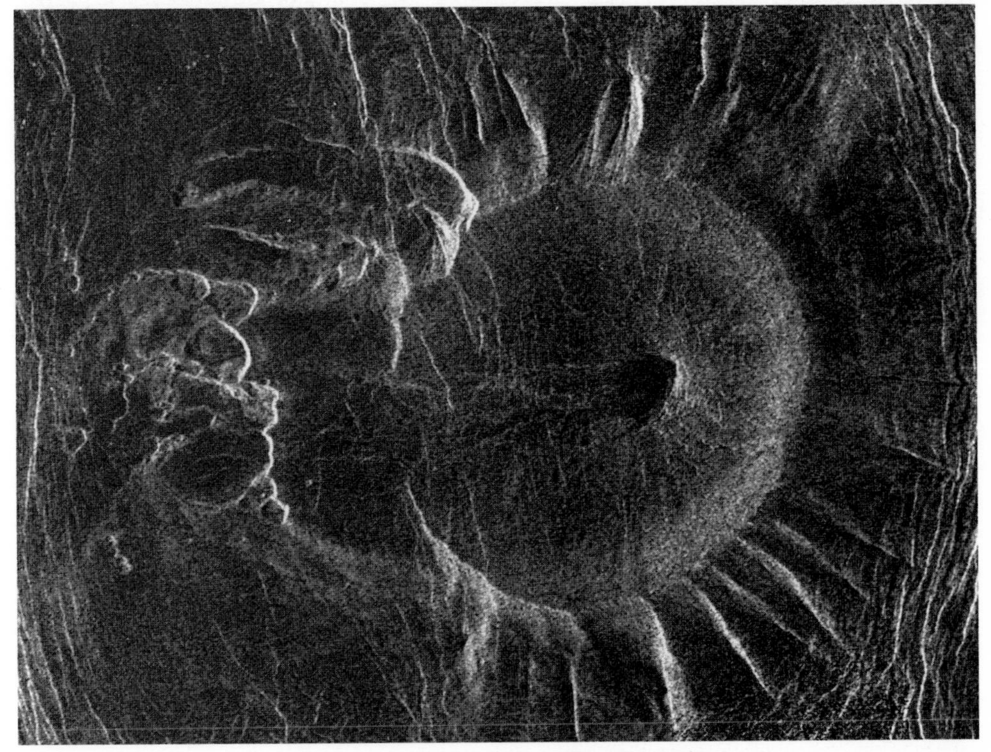

Figure 9.11 Dome with scalloped margin

This landform, situated northeast of Alpha Regio, has a 5 km diameter summit pit, a depressed central area surrounded by a narrow rim and flanks characterized by radiating spurs and troughs. Magellan image F-MIDR 20S003.

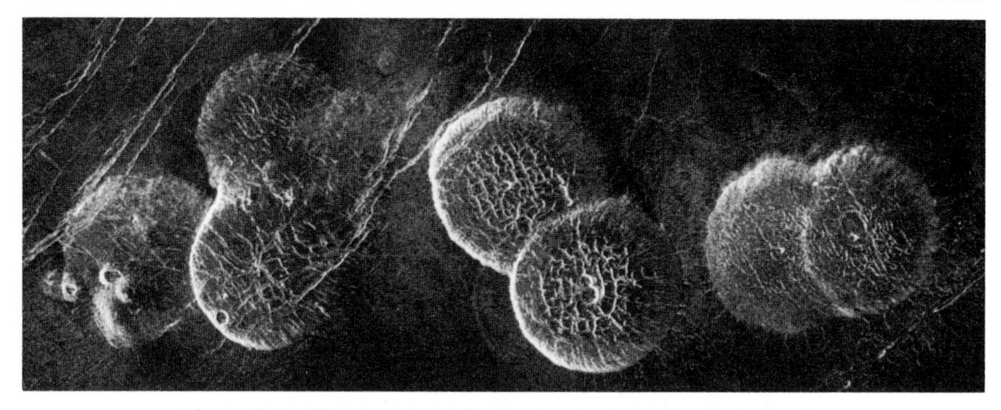

Figure 9.12 Circular steep-sided volcanic domes in Alpha Regio
Each of the seven circular domes is approximately 25 km in diameter. Each is characterized by a heavily fractured outer skin, steep margins and small collapse pit. Magellan image P-37125.

lapsed sectors. In one case, a breached skin with associated outflow of lava has been identified, presumably representing damage to the **carapace** by hot lava beneath (Guest et al. 1992). Such collapse events are not uncommon on the Earth and usually are a response to either oversteepening or explosion at dome margins, which may set off relatively small-scale avalanches of rocks and ash. However, the scale of such events on Venus - they may involve as much as 200 km^3 of material - is orders of magnitude greater than that of terrestrial ones, perhaps because the elevated ambient temperatures experienced on Venus allow only relatively weak carapaces to form.

Volcanic lava domes may occur singly or in groups, like small shields. Frequently they are located within zones of fracturing. Groups of domes are common and there may be some overlapping of individuals. Strangely, where two domes overlap, the relief of the (apparently) overriding dome is not disturbed by the relief of that underlying it. A possible explanation for this phenomenon is that the two individuals formed more-or-less contemporaneously, but one dome continued to grow laterally after cessation of expansion by the other, giving rise to overlap and a younger aspect (Fig. 9.12).

There is little doubt that domes represent the expression of relatively viscous lavas from central conduits that punctuate the plains. The high degree of circularity that typifies them implies that the plains surfaces over which they flowed must have been horizontal. McKenzie et al. (1992), after studying the form and surface textures of a selection of such domes, convincingly argue that their morphology is consistent with the spreading of magma within the viscosity range 10^{14} and 10^{17} Pa s, which implies magma temperatures of 610–700°C in dry rhyolite magma. Such temperatures agree well with laboratory measurements of the solidus temperatures of wet rhyolite, from which it follows that dome growth would naturally follow surface eruption and degassing of viscous magmas generated by wet melting at depths of >10 km.

9.3.4 Larger volcanoes and related structures

During mapping as a result of the Venera 15/16 missions, about 800 structures within the diameter range 20–100 km, and about 50 with diameters of 100–350 km, were mapped. Head et al. (1992) have termed these intermediate and large volcanoes respectively, a terminology that does appear to have some significance in terms of formational processes. A substantially larger tally of such structures exists now that Magellan has achieved global coverage. So far, 274 intermediate and 156 large volcanoes have been recognized; in addition, 86 calderas (without appreciable relief), 176 coronas, 259 arachnoids and 50 novae are known to exist.

Some of the intermediate-sized structures have associated with them a sequence of radiating lava flows, usually with lobate margins that stand out from the surrounding dark plains by virtue of their radar-bright signature (Fig. 9.13). These have been christened volcanoes of the *anemone type*, owing to the petal-like flow terminations that are prominent in radar images. Many have summit depressions. However, Head and his colleagues note that, particularly within the size range 20–40 km, such flows are not discernible and a volcanic origin is solely implied by the presence of a central pit, a conical peak or association with other nearby volcanic edifices. This suggests that lava flows emanating from centralized foci have lengths either less than 10 km or more than 50 km; however it may be that intermediate-length

Figure 9.13 Intermediate-sized volcano in Atla Regio
Several 30 to 45 km diameter structures with summit depressions at the centres of radar-dark regions can be seen. Each is surrounded by radar-bright lobate flows that extend outwards for a distance of 25 km, giving a petallate pattern. Magellan image F-MIDR 10S200;1.

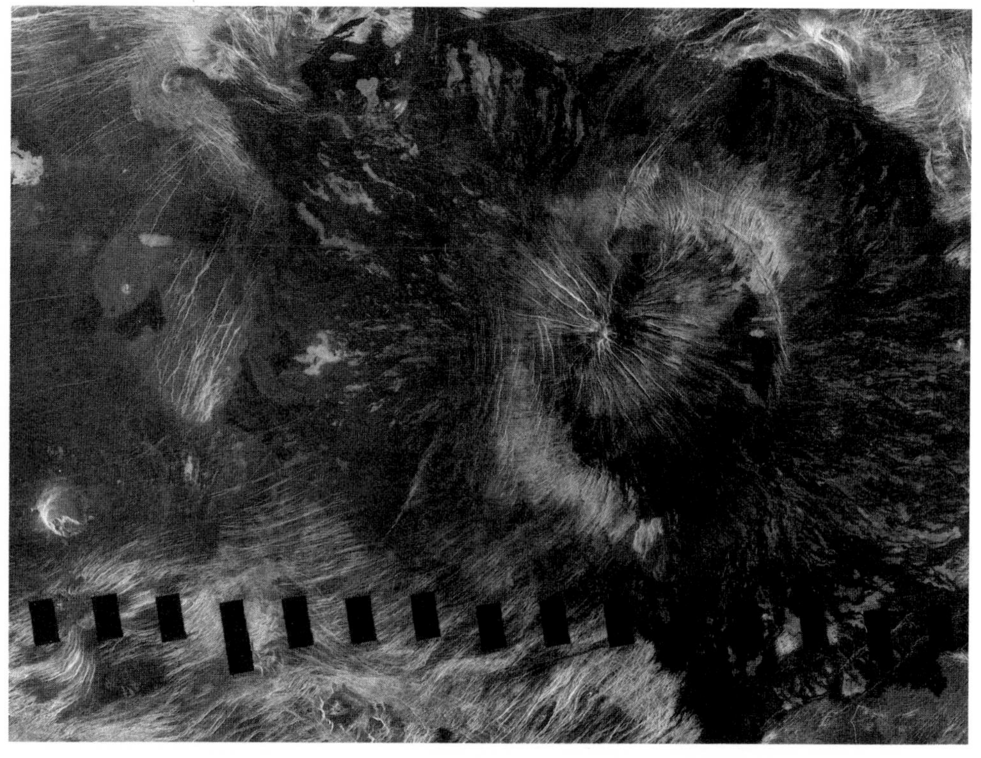

Figure 9.14 Large shield volcano southeast of Atla Regio
This 350 km × 450 km structure has a nova-like summit region characterized by radiating graben but also with some corona-like arcuate faults (NE flank). The associated, generally radar-dark, volcanic units that bury the underlying fractured plains have a plethora of mottled to radar-bright lobate flows superimposed upon them. Magellan image F-MIDRP 15S214;1.

flows do exist but that their surface texture does not give rise to prominent radar-bright signatures.

The anemone-type centres are particularly distinctive, some having a more-or-less equidimensional summit caldera, others showing elongate or even fissure-like depressions, which clearly are related to regional fault patterns. Their distribution is definitely patchy, there being concentrations in the Beta–Atla–Themis–Imdr region and on the flanks of the equatorial highlands. Few outcrop in the lowlands, a pattern that may be due to burial in low-lying regions or a response to the paucity or even absence of neutral buoyancy zones near or below MPR. Also within this size category are steep-sided domes, several of which show scalloped margins.

Larger volcanoes are often very distinctive and usually are characterized by fields of radiating lava flows and considerable positive relief, and often have a summit depression of considerable size. Sometimes the summit circular structure is not so much a depression as a corona-like structure, which may account for one-half of the radius. In other instances, both radiating and concentric fractures are attendant, giving the edifices an aspect intermediate between novae and coronas (Fig. 9.14). The flows associated with these larger structures typically have distinctive radar-bright lobate terminations and shield margins that contrast sharply with the surrounding fractured plains.

When the size–frequency distributions of intermediate and large volcanoes are plotted, it can be seen that there is a strong peak in the 20–40 km diameter range for intermediate structures, while the histogram for large edifices shows a broad flat top and a relatively small number of volcanoes with diameters exceeding 700 km (Fig. 9.15). Very large topographic highs, which are measurable in thousands of kilometres but which have attendant volcanic structures upon them – for instance the western part of Eistla Regio with the volcanic summits of Sif and Gula Montes – are classed not as "volcanoes" but as *rises*, since processes other than volcanism may have contributed to their growth. As far as the global distribution of the larger volcanoes is concerned, they tend to cluster within the upland regions of Atla–Beta–Phoebe–Theis Regiones, and along Eistla Regio, and generally are absent from complex ridged terrain and exist only sparsely on the lowland and rolling plains. This concentration in the more elevated regions suggests that some primary formational factor is responsible; certainly there is no evidence that large volcanic structures have been buried in the lower-lying regions, and as yet little-understood altitude-related magmatic phenomena probably are responsible.

One prominent example of a large volcano is Sif Mons, located on the broad rise of western Eistla Regio at 352°E, 22°N. A geological sketch map is shown in Figure 9.16. Sif is a 300 km diameter volcanic mountain that reaches 3.7 km above MPR, that is, 1.7 km above the general level of the Eistla rise, itself an elevated region of interdigitating volcanic plains. Near to the volcano there is a strong concentration of northwest-striking fractures that occupy a zone approximately 125 km wide; farther from the focus, in the northwest sector of the volcanic zone, fracturing is radial. This phase of fracturing evidently preceded much of the volcanism at Sif, since a field of 15–30 km wide radar-bright flows can be traced for at

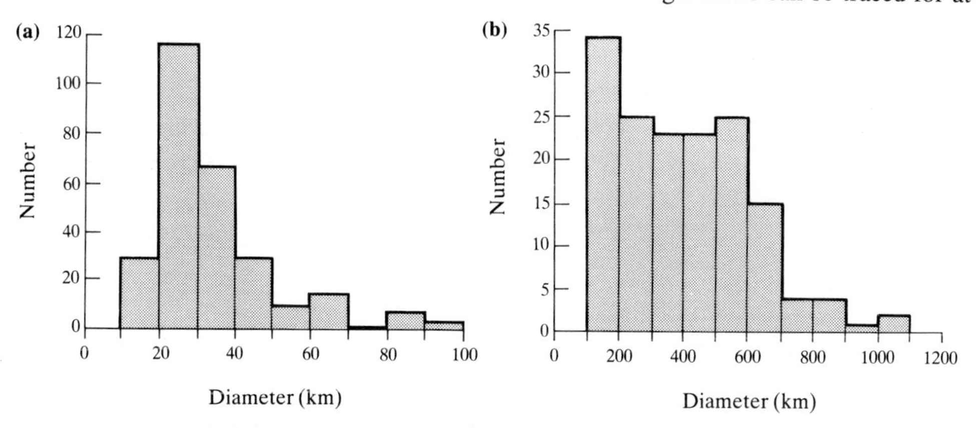

Figure 9.15 Size distribution of intermediate and large volcanoes
(a) Intermediate volcanoes, (b) large volcanoes. After Head et al. (1992).

least 250 km down the northern flank, partly inundating a cluster of small shields at lower altitude. Another series of flows spread down the same flank for at least 400 km from their source, high on the northern slope of Sif, before entering a fault trough and turning through

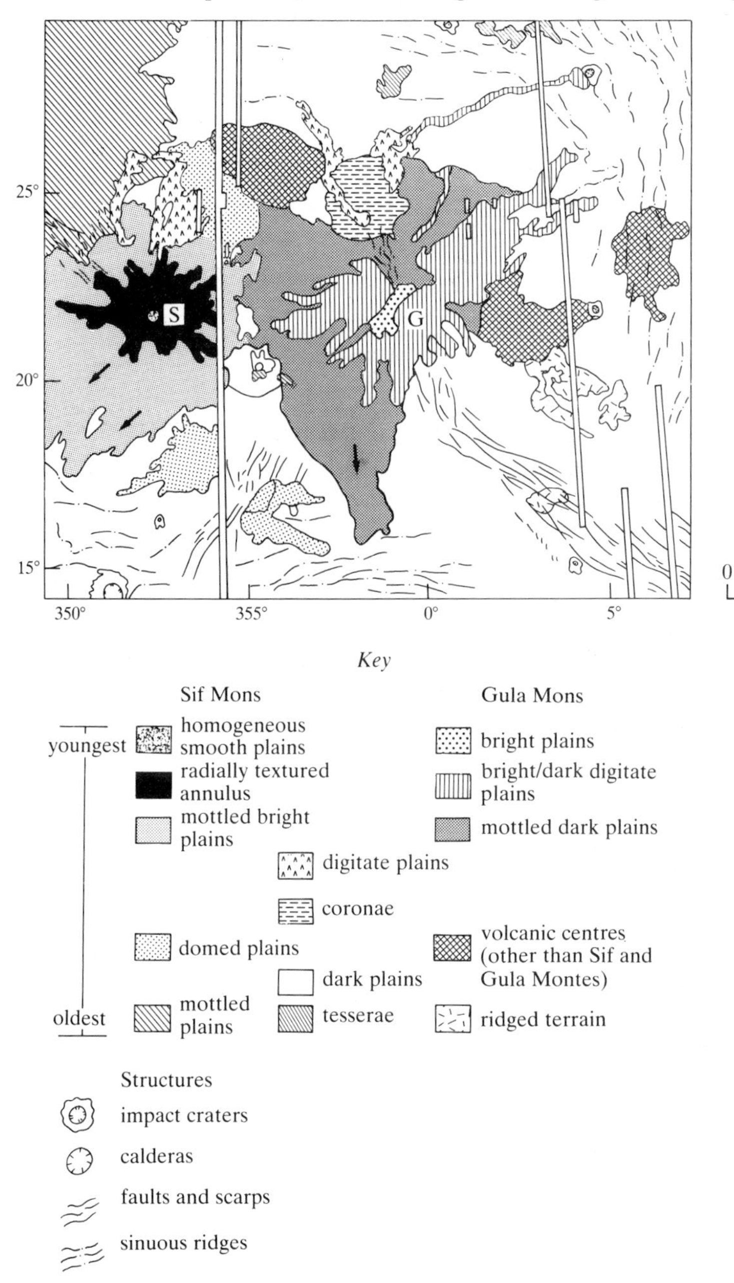

Figure 9.16 Geological sketch map of the region of Sif Mons volcano
After Senske et al. (1992).

(a)

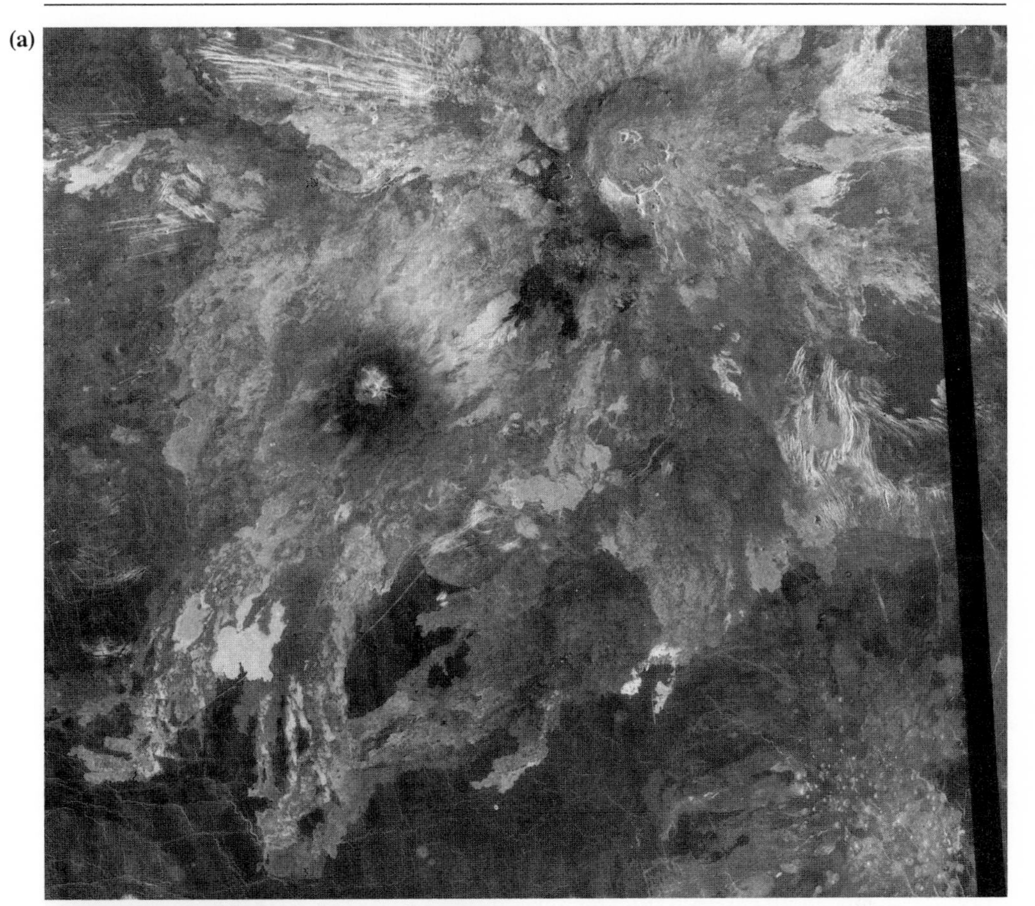

(b)

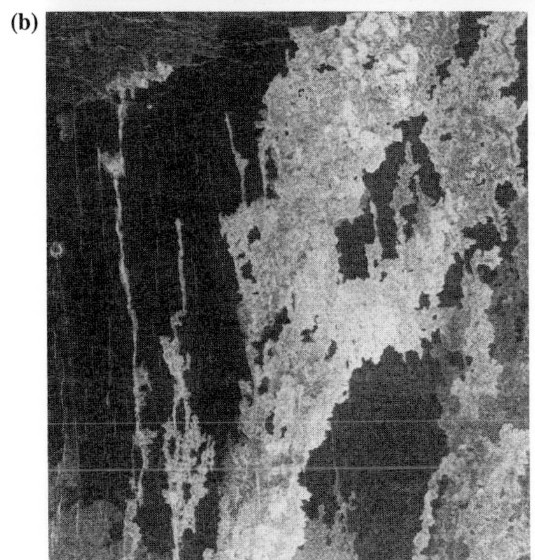

Figure 9.17 The volcano Sif Mons
(a) Central and southern part of Sif Mons volcano; the radar-bright flows overlie radar-dark fractured plains. Note the shield-field towards the bottom right-hand corner of the image. Magellan image F-MIDRP 20N351;1. (b) Northern flank: complex lava flows, several of which occupy narrow troughs formed by long fractures. Frame width 140 km. Magellan image P-38169.

90° of arc. Further flows emerged from the summit region, spreading east and southwards; these also have radar-bright signatures (Fig. 9.17); some may have been the products of flank eruptions, in much the same way as typifies Hawaiian-style activity. Such lengthy flows imply either high mass-eruption rates, or low-viscosity magma, or both.

The summit itself is marked by a caldera structure about 50 km across inside which are smooth flow units of variable back-scatter, which, in places, appear to have completely filled the depression and flowed over the lip. A plethora of lobate flows with generally bright radar signatures surrounds the caldera, giving rise to a prominent apron of lavas between 100 and 150 km across. Also within the depression and down the southeast flank are chains of smaller depressions 3–10 km in diameter that reflect the regional NW/SE structural pattern associated with western Eistla Regio. These presumably formed in response to flank eruptions and possibly activity associated with dykes injected along the flank rift zone. The very close similarity between Sif and Hawaiian shields such as Kilauea strongly suggests that volcanism occurred as a result of mantle plume or hotspot activity beneath the western end of Eistla Regio, fracturing the crust both radially and along a NW-trending zone before voluminous eruptions associated with the plume head emerged at Sif Mons itself.

Multiple collapse of volcanic depressions is also evident. Such a process is revealed by centres such as Sachs Patera, shown in Figure 9.18 and located south of Lakshmi Planum. Here, a slightly elliptical depression 40 km across and 130 m deep is surrounded by arcuate fractures and has associated with it an eccentric family of similar fractures to the north. Such an arrangement is suggestive of at least two stages of collapse and, indeed, a smaller arcuate zone to the southwest may represent a further event. This implies small-scale movement of the focus of eruption with time. A fan of prominent radar-bright lobate flows outcrops to the north, extending for up to 25 km from the caldera walls. However, there is little relief across

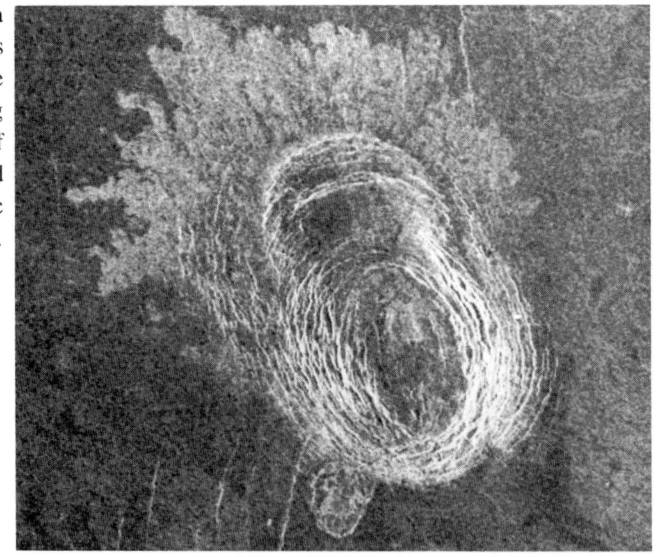

Figure 9.18 Sachs Patera This elliptical depression is located in Sedna Planitia. The pattern of arcuate fracturing suggests multiple episodes of collapse. Note the associated lobate radar-bright volcanic flows. Magellan image P-38284.

Sachs and it may be that collapse of the central area was accompanied by effusion of lavas from rift zones situated along its flanks.

9.3.5 Coronas, arachnoids and novae

Coronas were first identified on Venera 15/16 images and these, together with related arachnoids and novae (Janes et al. 1992), range in diameter from 60 to over 2000 km (Stofan et al. 1992). They have a range of morphology but most coronas have an associated annulus of fractures that ranges from 10 to 150 km across. Novae tend to lack annular faults but to be dominated by a central dome with radial fracturing, while arachnoids are characterized by concentric fracturing beyond which strong radial faulting may extend for several radii. Head et al. (1992) identified 259 arachnoids and 50 novas. There is some confusion as to whether or not the three types represent an evolutionary sequence generated by mantle upwelling.

Coronas are among the most widespread of Venusian landforms and have been studied by several different groups, not least because an appreciation of their origins is fundamental to an understanding of the internal workings of Venus (Barsukov et al. 1986, Pronin & Stofan 1990, Stofan & Head 1990, Squyres et al. 1992, Stofan et al. 1992). Their distribution is not entirely random, there being a clustering of such features around longitude 250°E, many being located between the longitudes of Beta, Phoebe and Themis Regiones, and near Atla Regio, while a distinct chain of coronas concentrates along Parga and Hecate Chasmata, which extend eastwards from Atla Regio (Fig. 9.19).

On the basis of a preliminary study of Magellan imagery, Stofan and her colleagues recognize five main types: concentric, concentric-double ring, asymmetric, radial/concentric, and multiple. Of these, the first is the most common, 186 landforms (about 50% of the total) being of this kind. Concentric coronas also have the greatest size range (60–2600 km), although the upper figure refers to the atypical and very large structure, Artemis Chasma. They also exhibit a wide range in morphology.

Tamfana, situated on the plains south of Alpha Regio, is a *concentric corona* with a raised rim, the opposite crests of which are 400 km apart. This rises 700 m or more above the level of the surrounding plains from which it is separated by a continuous 500 m deep moat, seen as the darker, smoother region beyond the radar-bright rim in Figure 9.20. The rim has a steep and somewhat hummocky outer slope, but its inner face is gentler. Concentrically arranged extensional faulting is concentrated across it, the faults mainly being graben about 1 km wide. Beyond the moat are further extensional faults, with two distinct trends, one radial and the other concentric; these are well seen at the top right of the image shown. Most faults are paired to form long graben. The concentric faults do not extend beyond the perimeter of the moat, but the radial fracturing splays out over the surrounding plains well beyond the moat, while some of the radial faults transect lavas that have flooded the moat floor, indicating some

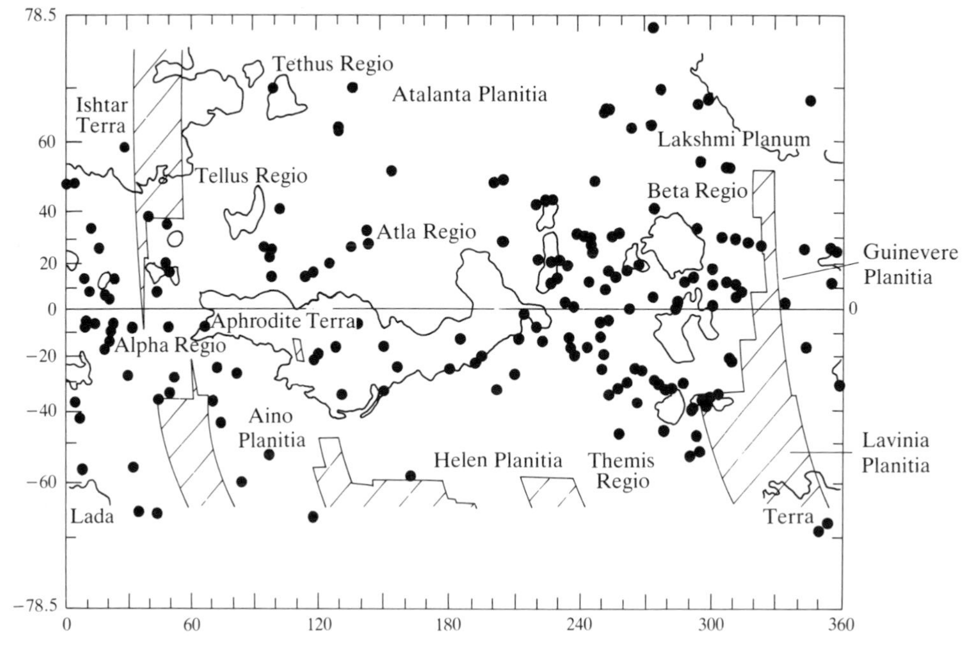

Figure 9.19 Distribution of corona and associated features on Venus
Coronae are shown by the polygonal symbol, corona-like structures by asterisks. After Stofan et al. (1992).

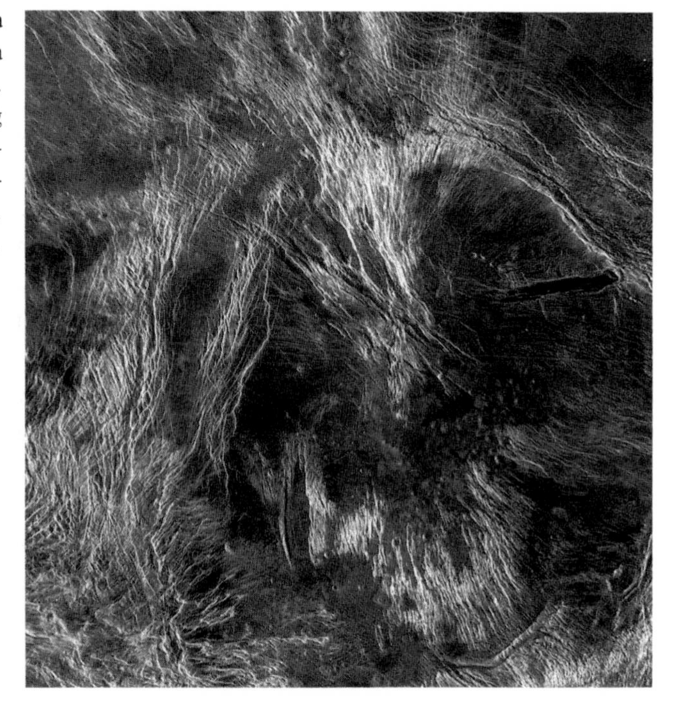

Figure 9.20 Tamfana Corona A 400- km-diameter corona located in Lavinia Planitia. Concentric and radial fracturing is clearly seen, while a well-developed moat almost completely surrounds the structure. Magellan F-MIDRP 35S003;1.

overlap between radial faulting and effusion.

The corona interior has an area of approximately 450 km^2, which in places lies 500 m below the plains and on which outcrop ponded lavas that in places are superposed on the concentric fracturing. Lavas too have largely flooded the moat, lobate and overlapping flows being clearly discernible towards the left edge of the same image. Interestingly, the flows of the corona interior lie at a considerably lower level than those of the moat, a situation which, if the magmas had the same source, can most easily be explained by assuming some measure of post-extrusion deformation. Alternatively, if deformation is not invoked, the implications are that magma did not rise to the same hydrostatic level, and therefore conduits may not have been physically connected when lava extrusion occurred. Of the two, the former seems the more preferable. There may also have been a much earlier phase of extrusion that post-dated both the radial and concentric faulting exterior to the moat. This manifests itself in a series of approximately radial radar-bright lobate flows, which extend out to distances of 400 km from the corona focus. Best seen in the north and west sectors, these flows have been buried by younger effusives associated with neighbouring coronas in the south and east. Their outcrop pattern strongly suggests that these flows emerged from a locus either at or near the present corona centre, preceding both the concentric and radial fracturing.

Although some evidence may have been buried by later events, it appears that activity at Tamfana, like other coronas, commenced with uplift – and perhaps volcanic construction – of the rim and downwarping to form the encircling moat. The precise sequence of these events cannot be fixed. However, both phenomena generated concentric flexuring, which was followed by the extrusion of highly fluid lavas, both on the corona interior and within the moat. Radial faulting apparently pre-dated volcanism within the moat region but in places also continued later.

Concentric-double ring coronas share many of the characteristics of the first group but are encircled by two distinct annular ridges and/or moats. Generally speaking, individual 15–20 km wide rings are separated by smooth plains, such structures typically being 60–70 km across. Thus far, 38 such structures have been identified, the majority of which occur in chains, especially in the region of Parga Chasma–Themis Regio (Fig. 9.21).

Radial/concentric coronas, on the other hand, exhibit an intense radial fracture pattern focused at the corona centre, which is surrounded by concentric troughs and ridges. Seventeen of these features have been mapped. Selu, the structure shown in Figure 9.22, is a broad plateau-like uplift whose summit lies 645 m above the surroundings; it has a diameter of 350 km and is located south of Alpha Regio, on the plains of eastern Lavinia Planitia. It has a double annulus, the outer one spanning about 180° of arc and open towards the east and being dominated by troughs and grabens. The inner annulus is even less complete (90° of arc) and is characterized by outward-facing fault scarps. Individual graben are seldom longer than 100 km and typically are 4–6 km in width. In the east and north the rim is traversed by two intersecting sets of fractures, one set striking north–south and parallel to plains-crossing regional fractures.

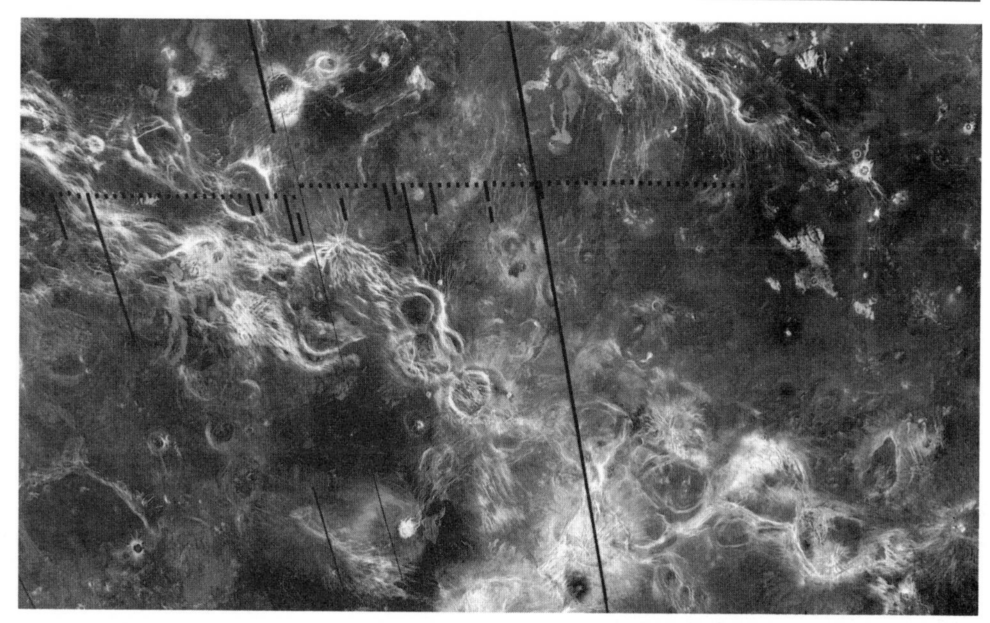

Figure 9.21 Chain of coronae running along Parga Chasma and Themis Regio
This region contains the highest concentration of corona on Venus. The mosaic is 4275 km across. Magellan C2-MIDRP 30S284;1.

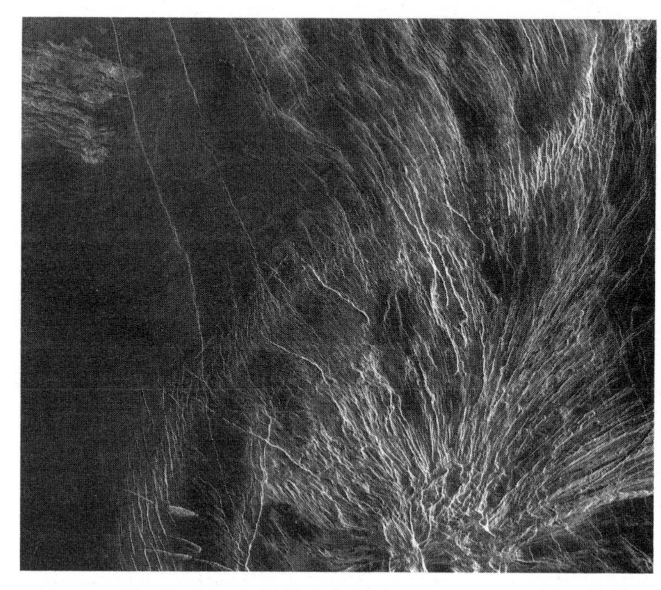

Figure 9.22 The corona, Selu
Situated south of Apha Regio, this 350 km diameter structure has a double annulus of fractures, open to the east. This iamte shows the north-west sector only, with the two annular fracture sets and plethora of radial fractures. Magellan C1-MIDRP 40S004;1.

The corona interior is extremely complex. The central region actually lies 500 m below the highest point (situated 50 km to the south of the focus) and is three-quarters encircled by higher topography. It is traversed by a plethora of cross-cutting fractures, pits and pit chains, and is partly covered by a small region of radar-dark plains. The radial lineaments comprise small ridges, troughs and scarps spaced about 1–2 km apart, while larger features include steep-sided troughs – seen towards the bottom left-hand corner of the image – which have rounded terminations and are very sharply defined; these frequently have been inundated by outward-flowing lavas. They appear to be volcanic collapse depressions, a supposition supported by their association with collinear chains of small pits with smooth floors.

Radar-dark plains outcrop at several locations within the interior of the corona, particularly between the two annuli, and between the inner ring and the centre. Such deposits generally embay radially fractured terrain but themselves are embayed by radially trending collapse depressions and by the concentric faults associated with the two rings. They have a rather featureless aspect. In contrast, well away from the central regions, on the south flank of Selu, radar-bright lobate deposits extend onto the adjacent plains and many inundate the concentric faults. These are presumed to be lava flows.

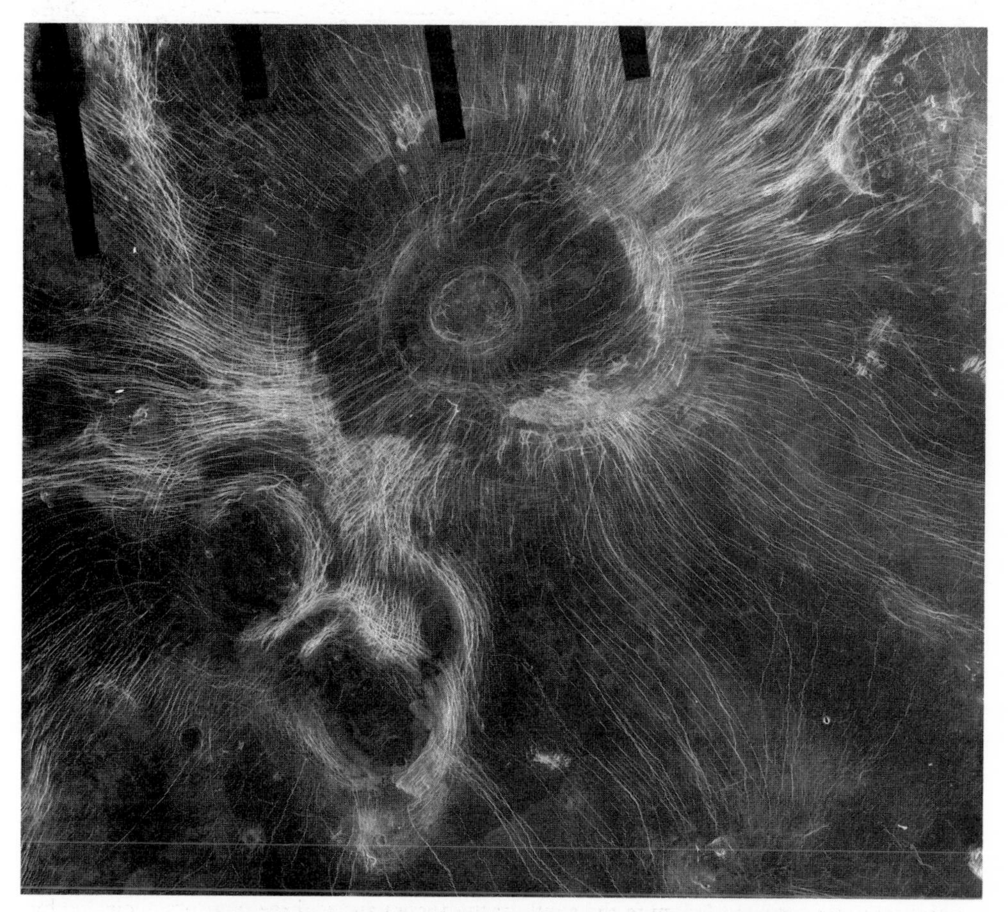

Figure 9.23 Multiple coronas located west of Themis Regio
This structure measures 200 km × 380 km across, and comprises two linked structures. A further, larger corona lies to the north-east. Magellan F-MIDRP 20S221;1.

Asymmetric coronas, of which 60 have been identified, are marked by an asymmetry of form that produces annuli that range from sinuous to angular. *Multiple coronas* consist of two or three linked structures that have a single encircling peripheral annulus (Fig. 9.23). Thirty-five have been mapped, several of which are located along ridge-and-trough belts to the east of Atalanta Planitia. Sometimes it is possible to establish an age sequence for individual members of a group; for instance, in the case of the multiple corona, Akajata, located north-west of Alpha Regio, the four corona-forming events evidently waned in strength with time, generating successively smaller structures.

9.3.6 Origin of coronas

The immense amount of detail revealed by Magellan imagery allows details of the evolution of the above structure to be formulated. The earliest tectonic event appears to have been west–east extension, which generated north–south faulting; a similar stress field appears to have prevailed over much of the lifetime of Selu. Plains deposits that embay these early faults are themselves cut by the radial fractures that characterize the central regions; these are presumed to have formed in response to the uprise of the central plateau. Any radiating troughs in this region probably represent volcanic collapse events mimicking pre-existing lines of weakness. The concentric faulting was largely subsequent to the above, the scarp-like nature of the innermost faults suggesting that uplift rather than subsidence affected their formation. However, some subsidence does appear to have affected the innermost part of the corona. The relations between plains deposits, lava flows, collapse depressions and fracturing all point to volcanism having occurred throughout the development of Selu, although its importance may have diminished with time.

Synthesis of all the geological relationships observed indicates that radial fracturing consistently was the first stage in corona development. Such a phenomenon would accompany domical uplift induced by a rising mantle **diapir** (Fig. 9.24a). Of course, not all coronas show such fracturing at their cores; many instead have volcanic deposits in these regions, a characteristic that may be anticipated, since melts would have risen into the crust above the rising diapir as it hit the base of the Venusian lithosphere (Squyres et al. 1992, Janes et al. 1992, Stofan et al. 1992). Thus, at several coronas volcanism clearly accompanied radial faulting, since radial fractures are both buried by flows and also transect them, while some flows evidently emanated from radially directed fissures. The discovery by Magellan of a small number of radially fractured domes (e.g. Carpo and Mokos) further supports this hypothesis; in fact, these landforms may represent the first stages of corona development.

As the geodynamical modelling of Janes et al. (1992) shows, the evolutionary stage ensuing upon uplift may comprise viscous relaxation to a more plateau-like form. Such a plateau will be isostatically uncompensated. Plateau formation can be accompanied by a central sub-

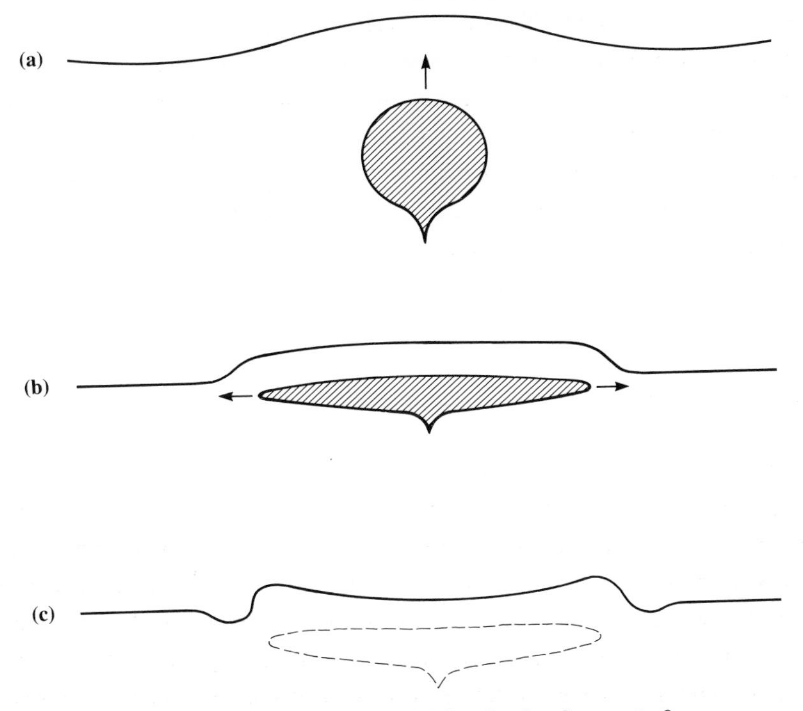

Figure 9.24 One conceptual model for the development of coronae
(a) A rising mantle diapir causes domical uplift of surface. (b) On impinging on the underside of the lithosphere, it spreads radially and flattens. (c) Finally, the diapir cools, allowing gravitational relaxation. After Squyres et al. (1992).

sidence, a topographic rim, an annular moat and concentric faulting. Volcanism may accompany all of this activity. Modelling shows that a plateau-like landform will develop as the rising diapir impinges against the lithosphere, spreading radially and flattening out (Fig. 9.24b). Concentric fractures are to be expected on the inward-facing slopes of a moat, and along the crest and outward-facing slope of a rim, as thermoelastic cooling and gravitational relaxation of the thermally supported plateau occurs (Fig. 9.24c). This sequence of events illustrates why dome and plateau formation precedes fracturing.

The later evolutionary stages of corona development involve continued volcanism, this often being so intense that it obliterates much or almost all of the evidence of early radial fracturing. The sunken interior regions of coronas and the floors of moats may also be flooded by lavas, while the former may see the growth of shields or shield fields, steep-sided domes and collapse pits. Because the later stages of diapir activity involve the cooling and thinning of the diapir, and concomitant relaxation of the topography, volcanism will decline and the youngest volcanic flows tend to pond in topographic depressions. Such a pattern is dissimilar to that of volcanic shields, where late-stage effusive activity is characteristic. Exactly how much topography will remain after this late stage will depend largely upon the strength of the local crust and any remanent thermal activity in the underlying mantle. It is likely to vary quite widely from region to region.

A few coronas approach the size of volcanic rises; however, most lack the cross-cutting rifts common to most such rises and have less pronounced topography. Artemis Chasma is something of an exception, and, in measuring 2600 km across, is intermediate between the two. This structure, located south of Thetis Regio, in Aphrodite Terra, has a structurally complex interior surrounded by a 150 km wide, circular annulus that shows strong deformation (Fig. 9.25). This annulus is characterized by graben, compressional ridges and other lineations whose origin is unclear. In places its floor lies 4 km deeper than the adjacent plains, while it is bordered by 4 km high ridges, which gives it a maximum height difference of at least 7.5 km. Complex troughs cross the interior, some of which are inundated by radar-dark volcanic flows. However, the interior does not lie at a higher level than the exterior plains – a major difference between it and a typical corona.

The origin of this enigmatic feature remains uncertain. Evidently little vertical uplift occurred during its formation, but equally there was considerable horizontal movement, both extensional and compressional. It has been suggested that some form of retrograde subduction might have played a part in its formation (Sandwell & Schubert 1992). Without doubt it is a region worthy of future close study.

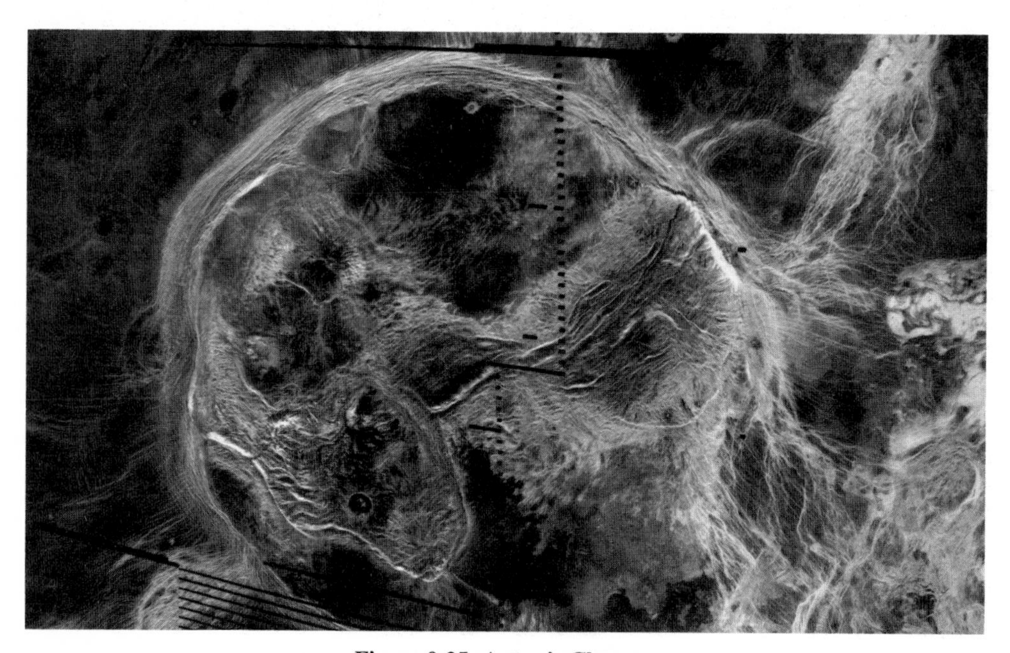

Figure 9.25 Artemis Chasma

This 2600 km diameter feature appears to be intermediate between a large corona and a volcanic rise. Magellan image P-38857.

9.3.7 Circular depressions and calderas

There are certain other landforms that share some of the characteristics of true coronas but which appear to have a different origin. Squyres et al. (1992) refer to these as *circular depressions*. Like coronas, they exhibit a concentric fault pattern, but unlike coronas, this encircles a simple depression whose floor may lie 500–600 m below the surrounding plains and some regions of the floors may have been covered by lavas. Occasionally a very low rim surrounds this. On the floor of Aramaiti, a typical member of this group, is a distinctive 100 km diameter dome that rises to the level of the exterior plain (Fig. 9.26). Squyres and his colleagues note that this structure essentially is the inverse of a typical corona, the raised rim taking the place of the depressed moat and the central dome taking the place of the central depression. While the sinking of the depression generated concentric faulting, subsequent uplift of the dome was accompanied by radial fracturing on the floor of the structure; again, the

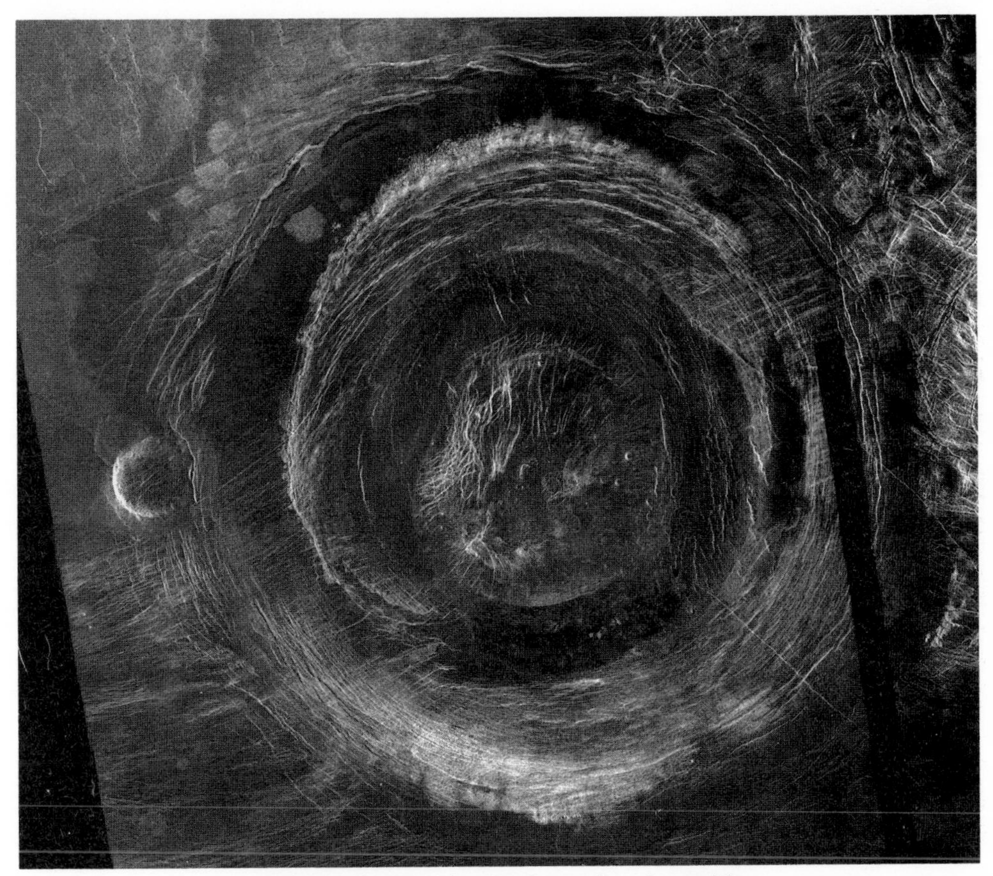

Figure 9.26 The circular depression, Aramaiti
Located in Aino Planitia, this large depression, 300 km across, has a narrow rim and a depressed floor on which is located a 100- km-wide dome. Magellan image F-MIDRP 25S082.

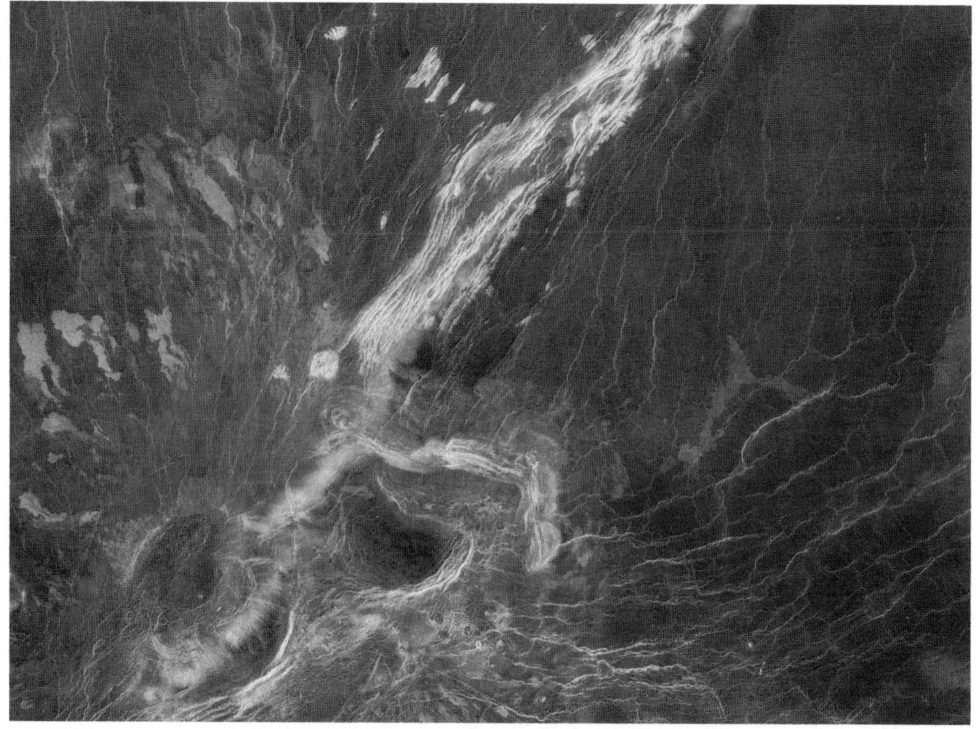

Figure 9.27 Caldera structures east of Bell Regio
This nested 90 km diameter caldera structure complex sits astride a prominent NE/SW tectonic zone. Note the adjacent radar-dark plains with sinuous ridges and the radar-bright flows associated with the main volcanic centre. Magellan image F-MIDRP 30N066;1.

reverse sequence to that typical of coronas. Such structures are more akin to calderas than coronas.

Calderas are circular to elongate depressions with little vertical expression of relief; their interiors generally exhibit outcrops, or a covering, of radar-dark lava flows. In a few cases a low rim occurs and there may be radial lava flows on the flanks. A typical example is Sacajawea caldera, situated amidst the plains of Lakshmi Planum. This measures 150 × 105 km, and is a nested depression surrounded by a zone of intense concentric faulting (Fig. 10.25). Its depressed floor is largely covered by relatively smooth, radar-dark, lava flows. Like similar structures located at the summit of unequivocally identified volcanic shields and paterae, these negative-relief features appear to have a volcanic origin and to be a manifestation of magma withdrawal and crustal subsidence perhaps following both intrusive and extrusive activity. A similar structure is illustrated in Figure 9.27

9.4 Summary

The surface of Venus is dominated by landforms and larger structures that are the result of mantle upwelling. Certain chains of coronas and corona-like structures may be similar to hot-spot chains on Earth, but the lack of clear age differences along such chains tends to militate against this idea. There appears to be a preference for the notion that such chains are a manifestation of integrated extension–upwelling activity. The fact that an evolutionary sequence can be identified for coronas, and that a small number of radially fractured domes appear to be coronas in their infancy, implies that upwelling continues at the present day, and consequently that coronas are still developing on Venus.

The broad global distribution of volcanic landforms over the Venusian surface contrasts greatly with the strong polarization of terrestrial volcanism along lithospheric plate boundaries, that is, along narrow, globally contiguous, linear zones. In particular, the absence of chains of major volcanic shields precludes the large-scale lateral movement of crustal material over mantle hotspots. However, the distribution of volcanic structures is not totally random, since more than 70% of all shield fields, coronas and large volcanoes fall within one-half of the surface (Head et al. 1992). Furthermore, apart from sheet lava floods, there is a deficiency of constructional landforms in the lowland regions.

This latter fact strongly suggests that there has been extensive flooding of topographically depressed regions by lavas that have emerged both from centres at higher elevation and also possibly from fissures within the lower-lying ground. Plots of major lava channels and flow orientations support this idea. A further rôle may have been played by altitude-dependent factors, which have been predicted to control volcanic style at different elevations (Head et al. 1992). For instance, the lowland regions should be preferentially endowed with high-volume effusions of lava, while at higher elevation, owing to the generation of neutral buoyancy zones, a greater profusion of shield volcanoes, calderas and domes would be expected. The ultimate distribution pattern undoubtedly is a function of both factors. Finally, there can be little doubt that relatively recent (0.5–1.0 Ma) volcanic resurfacing preferentially affected the equatorial latitudes, where the greatest incidence of modified impact craters occurs (see Ch. 8).

The very large proportion of volcanic structures that apparently are related to mantle instability over areas a few hundred kilometres across suggests that most shallow melt generation inside Venus is related to pressure-release melting associated with plumes or hotspots rather than to any form of globally widespread shallow melting process. It should be noted, however, that estimates of the volume of extrusive material embodied in the observed volcanic landforms falls significantly short of volumes predicted by some theoretical serial-volcanism models. The implication here is that processes other than those presently identified and understood may have played a part in the volcanic history of the planet.

10 Venus: the global view

10.1 Introduction: towards a tectonic map of Venus

The surface of Venus is endowed with a diverse array of tectonic features, from small-scale fractures to deformed belts many thousands of kilometres in length. Owing to an effective lack of weathering and erosion, these are excellently preserved and, from the styles of lithospheric deformation seen, they enable geologists to deduce their implications for the planet's tectonic history. However, it should be remembered that, at the time of writing, detailed structural analysis of Venus can, at best, be said to be in its early youth, and ideas and conclusions will doubtless change as time passes.

Pre-Magellan studies, based on Pioneer and Venera 15/16 data, saw several workers collating relevant data and presenting preliminary tectonic charts (e.g. Marchenkov et al. 1990, Nikishin 1990). Figure 10.1, a compilation by Nikishin (1990), shows the principal tectonic features of the planet identified at that time, the main structural units recognized being plains (1), dome-like uplifts (2), coronas (3), ridgebelts and fold zones (5) and "supercoronas" (6). The global structure is also depicted on USGS Maps I-2059 and I-2041 (1989). In his paper, Nikishin discusses three types of planetary belts:

- uplifted and weakened belts with an abundance of mantle hot-spot structures,
- a fan of ridgebelts traversing northern latitudes, and
- belts of low-altitude volcanic plains.

In particular he stresses the existence of an equatorial system of *weakened* belts, which terminates eastwards in Atla Regio, from which the meridional belt of large-scale dome-like uplifts (Beta–Phoebe–Themis Regiones) strikes northwards and crosses the north polar region. Nikishin comments that there must be a relationship between the origin and formation of the two kinds of global-scale structures, and he envisages the tessera (CRT) blocks as representing more ancient structures, possibly palaeocontinents. He bases this speculation on a number of facts, including the relatively young age of Beta, Atla and Eistla Regiones – indicated by correlation of their gravity and topography – and the observation that CRT/tessera blocks not only look anomalous against the background of global belts, but also, where partially buried by younger plains deposits in the low-lying areas of the planet, are older than these plains.

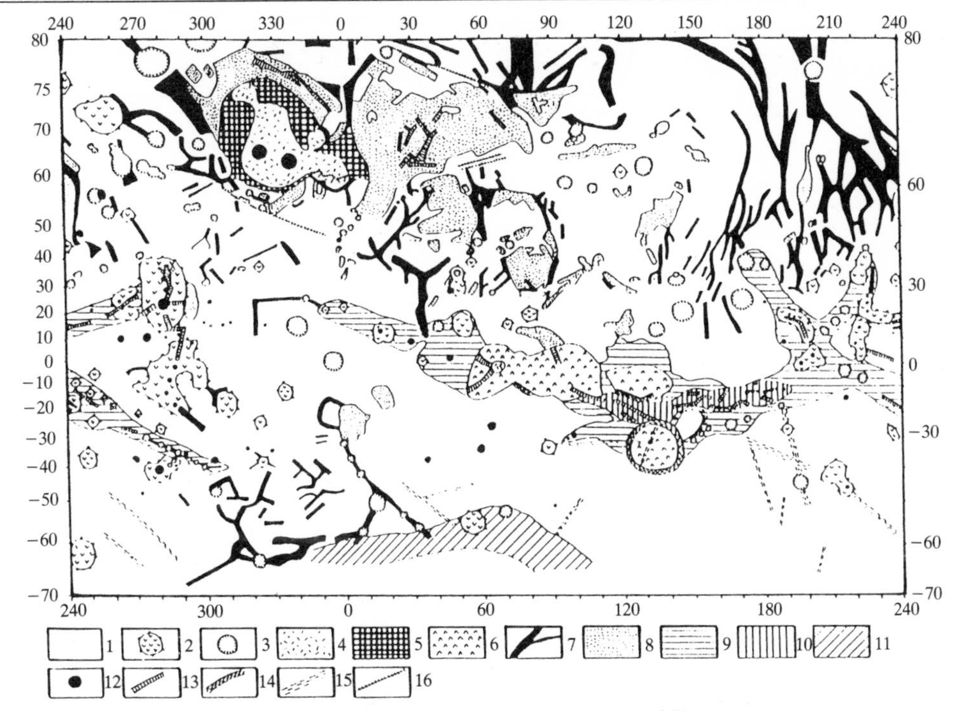

Figure 10.1 A preliminary tectonic map of Venus

This map, compiled by Anatoly Nikishin (1990), uses Pioneer Venus, Venera 15/16 and Arecibo data. 1 – plains; 2 – dome-like uplifts; 3 – coronae; 4 – lava plains of supercoronae; 5 – mountain ridge frame of Lakshmi Planum; 6 – supercoronae (undivided); 7 – ridge-belts; 8 – tesserae; 9 – large belts of domelike uplifts, coronae, rifts and probably ridge zones (uplifted "weakened" zones); 10 – Dali-Chasma rift zone; 11 – upland of Lada Terra; 12 – large volcanoes; 13 – rift-like structures; 14 – trench-like depressions; 15 – lineament zones; 16 – large faults.

Kozak & Schaber (1989) extended earlier speculations by Schaber (1982) concerning the existence of a global network of deformational and volcanically active zones. Using Venera 15/16 imagery, they discussed the north–south trending zone of tectonic disruption extending from near Eistla Regio and continuing over the north pole to rejoin the equatorial highland belt near Atla Regio. They concluded that a global-scale system of rift zones characterizes Venus and that fold/thrust belts are often associated with rifting, particularly where rifts were in close proximity to tesserae massifs. Although the scale of such a system is similar to the tectonic zones of Earth, they pointed out that the nature of the Venusian deformational zones was very different from terrestrial spreading centres. More recently Zuber (1990) reiterated the view that an understanding of the ridge-belt fan is fundamental to an understanding of mantle dynamics within Venus.

Further studies based largely on Venera 15/16 imagery, and concerning the relationships between tessera blocks and regional ridge belts, were completed by Raitala & Törmanen (1989, 1990) and Raitala & Kauhanen (1991). In noting the crustal shortening associated with the fan of global-scale ridge belts, they pointed to the smaller-scale make-up of ridge belts, which are composed of smaller components closely related to more local tectonic stresses. Furthermore, in describing the occurrence of small planitiae and tessera massifs bounded by ridge belts, they conclude that these units define Venusian mini-plates, and open up the whole argument associated with plate tectonics on the planet. In particular, within the region of Eastern Ishtar Terra–Allat Planitia–Fortuna Tessera, they conclude that structural relationships indicate small-plate movements against the highland massif of Ishtar–Fortuna. Similar arguments are used for the region of Ganiki Planitia, a part of the global belt of deformed rocks that follows the 200° meridian.

Magellan studies have built on this earlier work, which was admirably summarized by Solomon et al. (1991). Investigators using all available Magellan data have tended to use those regions first imaged by Magellan as "type areas", i.e. Ishtar Terra, western Eistla Regio, Alpha Regio and Lavinia Planitia. Currently, planetary geologists are applying what they have learned from focused studies of these few regions to Venus as a whole. The complexities of the highland regions, in particular the tesserae, have attracted the attention of many scientists, resulting in competing ideas as to how the highlands evolved.

As a basic concept, it is now reasonable to identify three broad kinds of highland terrains on Venus:

- *Beta-type highlands*, which best can be described as major volcanic rises;
- *Ovda-type highlands*, which are steep-sided highlands dominated by tessera massifs; and
- *Ishtar Terra*, which stands alone in having components of both of the above.

The highlands exhibit a wide variety of morphological and tectonic features. Magellan also has shown that tesserae (CRT) cover a much larger area than had been anticipated in the light of Pioneer data, and can be shown to share a close relationship with the other highland areas.

Major volcanic rises characterize the equatorial highland zone and extend also, with a meridional trend, into Beta Regio. They are characterized by topographic highs and substantial positive gravity anomalies. The general view is that they are associated with mantle upwelling. A part of Ishtar Terra may have a similar origin. The rises within Aphrodite Terra are spatially associated with "Ovda-type" highlands, that is, steep-sided massifs dominated by tesserae.

10.2 Highland rises and massifs of Aphrodite Terra

The earlier mapping by Pioneer-Venus showed that Aphrodite is by far the largest region of highlands on the surface of Venus, extending from 45° to 210°E, a distance of around 23 000 km. However, it is now clear that, far from being one contiguous highland massif, there are wide variations in topography, structure and gravity along its length. Western Aphrodite comprises the plateau-like highland massifs of Ovda and Thetis Regiones, which

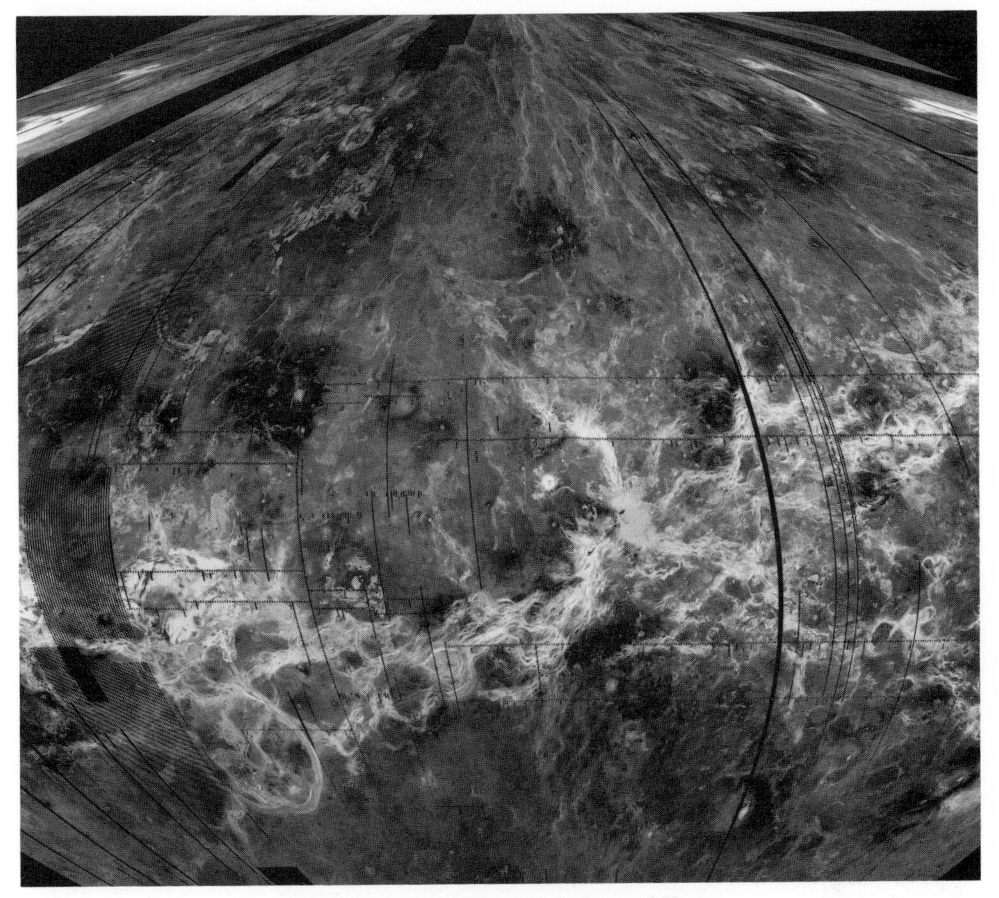

Figure 10.2 Aphrodite hemisphere of Venus

Compressed Magellan image showing complex structure of Aphrodite Terra. Magellan image C3-MIDRP 14N180;1.

rise between 3 and 4 km above MPR; eastwards of latitude 140°E is a much lower zone of troughs and elongated ridges, including Diana and Dali Chasmata, which extends as far as 190°E, a distance of around 5000 km. Eastern Aphrodite is built largely from the massif of Atla Regio, whose highest points (the summits of the volcanoes Ozza and Maat Montes) attain elevations of 9 km above MPR. Aphrodite is surrounded by the extensive equatorial plains of Aino, Niobe and Rusalka Planitiae.

Magellan imagery reveals the structural patterns within this region to be complex, and throws some doubt on the notion that Aphrodite should be considered as a single geological entity. Individual massifs, broad domes and many cross-cutting fracture lines can be clearly recognized in Plate 6. The Magellan altimetric map (Plate 7) and the preliminary geological map of Saunders nicely show the basic geological elements that build this complex region (see Fig. 10.3 and Table 10.1).

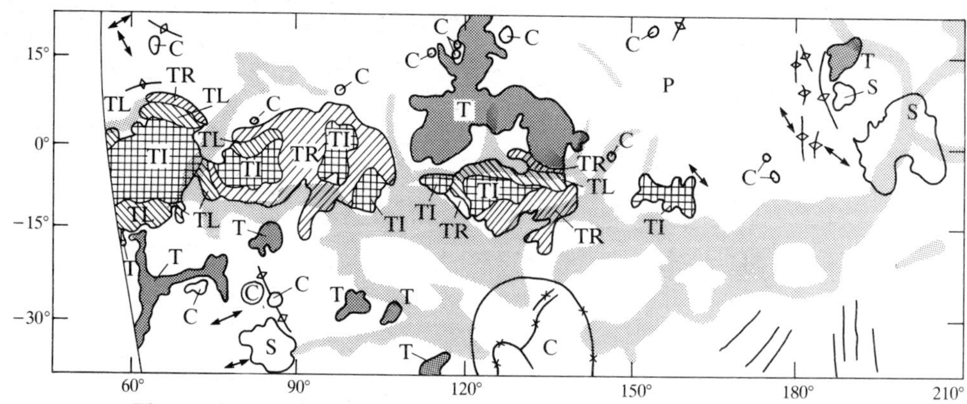

Figure 10.3 Preliminary geological map of the Aphrodite Terra region

The map was drawn by Stephen Saunders. Units delineated are keyed in Table 10.1. Double ended arrows indicate the general trend of lineations in a given area. Lines with diamonds = ridges; lines with crosses = troughs; unmarked lines = lineations. After Solomon et al. (1992).

Table 10.1 Definition of map units in Aphrodite Terra.

Symbol	Unit	Definition
TI	Interior blocking terrain	Irregular raised blocks, 15–75 km across, separated by curvilinear troughs
TR	Linear ridge terrain	Linear rounded ridges cross-cut by narrow troughs, probably graben
TL	Linear trough terrain	Cross-fractured, with at least one system of straight parallel troughs
T	Other tesserae	Similar to TR and TL but low-lying and not so radar-bright
S	Shield volcano	Topographic domes, radar-dark planar surfaces. Presumedvolcanic
C	Crater material	Craters, walls, floors and ejecta
P	Plains, undivided	Smooth, radar-dark, planar units
F	Festoon	Volcanic flow of thicker-than-usual dimensions. Viscous lava feature
	Fractured terrain	Swarms of radar-bright lineatios interpreted as extensional faults

10.2.1 Ovda and Thetis Regiones

A high degree of deformation characterizes the rock units that outcrop within both Ovda and Thetis Regiones. The core regions of both massifs are built from a unit characterized by irregular raised blocks with a more-or-less equidimensional form, and between 15 and 75 km across, clearly seen in Figure 10.4. These are separated by curvilinear troughs. The general

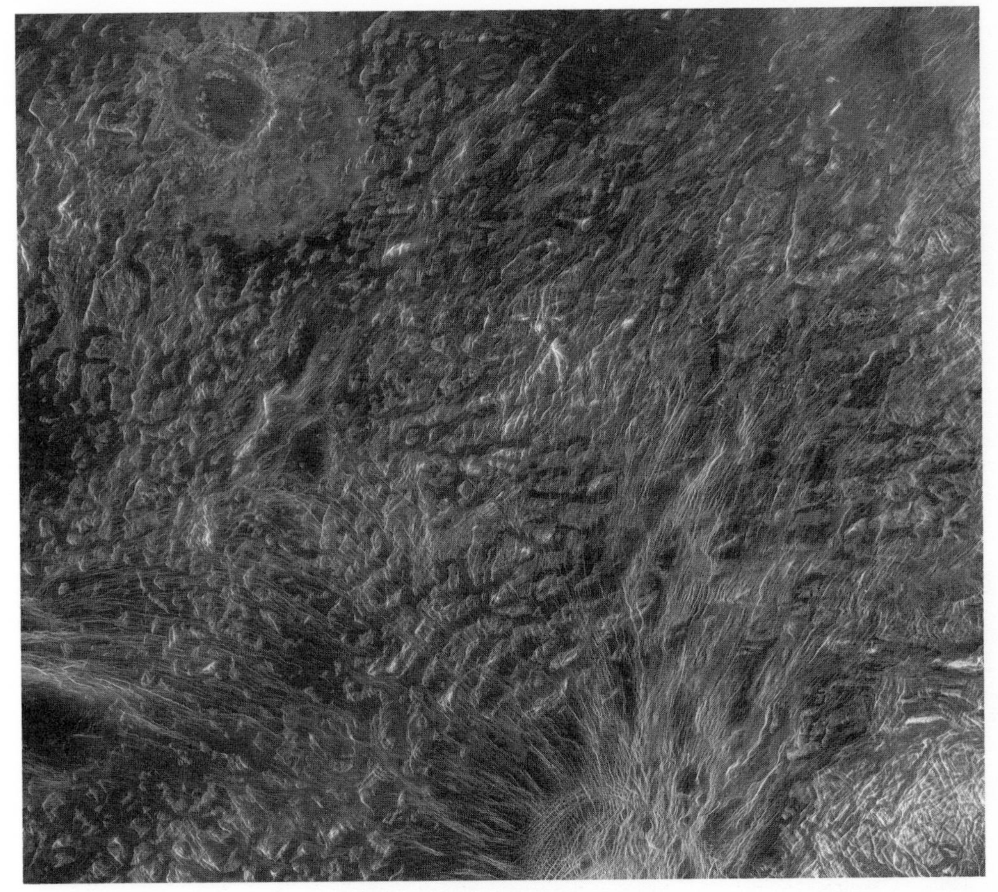

Figure 10.4 Eastern Ovda Regio

Interior block terrain. The oldest features are irregular broad domes and ridges about 20 km across. The intervening valleys were subsequently flooded by radar-dark lavas. An extensive radial fracture system associated with the feature at bottom right postdates domes, ridges and flooded valleys. The image is 600 km across. Magellan image F-MIDRP 05S070;1.

dome-like form and the presence of radar-dark (presumed) volcanic flows between the ridges give a distinctive signature to these areas. Such *interior block terrain* is accompanied by *linear ridge terrain*, in which ridges, which often are continuous for upwards of hundreds of kilometres and spaced 10–20 km apart, are often transected by fractures well developed in the right-hand part of Figure 10.5. Linear ridge terrain is widespread in central Ovda and also in Thetis Regio, although less extensively developed there. In places the ridges are cross-cut by fractures, generating smaller segments, these perhaps being the first stages in the development of block terrain.

The above tectonic patterns all indicate an origin in crustal shortening; the lines of long ridges, in particular, are reminiscent of terrestrial fold and thrust belts. The other terrain type

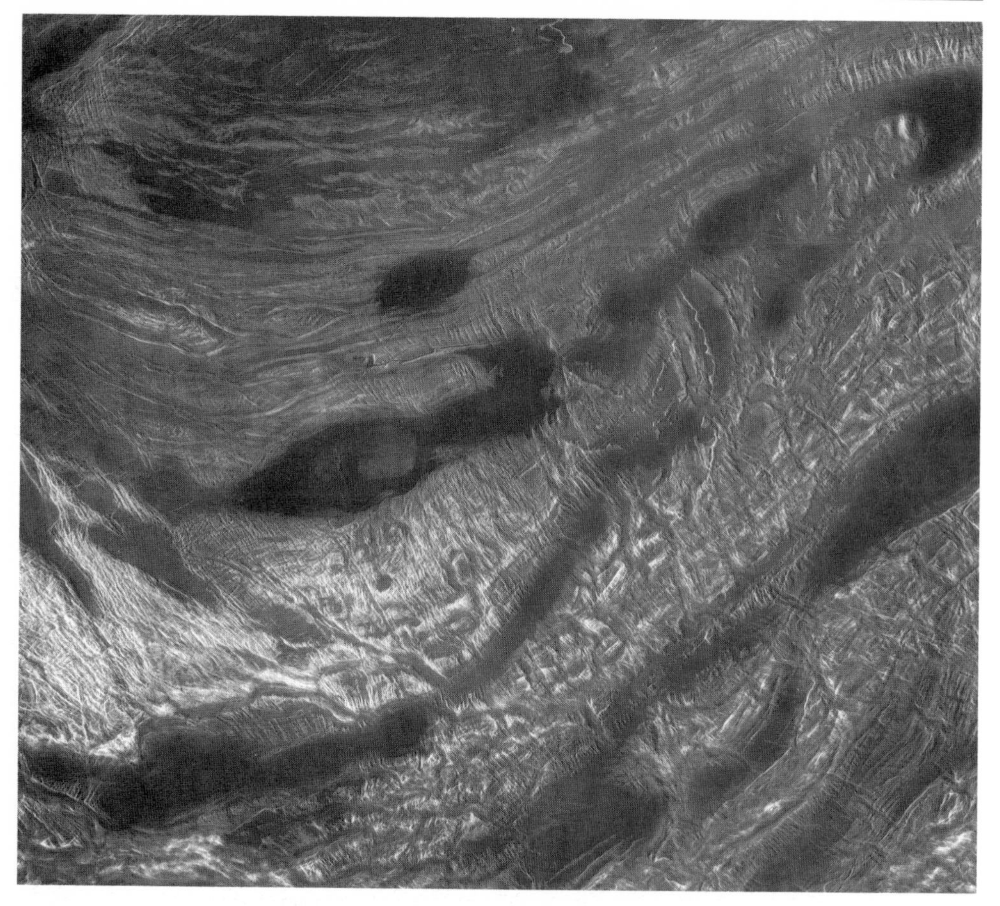

Figure 10.5 Northern boundary of Ovda Regio
The dominant features are low-relief, rounded, linear ridges, 8–15 km wide and 30–60 km in length. These constitute linear ridged terrain. Radar-dark material infills regions between some ridges. The surface slopes from south to north, with a gradient of 1 in 60. Magellan image F-MIDRP 00N082;1.

observed in both regions is rather different: this *linear trough terrain*, generally located at the periphery of the massifs, is typified by narrow graben just a few kilometres in width and tens of kilometres long. North–south trending graben intersect broader valleys with dark floors that strike approximately SW/NE. The downfaulted valleys frequently are embayed by radar-dark smooth units, presumed to be lava flows. Such units generally represent a later phase of deformation involving extensional stress.

The margins of both Ovda and Thetis Regiones are cut by fracture sets that extend outwards across the adjacent low-lying plains. Such 1 km wide trough faults are often continuous for hundreds of kilometres and are the youngest tectonic features present in the region; their formation appears to have overlapped the effusion of lava flows observed at both locations.

10.2.2 The zone of arcuate chasmata

Southwards and eastwards from the Thetis highland are some arcuate troughs that are among the deepest on Venus. Of these, Dali and Diana Chasmata are the most prominent. Up to 1000 km long and 100 km wide, each trough is at least 1 km deep and typically is bounded

(a)

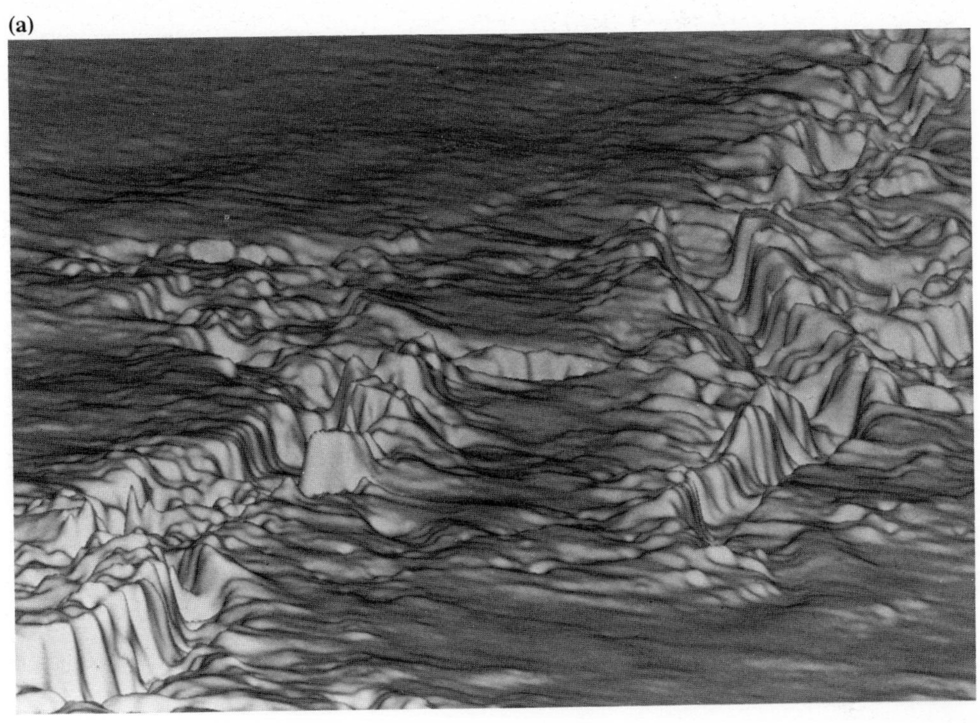

(b)

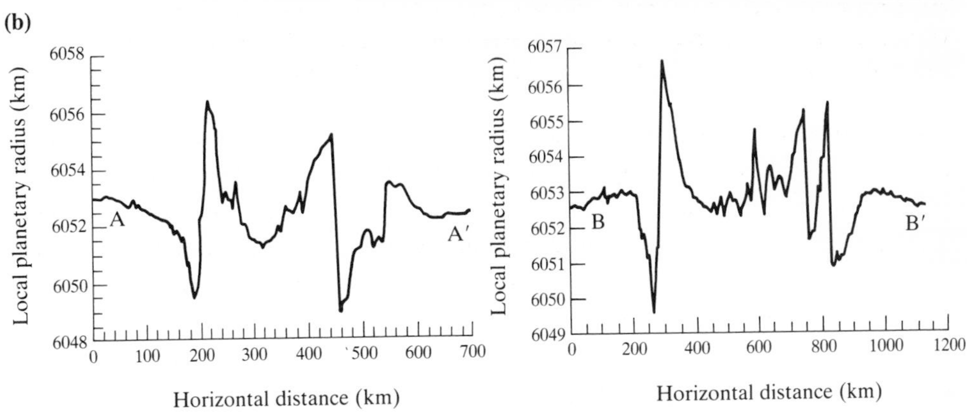

Figure 10.6 The region of Dali Diana Chasmata

(a) Perspective view of the topography. Magellan image P-38452. (b) Topographic profiles across the chasmata, from $A–A_1$ to $B–B_1$. After Ford & Pettengill (1992).

on one side by a ridge whose height is comparable with the depth; the opposite rim is much lower. The height difference between a trough floor and the summit of the highest point on its rim may be as much as 7 km (Fig. 10.5). A high degree of deformation typifies this zone of trough faulting, most of the smaller-scale deformation manifesting itself as fractures, often paired to form graben, and trending in an ENE direction along a line joining east Thetis Regio to Atla Regio (Fig. 10.6). In places fractures may radiate out from the foci of coronas, while long compressional ridges also occur. At one location – seen on the left-hand side of Figure 10.7 – a series of ridges run parallel to a north–south trending double trough, which appears to join Dali Chasma with the western end of Diana Chasma.

One interpretation of Diana Chasma is that it is a transform zone that has accommodated

Figure 10.7 The volcanic rise of Atla Regio

This image, measuring 1600 km by 1800 km, is centred on the equator at 197°. Ozza Mons is situated within the radar-bright region near the image centre, while Maat Mons is located to the southwest. Note the radar-dark summit regions of both. Ganis Chasma trends northwards from Ozza Mons, while Parga Chasma extends eastwards. Magellan C1-MIDRP 00N197;1.

strike-slip movement. McKenzie et al. (1992a) draw attention to the similarity between this part of Aphrodite, where curved trench-like features have high topography on their concave side, giving an asymmetric profile, and the area of the East Indies between New Guinea and Fiji, where active subduction is taking place and, of course, there is active magmatism. In the East Indies island arc location, the complex geometry of the trench pattern is a reflection of motions between several small lithospheric plates, rather than of the two main (Pacific and Australian) plates. The complicated faulting seen on Magellan imagery of this region may be a reflection of flexure within Venusian plates being subducted. In suggesting that along this part of Aphrodite subductive movements may be taking place on Venus, they explain that the absence of active volcanism (a characteristic of subduction zones on Earth) may be anticipated on Venus, since there can be no lowering of melting point within the subducted wedge by hydrated sediments being carried down with it, and therefore no crustal melting to produce subduction melts.

Suppe & Connors (1992), on the other hand, prefer to interpret both the Dali–Diana and Artemis Chasmata as regions where fold-and-thrust belts have developed by a process of *taper wedge mechanics*. Such a process is considered likely to have occurred in terrestrial locations where fold-and-thrust belts and accretionary wedges have been recognized. It involves plate motion but not necessarily plate tectonics; thus fold-and-thrust belts are known to exist at the base of sliding continental margins and deltas (e.g. offshore Texas) and on the flanks of growing volcanoes (e.g. Earth and Mars). Such a process – which Suppe & Connors liken to the mechanics of a wedge of soil pushed along by a bulldozer – will produce a flexural foredeep, a toe of less intense folding, a 10–30 km wide steep zone with gradients of between 5° and 20°, and a flat crest.

Their interpretation, from analysis of the structural geometry of such zones on Venus, is that near-surface deformation on Venus is brittle, and dominated by cohesive strength in the uppermost 1–2 km of the crust. Applying known models of taper wedge mechanics, they predict that, under Venusian conditions, brittle wedges should have maximum surface slopes of between 10° and 20°. All of these features can be recognized as elements of the major troughs in this region and their gradients are of the predicted order. However, a plane of decollement is also a characteristic of terrestrial examples, and the viability of this idea largely depends upon the interpretation of features, particularly in the Artemis Chasma area, that may represent such lines of decoupling on Venus. It will be interesting to see what further clues may emerge after Magellan stereo imagery is processed. In the meantime, it should be noted that Suppe & Connors estimate, from measurements made on the toe associated with Artemis Chasma, a shortening of around 20% in this area, produced by fault-bend folding above a plane of decollement estimated to lie at a depth of approximately 1.5 km.

10.2.3 Atla Regio

Atla Regio lies at the eastern end of the Aphrodite highland zone (see Plates 6 & 8, and Fig. 10.2). It takes the form of an elongate upland measuring approximately 1500 × 2500 km, most of which lies between 2 and 4 km above MPR but rising to nearly 6 km in some places. It is by nature a broad highland dome, with several major volcanic centres and rifts (Fig. 10.7). The coloured Magellan topographic map of Atla (Plate 9) shows a series of chasmata radiating from the elevated central zone, one of which extends northeastwards towards Beta Regio. Like Beta Regio, it is a kind of tectonic junction. This characteristic is clearly shown on the plate, the three main rift zones – Ganis, Parga and Dali Chasmata – radiating out from the centre, the location of Ozza Mons.

Geoid and gravity anomalies for Atla Regio both are high (120 m and 130 mgal respectively) and are centred over Ozza Mons which, like Maat Mons, stands out clearly on the same plate. Estimates of the depth of compensation for this area range from 200 to 250 km (Smrekar & Phillips 1991, Phillips et al. 1991).

Maat Mons is the most central and also the highest of Atla's volcanoes. Rising to nearly 9 km above MPR, the shield is 300 km in diameter. Lava flows and radial graben extend for hundreds of kilometres from the summit. Many episodes of volcanism and faulting are apparent, some flows infilling, others embaying older fractures, and some faults transecting volcanic flows. The summit region of Maat Mons is marked by a region of radar-dark material 100 km in diameter, within which can be seen pits, flows and small domes. It is bounded to the south and west by concentric graben, 5–10 km wide. A striking field of small domes is situated north of the summit itself.

Ozza Mons rises to 7.5 km and is located northeast of Maat; it shares many of Maat's characteristics. However, although a few NNE-trending faults intersect the caldera walls, fracturing is by no means the dominant characteristic of this volcano (Fig. 10.8). There is no summit caldera, rather a radar-dark plateau feature measuring 50 km × 100 km, and rising 1.5 km above its surroundings – seen just below the centre of the image. A cluster of small volcanoes and associated lava flows that lie north of the summit are responsible for partially burying Ganis Chasma, giving a clear indication of the age relationships in this area. To the south lies a further subsidiary volcano, which also is elongated along a NE-striking line. Volcanic flows and fractures are both abundant here. The fractures, although many are radial, tend to be aligned northeast and are continuous with the family of fractures that connect Atla with the region of chasmata to the southwest. A close connection between the siting of volcanic centres and the regional extensional faulting is very clear.

In addition to these major edifices, there are several pancake domes, small domes, pits and lava flows in the environs. Sapas Mons, another large shield volcano (Plate 12), is situated close to the western margin of Atla Regio, while several impact craters show modification by volcanism, as described in Chapter 8. The radar-dark plains surrounding Atla contain

(a)

(b)

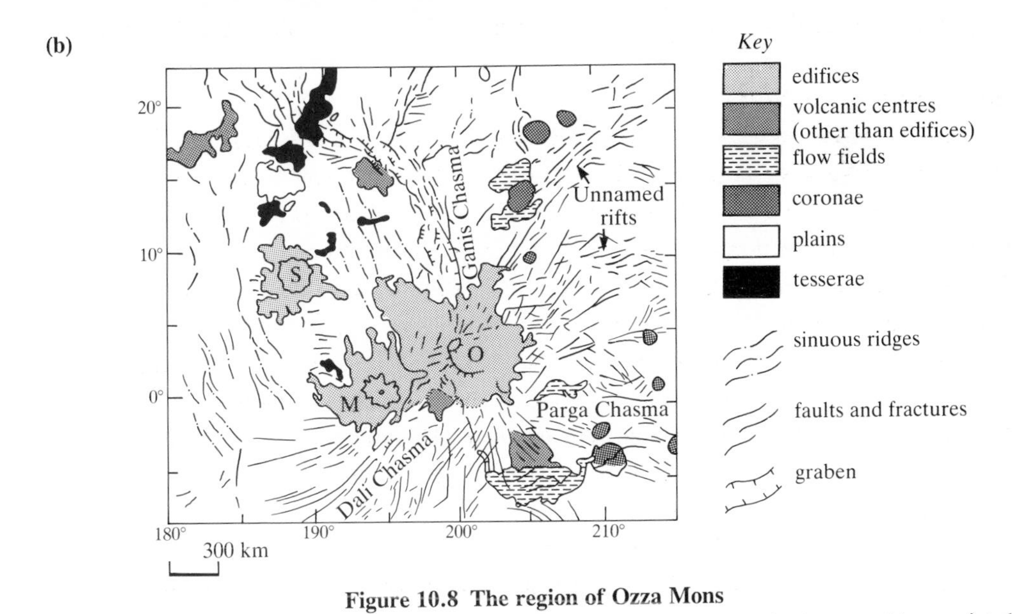

Figure 10.8 The region of Ozza Mons

(a) Magellan image of the volcanic summit, showing the radar-dark summit plateau, with associated fractures and flows. Magellan image F-MIDRP 05N200;1 (b) Geological sketch map of the same region. After Senske et al. (1992).

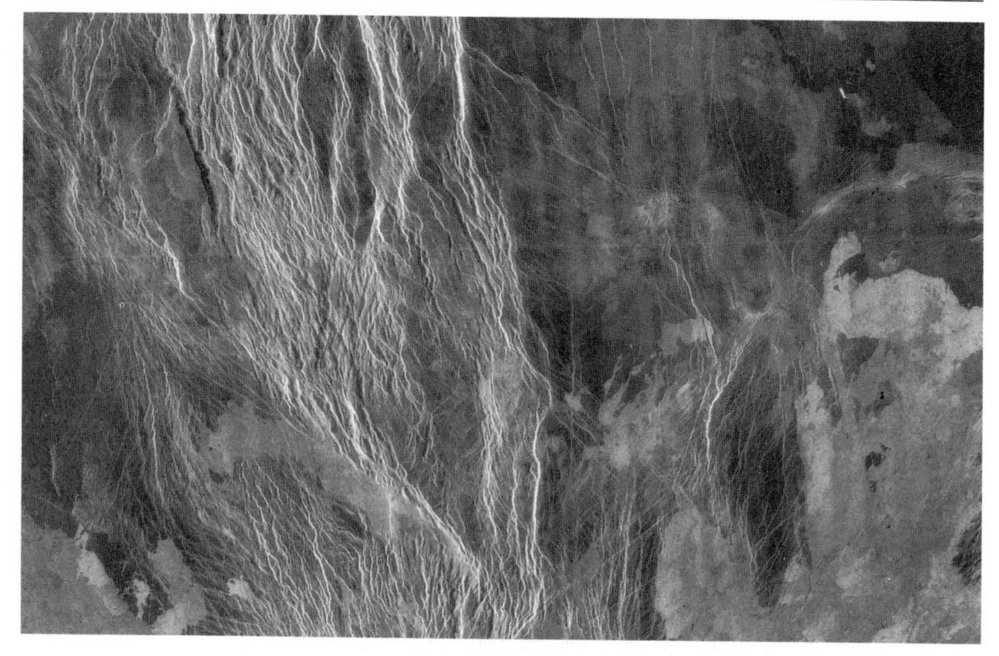

Figure 10.9 Part of Ganis Chasma rift system
The 400 km wide image shows a series of overlapping graben, 1–10 km wide. These constitute the Ganis rift system which attains a width of 300 km and extends for 1000 km. The main trend is north–south. Note the lobate volcanic flows with differing radar signatures to the east. Magellan image F-MIDRP 10N200;1.

several highly embayed tessera blocks. This is true also of the region surrounding the northern termination of Ganis Chasma. These areas of CRT appear to be the local basement, a relationship they hold also in the vicinity of both Ishtar and Beta Regiones. However, older CRT units are much more widespread within Beta Regio than at Atla.

Despite the abundance of volcanism, fracture systems are the most prominent features of Atla and extend outwards over the surrounding area. Ganis Chasma connects Maat Mons with Nokomis Montes to the northeast. This 1.5 km deep feature is approximately 1000 km long and as wide as 300 km in places. Figure 10.9 clearly shows Ganis Chasma to be a major rift zone consisting of sets of overlapping graben. Volcanic flows both cross-cut and are transected by the fault scarps. The broad similarity between Atla and Beta is obvious, a feature that has led several groups to consider Atla as a region of mantle upwelling (Senske et al. 1992, Solomon et al. 1992, Bindschadler et al. 1992a).

10.2.4 Eistla Regio

Eistla Regio lies astride the WNW extension of a line continuing the trend of western Aphrodite.

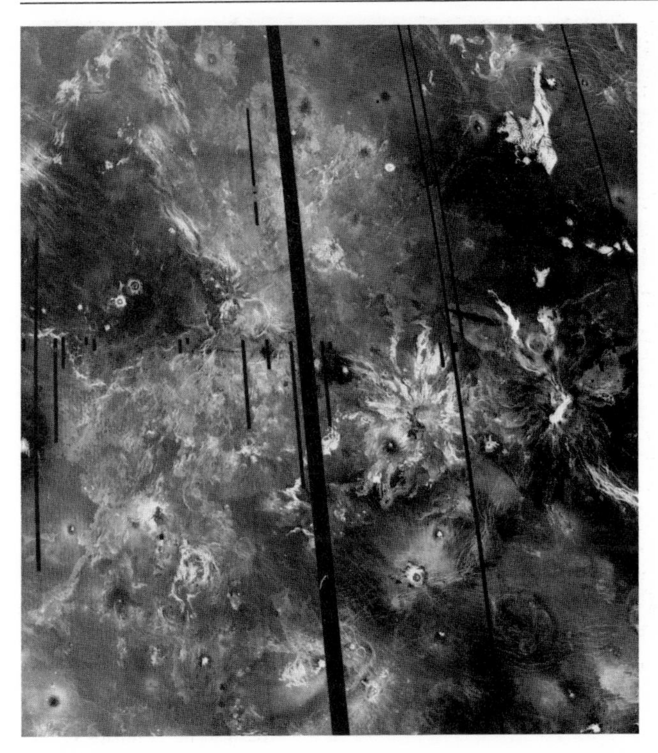

Figure 10.10 The region of Western Eistla Regio
The mosaic is 1500 km wide. The summits of Sif and Gula Montes, Guor Linea and the two coronae, Nissaba and Idem-Kuva, are shown. Magellan C2-MIDRP 30N335;2 (part of).

By nature it is a series of broad crustal rises, each several thousand kilometres across, which trend WNW/ESE and extend between 340°–55°E and 10°–25°N. It was the first of the equatorial highlands to be imaged by Magellan and its general aspect is shown in Figure 10.10. Western and central Eistla are both characterized by extensive areas of radar-dark and mottled plains, and are broad volcanic rises traversed by rift zones, with large volcanic structures occupying the higher ground; these tend to have high radar back-scatter (Senske 1990, Senske et al. 1992). The region is marked by strong positive gravity anomalies, and estimated compensation depths of between 100 and 200 km have been reported (Smrekar & Phillips 1991, Grimm & Phillips 1992). Eastern Eistla, which was not imaged during the first Magellan mapping cycle, is dominated by the 252 km × 370 km corona structure, Pavlova. The rises are surrounded by the volcanic plains of Sedna, Guinevere and Tinatin Planitiae. Fracture families present within the bounds of Eistla extend onto these low-lying regions, particularly in the west.

The highland rise of western Eistla measures 3200 km × 2000 km and is dominated by the large volcanoes of Sif and Gula Montes; Gula Mons is the larger of the two, and is approximately 400 km × 250 km, being broadly elliptical in planform; it rises 4.6 km above MPR and 3.2 km above its surroundings. The flanks slope away from the summit at angles of between 0.25° and 1.4°; thus it has the profile of a large shield volcano. Volcanic flows radiate down its flanks and extend outwards for up to 300 km. Their radar signature ranges from radar-dark to radar-bright, the latter contrasting strongly with the surrounding plains shown on the above image. Sif Mons is smaller, and is an isolated mountain that rises 3.4 km above MPR with a local relief of around 2 km and a diameter of 200 km. It has a prominent summit caldera, southeastwards from which run a line of 3–10 km diameter collapse pits, seen towards the left-hand side of Figure 10.11. North of both Sif and Gula Montes and roughly midway between them is a further volcanic edifice, Nissaba – the roughly circular feature with intermediate radar signature due west of the most northerly of the Gula flows. This is a broad volcanic dome and transection relationships suggest it to be older than the corona, Idem-Kuva, located to its east and between the two bright flows north of Gula Mons.

Both shields are the sources for a plethora of volcanic flows that extend outwards for at least 300 km. Mapping of these by Senske et al. (1992) shows that both Sif and Gula expose three assemblages of lava flows: to the north of Sif, two phases of eruption are separated by an episode of extensional fracturing. The earlier phase is dominated by flood eruptions that generated radar-dark flows whose source cannot be discerned; these are overlain by mottled flows that may embay earlier volcanic landforms. The post-fracturing volcanism saw the effusion of mottled and radar-bright flows that form well defined flow fields whose sources can often

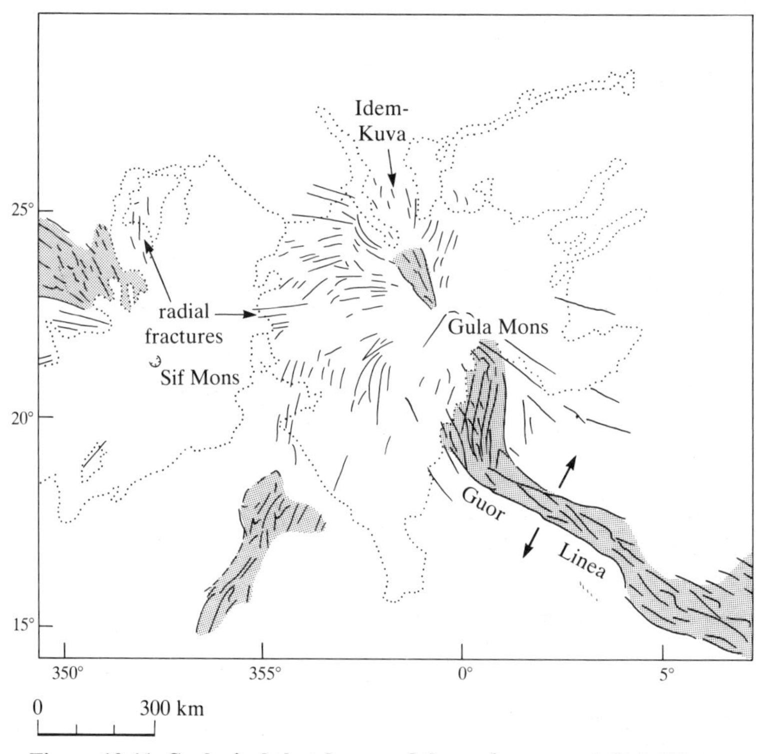

Figure 10.11 Geological sketch map of the region around Gula Mons
The volcano has no summit caldera, rather a 150 km long linear fracture zone. It lies at the focus of intersecting rift systems. After Senske et al. (1992).

be tracked down. Thus, the westernmost flow field appears to originate in a large dome. Other late-stage flows, particularly those to the east of Sif, exhibit flow channels and may inundate graben. The longest of such flows is 110 km in length and less than 1 km wide. An even younger sequence of smooth mottled volcanic units embays the previous unit and is interpreted by Senske et al. (1992) to represent late-stage flank eruptives associated with shield summit construction. This type of sequence is very reminiscent of several Martian shields and paterae.

Both shields, but Gula Mons in particular, are also the focus of an array of extensional fractures (Fig. 10.11). The latter cross the volcano's flow apron and extend onto the adjacent plains. Fracturing is also widespread to the northwest of Gula Mons, where two prominent coronas, each 250 km in diameter, have associated families of both concentric and radial fractures. This is taken as an indication of lithospheric extension to the northwest of the Gula structure. The most prominent fracture set in this region is that of Guor Linea, which extends southeast of Gula Mons for at least 1000 km (Fig. 10.12). This feature lies astride a topographic ridge that joins western and central Eistla. Many of the Guor Linea faults are located along the walls and floor of what essentially is a linear 50–75 km wide trough running along the axis of a broad linear rise. Such an association clearly shows that Guor Linea

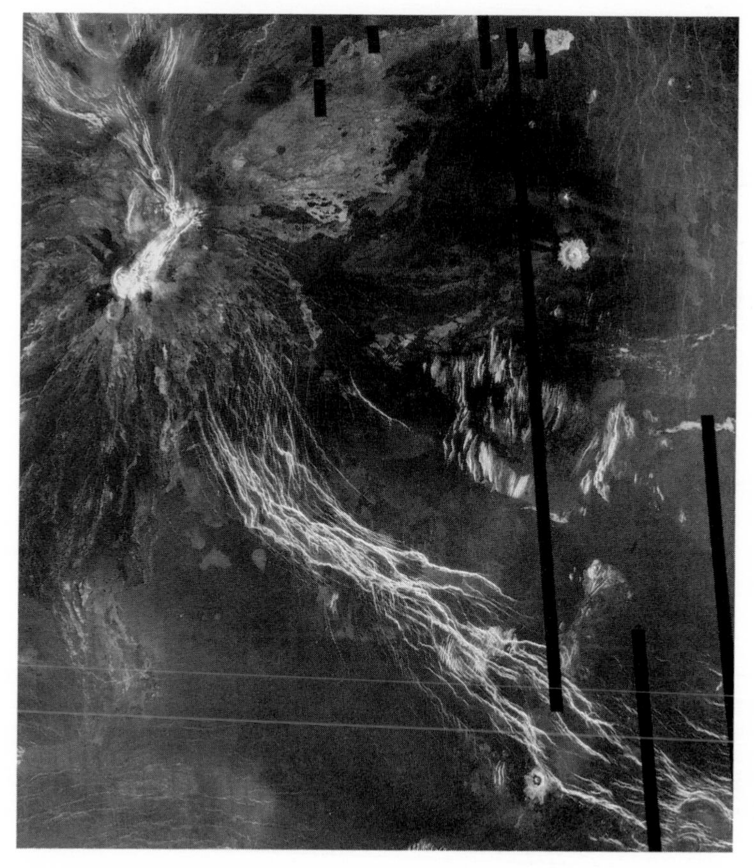

Figure 10.12 The Guor Linea rift system This structure – seen on the left side of this mosaic – extends southeast of Gula Mons in Western Eistla Regio. It is located within an area of thinner than average crust. The image is 700 km wide. Magellan C1-MIDRP 15N009;1.

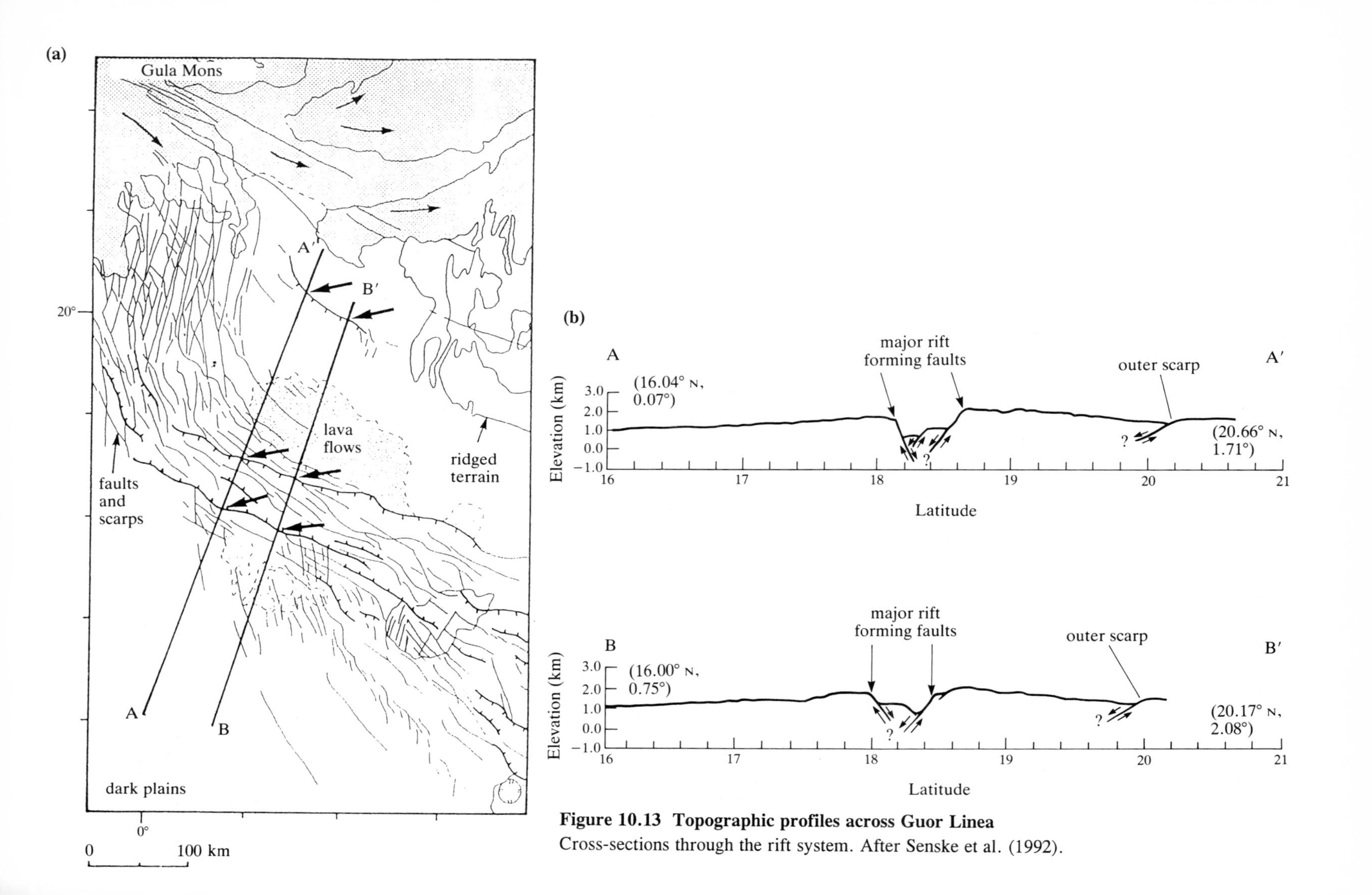

Figure 10.13 Topographic profiles across Guor Linea
Cross-sections through the rift system. After Senske et al. (1992).

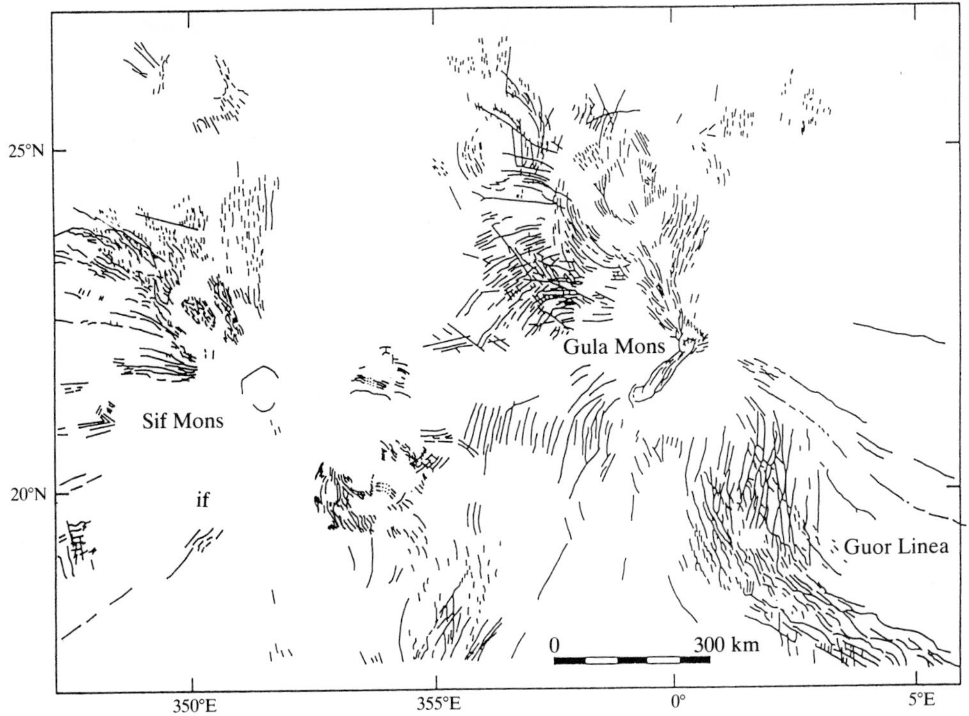

Figure 10.14 Tectonic features of a part of Western Eistla Regio
Showing the fractures in the region of Sif and Gula Montes and Guor Linea. After Grimm & Phillips (1992).

is a rift system that has developed in response to local extension and thinning of the lithosphere (Solomon et al. 1991, Grimm & Phillips 1992). The topographic profiles shown in Fig. 10.13 clearly show the rift-like structure.

The most pervasive of the fractures extend for 300 km or so, and Magellan imagery has revealed these to be graben faults. Most are far shorter; however, owing to the observed widespread local negative relief, these also are taken to be graben and confirm the rift-like nature of this area. Grimm & Phillips (1992) have conducted a detailed analysis of the faulting in this region, and their work, represented here by Figure 10.14, indicates that there are at least three major fracture families, the most recent of which strike nearly west–east and are associated with a deep trough that transects the earlier NW/SE faults that are the principal structures of the rift. Some of the latter structures well away from the rift valley occur as en echelon sets consistent with a component of right-lateral shear superimposed on the overall regional extensional strain. The age relationships between these two sets, as indicated by superposition/transection relations, suggest that approximately north–south extension was the most recent extensional strain to prevail, while earlier extension was mainly oriented NE/SW. There is, however, the possibility that the earlier faults may have been originally more nearly west–east but subsequently were rotated clockwise. The third family of faults is much less prominent and strikes approximately north–south. Field relations suggest that this set formed early on, providing mechanical boundaries for younger NW-trending faults, but was later reactivated and now transects both NW-striking fractures and Gula Mons volcano.

The main line of Guor Linea is interrupted by the massive edifice of Gula Mons. This implies that, in general terms, volcanic construction post-dates the main rifting; however, some fractures transect Gula flows, indicating that extensional deformation was also operational post-volcanism. The main fractures associated with Gula are radial and most are found to the southwest, northwest and east of the volcano. What happens in the northwest, incidentally, is that the radial faults merge into a set of fractures concentric and radial to the Idem-Kuva corona. Sif Montes-related radial fractures are found to the north, northwest and west of Sif, but are less prominent than those of Gula Mons.

Extending around the rise are extensive regions of *domed plains*, several areas being characterized by large numbers of small volcanic domes. Grimm & Phillips (1992) quote frequencies of 400 per million square kilometres. The emplacement of such plains appears largely to have pre-dated volcanism associated with Sif Mons, but some clusters of domes are superimposed on later fracturing. The formation of such units implies a strong local concentration of volcanic activity but with only limited surface access of partial melts to the Venusian surface.

Central Eistla Regio was originally called Sappho, a prominent radar-bright feature seen on earlier Pioneer and Arecibo images. Later imagery, particularly that of Magellan, reveals that this region is actually built from two volcanic structures that sit atop a broad volcanic rise. Sappho Patera is a 300 km diameter volcanic edifice marked by concentric fractures and a 200 km diameter summit caldera. This makes it one of the largest calderas on the planet. To its south is another volcanic structure, Anala, which is less well defined but rises 1 km above MPR. Flow channels have been identified on the high flanks of the latter. Major flow fields extend outwards for 900 km from these two centres, that of Anala overlapping parts of Sappho's, giving a clue to the broad age relationship. There are also several partially buried coronas flanking the main rise.

Fractures on this rise have a curvilinear pattern but strike mainly north–south, whereas those to the east and west strike perpendicular to these. NNW-trending fractures are cut by the flow apron, but faults striking just east of north extend all the way to Sappho. At Sappho itself, the fractures are deflected around the periphery of the caldera, and NNE- to NE-trending fractures join the two volcanic edifices. Another family of fractures trends approximately west–east and is buried by Sappho (and may also be offset beneath it). The Guor Linea rift stops some 900 km west of Sappho and is truncated by a corona on the western flank of the central Eistla rise.

In terms of the age relationships of the four major volcanic centres, it appears that Sif is younger than Gula, while Anala is younger than Sappho. Although no direct evidence can effect correlations between west and central Eistla, Grimm & Phillips (1992) note that volcanic styles in central Eistla appear to be more mature than those in the west, suggestive of greater age there. While Senske et al. (1992) feel that volcanic construction largely post-dated uplift, it does seem that the fracture systems that typify the area both cross-cut and are inundated by volcanic deposits on both rises, indicating that volcanism and tectonism were broadly coeval. The surrounding domed plains may pre-date this centralized volcanism.

Pioneer-Venus gravity data have been utilized by Grimm & Phillips (1992) to determine the two-dimensional gravity anomaly for this region. From the gravity and topography they derived mass anomalies on two internal horizons: the first lies at 20 km depth and is a manifestation of the density contrast experienced at the crust/mantle boundary; a deeper boundary was found at 200 km, and is presumed to lie where lateral density variations attributable to mantle circulation are greatest, i.e. beneath the thermal boundary layer at the base of the lithosphere. They interpreted the deeper set of anomalies and then solved for the mantle flow that would be driven by such anomalies. On the basis of their chosen flow model, both western and central Eistla are seen as sites of strong mantle upwelling, this being stronger beneath Gula Mons than Sif. The zone of upwelling also extends to the southeast beneath Guor Linea as a less intense *saddle* with about one-half to one-third the strength of the main plumes. This implies that, rather than indicating simple passive extension beneath Guor, the stresses are being applied by the actively convecting mantle. The solution for the shallow set of anomalies suggests that the crust beneath western Eistla, particularly Guor Linea, is thinner than average for the region; however, crustal thinning is not indicated for Sappho Patera – indeed, some solutions suggest that it may have been somewhat thickened.

Using the working hypothesis that variations in apparent depth of compensation are, at least in part, a reflection of crustal thickness or shallow thermal structure, Grimm & Phillips suggest that western Eistla, Guor Linea and the plains to the north are related to crustal *thinning* (or shallow cooling), while central Eistla and the southern plains are regions of crustal *thickening* (or shallow heating). Interestingly, the same workers find that maximum crustal thickening/shallow heating occurs beneath Heng-O corona, situated to the south of western Eistla; however, various trial solutions for the gravity data usually produced no anomaly for the lower layer. This prompts the observation that, if coronas are sites of mantle upwelling, as many suspect, then Heng-O must no longer be active.

The large gravity anomaly of western Eistla indicates that the rise overlies a site of strong mantle upwelling and crustal uplift; this is where thinning has occurred. In contrast, the larger component of crustal compensation and somewhat weaker flow-related density anomaly at central Eistla may imply that the crust has been thickened magmatically, in which case mantle upwelling may presently be declining. Perhaps, therefore, central Eistla sits above a more mature plume than that underlying the western rise. Complementary loci of mantle downwelling are found beneath the plains of Guinevere and Tinatin Planitiae, to the south. Strong northerly flow also is predicted directly to the north of Gula Mons but is weaker elsewhere. The asymmetry of the large-scale mantle flow in this region probably is manifested in the regional tectonic pattern. One schematic interpretation of events in the region is shown in Figure 10.15.

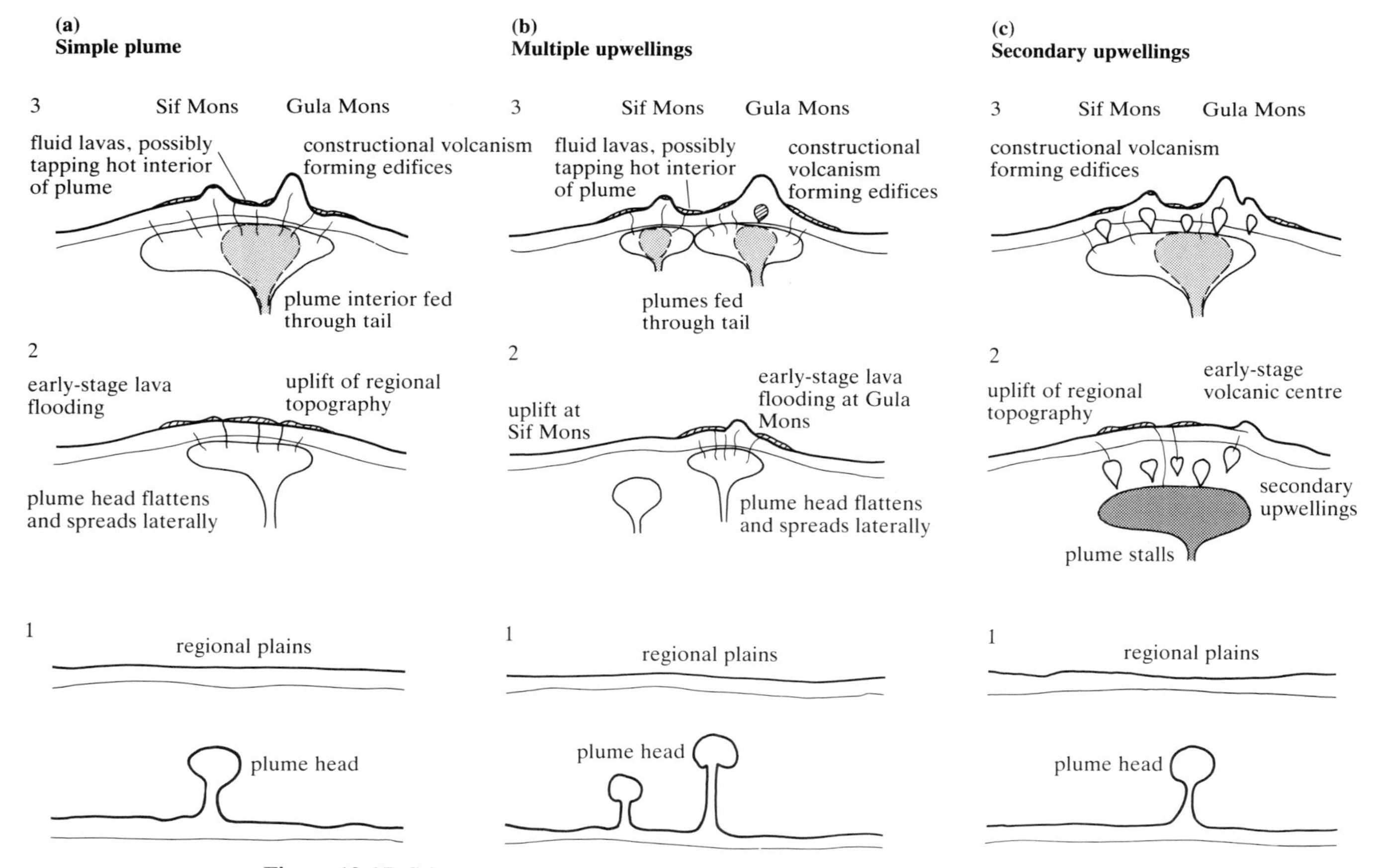

Figure 10.15 Schematic interpretation of the generation of Western Eistla Regio

Three models are shown: (left) simple plume model; (centre) multiple plume model; (right) single plume with secondary upwellings. After Senske et al. (1992).

10.3 The volcanic rises of Bell Regio

Bell Regio is a broad crustal dome 1500 km across situated to the north of Eastern Eistla Regio and, in many respects, is somewhat similar to Western/Central Eistla. It consists of two rises, the northern of which contains the corona Nefertiti and numerous volcanic flows; the southern rise contains the volcano, Tepev Mons, and several smaller volcanic centres. The two foci are joined by a family of north–south trending fractures which were interpreted as a rift valley from Venera 15/16 imagery (Jannle et al., 1987). Magellan data reveal Tepev Mons to rise 5 km above its surroundings and to consist of two foci elongated in a west–east direction. To the north of Tepev and located about 100 km away from the summit, is a family of narrow concentric graben. The eastern component of Tepev is a volcanic peak with a radar-dark signature; western Tepev has a shallow 40 km diameter summit caldera with a radar-dark floor. Numerous relatively radar-bright volcanic flows radiate out from both foci. Another smaller depression with a pancake dome lies to the southeast of the summit, together with prominent NW–SE fractures.

On the southeast flank of southern Bell Regio, roughly due east of Tepev Mons, is another gentle rise, which is a volcanic focus and with which are associated numerous radar-dark flows. Pit chains and other collapse features are to be observed here too. A small family of concentric graben is located farther south; this seems to be associated with the same eruptive focus. Northern Bell Regio is lower than its more southerly counterpart, and is the site of less intensive volcanism. The elongated 500 × 225 km corona, Nefertiti, has numerous radiating flows as well as being defined by concentric faults. South of this, a family of rather ill defined north–south trending fractures appear to delineate a minor rift zone (Jannle et al. 1988).

To the southwest of southern Bell Regio is a series of tessera blocks; further small blocks outcrop to the south. Within them are two main sets of lineations: one NW/SE, the other NE/SW; sometimes there is a third, west–east set. In all cases the NE/SW set is the youngest and, in places, the areas between the faults are inundated by volcanic flows associated with the volcanic rise. As elsewhere, this implies that CRT is the oldest terrain in the region, pre-dating volcanism, and possibly being the initial stage of deformation associated with the evolving mantle plume. Smrekar & Phillips (1991), from Pioneer gravity data, derive an apparent depth of compensation beneath Bell Regio of between 100 and 200 km, i.e. very similar to that beneath western Eistla Regio.

10.4 The volcanic rise of Beta Regio

Beta Regio is a broad rise located 6500 km west of western Eistla Regio. It is a region of extensive volcanism and rifting that Pre-Magellan work had already identified as being located at the intersection of several major tectonic lines. In this respect, Beta bears similarities with terrestrial rift-related volcanism that is plume-related (Dewey & Burke 1974), as a consequence of which, not surprisingly, it had been compared with the East African Rift Valley (McGill et al. 1981, Schaber 1982, Campbell et al. 1984, Stofan et al. 1989). While early work suggested that the region was predominantly a rifted volcanic rise, Arecibo high-resolution images revealed that extensive units of tessera outcropped in the neighbourhood of Rhea Mons and also on the eastern flank of the Beta rise. This prompted Senske et al. (1991) to suggest that Beta Regio, rather than being a simple volcanic rise, actually is a zone of mantle upwelling that has severely disrupted an older tessera block.

10.4.1 Devana Chasma

One of the more prominent characteristics of Beta Regio is that it is cut by a major rift, Devana Chasma, which strikes north–south (Fig. 10.16). The volcano Theia Mons is superimposed on this rift, and itself is transected by younger faulting. In the central region of the rift valley, particularly, faulting occurs over a broad region. Towards the northern end, the character of the surrounding terrain differs from that to the south. Thus, near the northern end of the rift, the surrounding region is composed of CRT, while in the central regions and to the south, the rift is surrounded by fractured plains with volcanic flows. The topography of the rift also shows a great deal of variation along its length: thus, while being a narrow 80 km wide, 2 km deep trough near Rhea Mons, it becomes a broader 130 km wide trough that has upstanding blocks of older crust in its centre. Furthermore, south of Theia Mons, in the belt between Beta and Phoebe Regiones, Devana Chasma is characterized by considerably greater relief (as much as 6 km) and more distinct flanking highs than are seen within Beta itself.

10.4.2 Tessera blocks

The tessera units in northern Beta are composed of NE-trending ridges and grooves spaced 1–2 km apart; to the east of Devana Chasma, in some parts of the rift zone the more prominent trend is nearly north–south. This dominant tectonic fabric is cut by grooves and lineations, which strike on a variety of directions. Normal faults and graben extending out of Devana Chasma transect the tessera fabric, and blocks of CRT can be identified on the rift floor. High-resolution images reveal that, in many places, the tessera terrain is embayed by smooth radar-dark plains, probably volcanic in origin.

Faulting within the rift valley varies in type from place to place. In some sections, such as the western rift wall, > 100 km long, relatively straight, faults with extremely steep walls are typical; elsewhere, for instance in the rift's central regions, more sinuous faults occur;

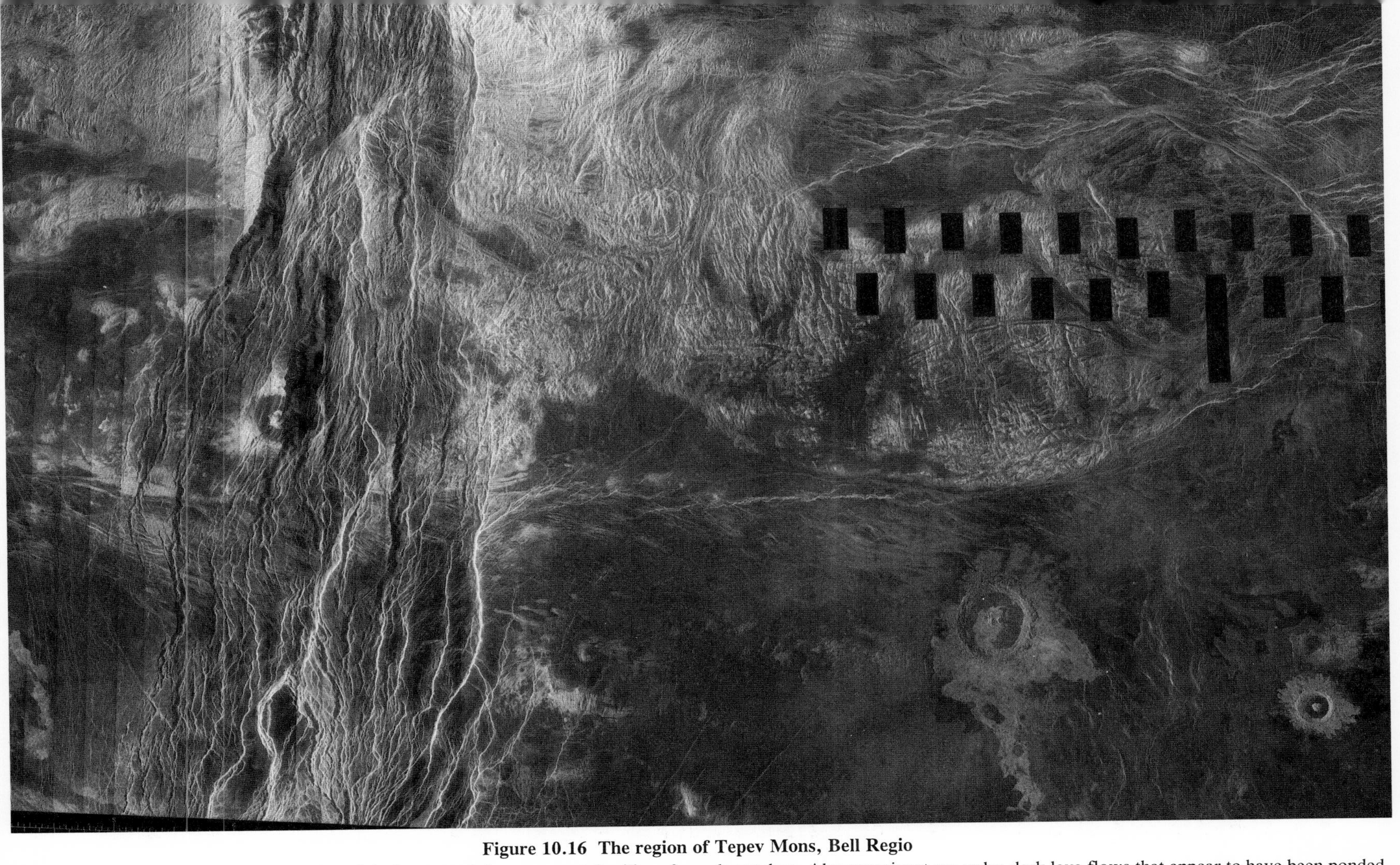

Figure 10.16 The region of Tepev Mons, Bell Regio

The volcano, Tepev Mons, is at the centre of the image and shows several families of annular graben. Also prominent are radar-dark lava flows that appear to have been ponded inside a flexural moat. The image is 450 km wide. Magellan image C2-MIDRP 30N026;1.

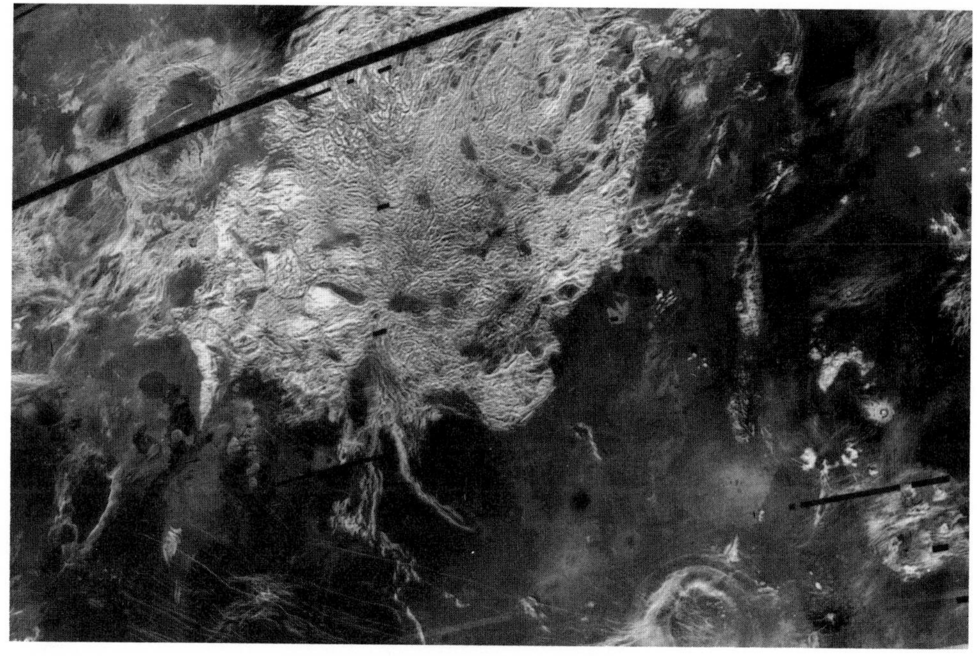

Figure 10.17 The region of Alpha Regio

The complex nature of this tessera plateau is clearly seen. It is surrounded by radar-dark plains, while the prominent corona Fatua lies to the northeast. This has a double ring structure. Magellan image C2-MIDRP 30S026;1.

elsewhere, again, are found mainly smaller graben. A useful indication of the amount of extensional strain experienced over this region comes from the 37 km diameter impact crater Somerville, which has been split apart by the rifting (Fig. 10.17). This, originally circular, crater has been extended in width by about 11 km and allows estimates of the average accumulated extensional strain to be made: between 20% and 30% extension is required to enable the crater to reach its present configuration.

10.4.3 Rhea Mons

Rhea Mons, which lies west of Devana Chasma, has had to be reinterpreted now that Magellan imagery has been returned. These images of Rhea Mons indicate that the topographic high, rather than being a caldera-topped volcanic construct with radiating lobate flows, is actually a region of disrupted tesserae. Although there are outcrops of smoother units around the summit region, which may be of volcanic origin, those lobate radar-bright flank features previously interpreted as flows (Stofan et al. 1989) are found to be lobate regions of CRT. Consequently, this mountain massif cannot now be considered primarily as a volcanic structure.

10.4.4 Theia Mons

There is no doubt that Theia Mons definitely is a major volcano. Lying at the intersection of converging rifts, it is surrounded on all sides by volcanic flows and plains. These extend for up to 1000 km from the summit. Beyond the latter, mainly to the east, west and north, isolated blocks of tessera occur, and were identified first on Arecibo imagery; Magellan images indicate that CRT also exists on the western flank of the highland massif. This latter tessera region is embayed by plains with shield fields and is transected by fractures related to adjacent coronas. The transection/superposition relationships in the region clearly show that the tesserae are the oldest crustal units exposed.

On the northern flank of Theia, where Devana Chasma intersects the edifice, the rift narrows from 200 km to a width of 50 km. To the east, radar-bright and mottled lava flows partially infill the rift floor and evidently are younger than the faulting. At the summit of the volcano is a high plateau feature, to the south of which is a radar-dark 75 km × 50 km caldera whose floor lies below the base of the mountain. As is the case with large Martian shields and paterae, the upper flanks of Theia are covered in relatively short, lobate, radar-bright flows that overlie longer flows that extend towards more distal regions.

10.4.5 Geological history

Prior to Magellan, several workers had used the apparent rifting of Rhea as evidence that volcanism had preceded rifting in Beta Regio, and that what had happened was that the thermal anomaly beneath this region had moved southwards with time. It now appears that rifting in Beta Regio occurred at the same time as the main phase of volcanism centred upon Theia Mons. On the basis of the Magellan imagery now available, it seems that Beta Regio may represent a rifted tessera block, disrupted by a mantle plume. Identification of CRT fragments west of Beta and the detailed characteristics of the tessera terrain in northern and eastern Beta Regio appear to support this hypothesis. No CRT has yet been identified in southern Beta Regio, but they may have been covered by younger volcanic units. Theia Mons sits at the junction of several tectonic trends, and in this respect is similar to major volcanic centres at Eistla and Atla Regiones. This geometry is also similar to terrestrial rift-related volcanism, where faults tend to propagate and link up between regions of hotspot-related volcanism. To the north of Beta Regio, a zone of faults and graben that connect a chain of coronas cuts faults extending from Devana Chasma and may represent a late stage of tectonic activity in this region.

The combined data from Pioneer-Venus, Arecibo and Venera 15/16 missions indicate that there is a large positive free-air gravity anomaly over Beta, with an apparent depth of compensation of at least 300 km. This is entirely consistent with the interpretation of this highland rise as lying above a region of major mantle upwelling.

10.5 Alpha Regio

Alpha Regio was one of the first Venusian uplands to be imaged from Earth, standing out prominently as an approximately circular radar-bright region. Centred at 25°S – midway between Phoebe Regio and the western end of Aphrodite – it is located at the intersection of the X-shaped lowland region that is such a distinctive feature of the planet's physiography. Immediately to its southwest lies the prominent corona, Eve. Alpha measures 1300 km × 1500 km, and attains heights of around 2 km above MPR. However, the core of the massif is lower than the periphery, lying at approximately MPR; it has a rugged character (Plate 10). The exterior plains are some 500 m lower. The highest terrain sits along the perimeter, rising as a series of peaks and elevated plateaux. Available gravity data show that a modest high (2–4 mgal) is located near to Alpha's eastern border, implying a relatively shallow depth of compensation, although the Pioneer-Venus data for this region were of rather low resolution (Bindschadler et al. 1992a).

Figure 10.18 Structural sketch map of Alpha Regio

This map, compiled by Bindschadler et al. (1992a) shows all structures ≥ 5 km in width with Alpha Regio.

10.5.1 Detailed structure

The interior of the massif comprises broad linear to arcuate ridges 10–20 km in width, finer-scale < 3 km wide ridges, graben, scattered radar-dark units and several prominent linear zones along which disruption has occurred. The latter appear to be indicative of shearing movements. The cumulative effect of this array of landforms is to give rise to a very complex fabric clearly discernible in Figure 10.17.

Bindschadler et al. (1992a) have conducted a detailed survey of the structures developed at Alpha and compiled a map of the orientations of those whose width exceeds 5 km (Fig. 10.18). Particularly widespread are fine-scale ridges and troughs, commonly spaced 3 km or less apart. These may be subparallel, sinuous or arcuate, and frequently adjacent ridges meet. By and large the ridges appear to be symmetrical in form but a proportion have asymmetric profiles, and may represent fault scarps. These fine-scale features generally share the same trend as broader ridges and troughs; indeed, they are superposed (Fig. 10.19). Most have a

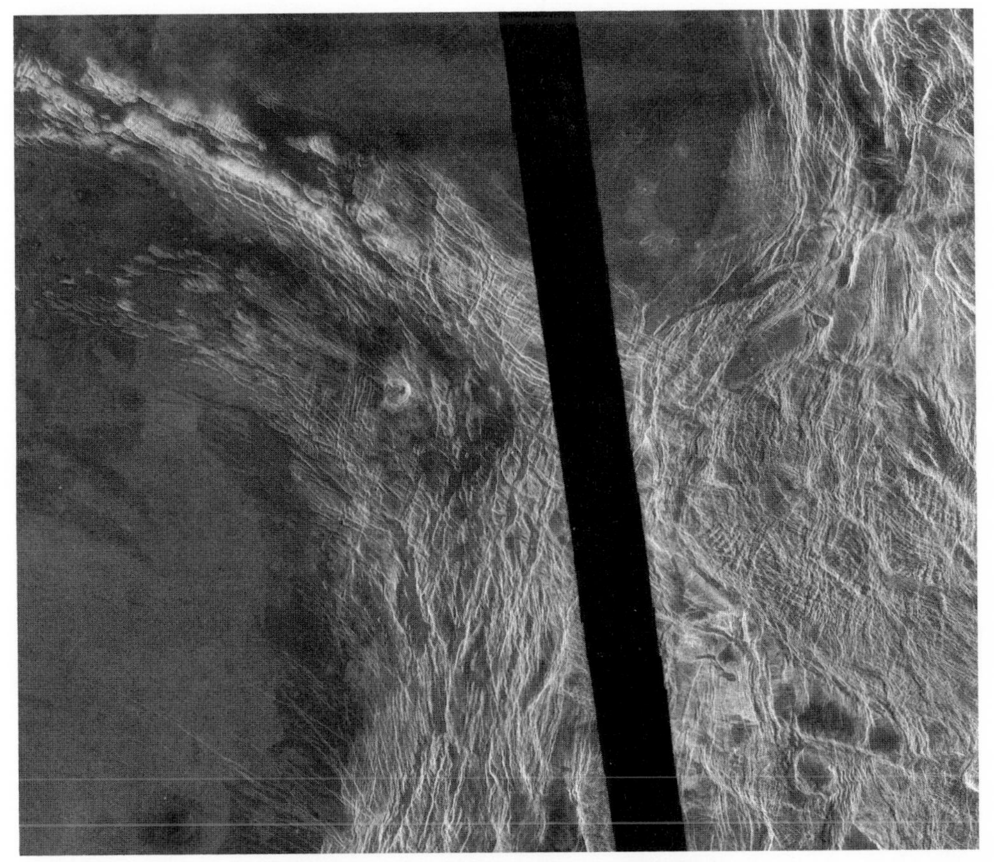

Figure 10.19 A part of southeast Alpha Regio
Fine scale ridges and troughs in a 250 km wide region of the plateau. Magellan F-MIDRP 20S357;1.

north–south strike, but orientations locally swing between NNW/SSE and NNE/SSW. In places, particularly where the structures have an arcuate form, a secondary family of fine-scale features may develop; again, these generally lie parallel to larger-scale structures. This close association implies a genetic relationship between the two scales of landform. The ridges, because of their fold-like character, are interpreted as compressional in origin. As Bindschadler and his colleagues note, if such is the case, their narrow spacing and widths suggest the possibility of rheological lamination on a scale of tens to hundreds of metres within the Venusian crust.

The broader-scale ridges and troughs range from linear to arcuate, with individual ridges varying between 20 and 150 km in length, and spaced 20–30 km apart. Those with a linear character tend to be rather rounded, suggestive of low slopes, although some are steeper-sided and asymmetric, particularly those developed along the western boundary of Alpha Regio, where flanking slopes of 15° to 20° are implied. Most strike north–south, but locally the strike swings to mimic the boundaries of Alpha. Where two sets of ridges occur, the points of intersection are represented either by dome-shaped hills or as kink zones. The overall morphology

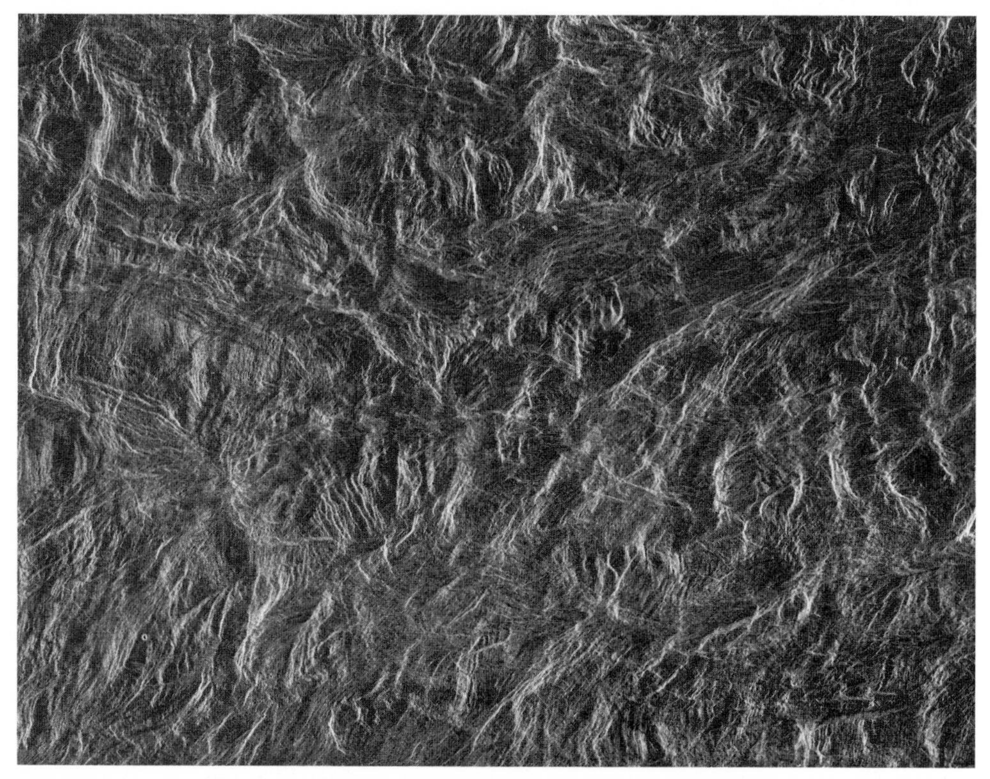

Figure 10.20 Linear disruption zones within Alpha Regio
This region is located just southeast of the centre of the plateau. Magellan F-MIDRP (number to be supplied).

and distribution of such features suggest that they are compressional in origin, the most likely interpretation being that they are folds.

Arcuate ridges are mainly confined to a small area in west central Alpha, where 30–40 km long ridge segments exhibit orientations that vary widely over quite small areas. Also observed in such areas are broad graben, scarps and other lineations with a dominant NW/SE trend. NE/SW-trending fine-scale ridges and scarps intersect these at right angles. In most cases, the NW/SE-trending lineations transect features with other orientations. The broad arcuate ridges may represent interference folding (Bindschadler et al. 1992a).

Arecibo images of Alpha Regio revealed broad belts of linear disruption that were interpreted by some workers as strike-slip offsets (Senske et al. 1991). Over large areas these disruption zones trend WNW/ESE or ENE/WSW. On Magellan imagery such features exhibit a variety of radar signatures, ranging from narrow troughs to wider zones with radar-bright signatures. Where other structures approach these zones, they are frequently deflected so that individual trends merge (Fig. 10.20). These characteristics suggest that, rather than being discrete shear zones, they represent broad zones of tectonic accommodation that absorbed strain during compressional movements.

Of all the features observed within the boundaries of Alpha Regio, steep-sided, flat-floored graben are the youngest. Approximately 5 km wide, they are spaced 1–20 km apart and range in length from tens to hundreds of kilometres. Their orientation is highly variable. On the northern border of Alpha, graben cross onto the adjacent plains, indicating them to be post-plains in formation (Fig. 10.21). They are clearly extensional in origin, and measurements of their width and downthrow imply a regional strain of between 3 and 11%. The spacing of the fault sets and the relatively modest heights of the fault scarp walls (a few hundred metres) imply that a relatively weak layer exists at depths beneath the surface perhaps as shallow as 1–5 km.

10.5.2 Formation of Alpha Regio

What can be deduced about the origins of Alpha from a synthesis of these structural data? It is an irregular plateau with an elevated rim that corresponds to regions of high radar backscatter. These radar-bright areas are characterized by broad ridges with an arcuate or dome-like form. Discontinuous 100 km wide troughs transect the higher ground, and it is within these low-lying zones that plains units are located. The troughs have a broadly radial arrangement about the core of Alpha Regio, an arrangement they share with many of the smaller graben faults that run through the tessera units. Bindschadler et al. (1992a) speculate that such a development is a manifestation of gravity-driven extensional deformation. They also note that such a strain regime can develop in response to either mantle downwelling (Bindschadler et al. 1992b) or relaxation of a plateau compensated principally by variations in crustal thickness (Bindschadler & Parmentier 1990, Stofan et al. 1991).

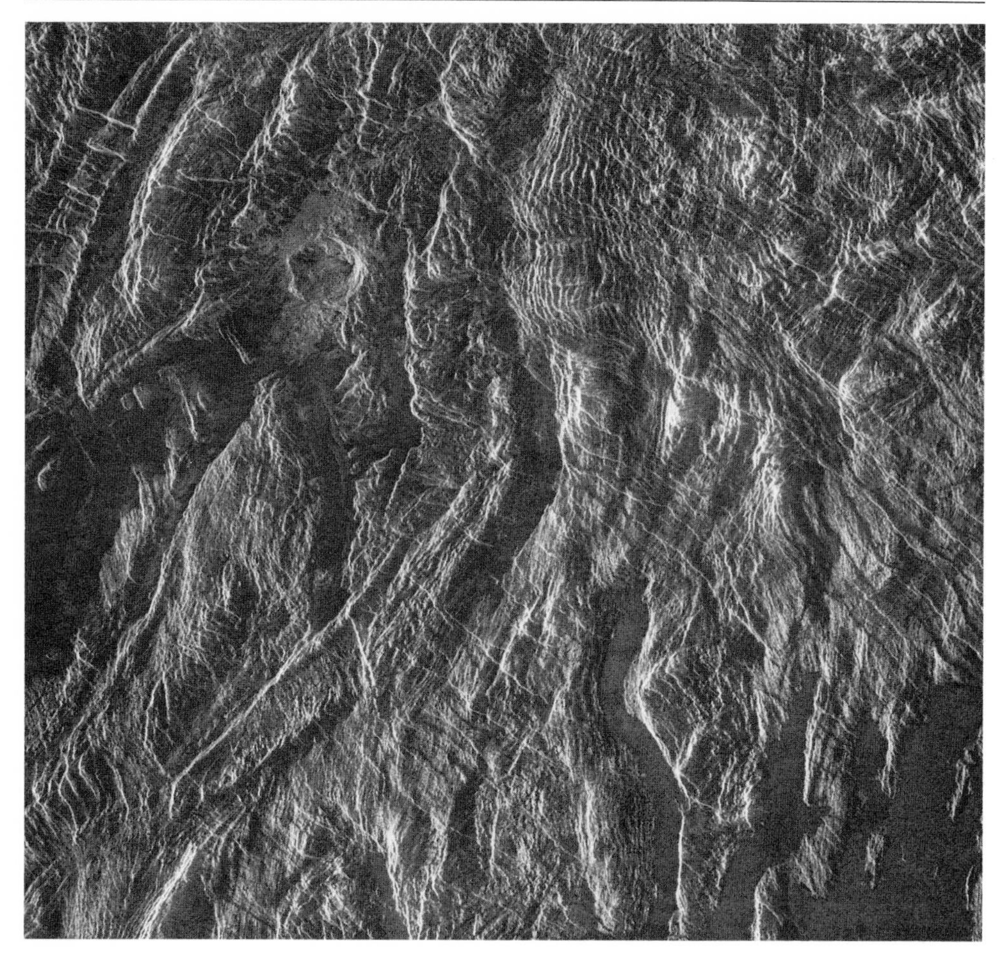

Figure 10.21 Alpha Regio: northern boundary
Graben cross from the tessera massif onto adjacent plains.

As the observed structure is one of the most closely studied areas of CRT, it is important to see what interpretation can be put on it in the light of current models for tessera formation. Two principal models are current: *hotspot* or *mantle upwelling* (plume) activity, and *coldspot* or *mantle downwelling* activity. There is also a third model, which is still largely unexplored, that considers that all crustal plateaux on Venus will suffer major deformation owing to regional stress, simply because they are zones of weak lithosphere (Solomon et al. 1991).

Hotspot (mantle upwelling model)

The hotspot model predicts that deformation and volcanism will affect the lithosphere when a major (1000 km scale) plume approaches the thermal boundary layer of the upper mantle (Herrick & Phillips 1990, Phillips et al. 1991). In the early stages of the sequence of events, when the plume is still relatively deep, say 100 km or more, the crust will be uplifted, stretched to form rifts, and be the site of shield volcanism. Regions of Venus believed currently to be at this stage are Beta and Atla Regiones. The continuation of extrusive and intrusive basic volcanism will form a region of significantly thickened crust. As a result of gravity spreading along the perimeter of the newly formed upland, a girdle of thickened crust will develop at the edge of the uplifted region. At the same time, there would be extension within the interior of the plateau, but shortening towards the exterior. Subsequently, as dynamic support for the plume wanes, the plateau would experience gravitational relaxation (a feature it shares with the coldspot model, by the way). Therefore, the complete sequence involves,

1. gravity spreading
2. membrane stresses
3. thermoelastic stresses
4. gravitational relaxation.

Coldspot (mantle downwelling) model

In the coldspot scenario, a CRT plateau begins to rise as the crust is flexed over a downwelling in the underlying mantle. At this stage, strains are anticipated to be small and to be dominated by brittle and/or flexural response of the lithosphere; this results in a largely radial pattern of compressional structures. Some workers have suggested that regions such as Lavinia and Atalanta Planitiae may be at this early stage (Bindschadler et al. 1992b, Squyres et al. 1992). Because of horizontal shear stresses and pressure gradients, the lower crust, being relatively ductile, is pulled in towards the focus of downwelling, initiating a region of crustal thickening. As this thickening process continues, compressional structures begin to be generated and have an azimuthal orientation; this happens because, at this stage, radial compression dominates over azimuthal compression. The growth of thickened crust over the downwelling mantle eventually presents a barrier to convergent, horizontal flow within the mantle, and a relatively weak zone within the overlying lithosphere. At this stage it is predicted that compressional deformation becomes concentrated at the perimeter of the plateau (Lenardic et al. 1992). The result is that a rim of relatively elevated topography develops, accompanied by annular compressional structures. However, within the interior, since the wedge of thickened crust sags under its own weight, the earlier compressional regime gives way to extension; thus, radial graben develop.

Which scenario?

Whether Alpha Regio is considered to have a hot- or coldspot origin, either way its structure is consistent with its being at a mature stage in its evolution. The overall topography – plateau with peripheral highs – appears to be more in line with the coldspot model. A suggestion made by Phillips et al. (1991) – that a ring of thickened crust may develop owing to gravity spreading of a plateau formed by hotspot activity – is ruled out by Bindschadler et al. (1991)

on the grounds that gravity spreading would be expected to decrease topographic slopes, thereby implying that original gradients were much steeper than those now observed; and the current 1 km of relative relief would necessitate at least 10 km of crustal thickening. However, they do note that quantitative data concerning gravitational relaxation are not available, and that the idea therefore cannot be disproved.

The structural features are perhaps more diagnostic. The upwelling model predicts a distinctive pattern of extension on regional topographic highs and compression in lows; in fact, compressional structures are found at all elevations, while extensional landforms mainly post-date compressional ones. Bindschadler and his colleagues doubt that membrane stresses could

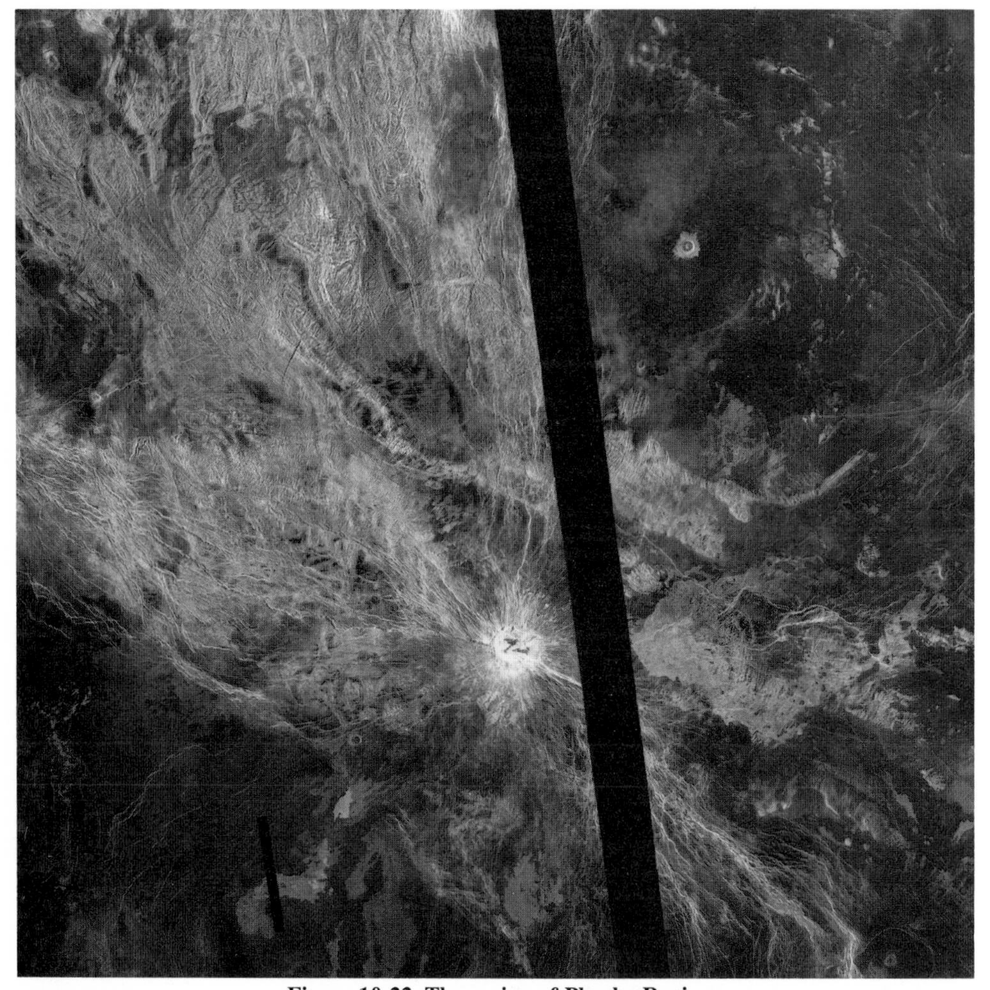

Figure 10.22 The region of Phoebe Regio

The bright signature of this region is mainly due to metre- to submetre-scale surface roughness. The brightest area is where a volcano has erupted onto the CRT. Image width 3700 km. Magellan C1-MIDRP 15s283;201.

account for the fine-scale ridges or the observed pattern of radial graben, and place their bet firmly with the mantle downwelling scenario.

Another aspect of the issue concerns the development of volcanism during plateau formation. At Alpha Regio volcanism appears to have been confined to the later stages of plateau formation, whereas the hotspot model predicts that early and extensive shield volcanism would develop. It is possible, of course, that subsequent tectonism has rendered earlier shield-style volcanic structures (which would not necessarily have had significant relief) unrecognizable; therefore closer analysis of stereo imagery may be necessary to settle this issue. At the present time, it would appear that the downwelling model may offer a more satisfactory solution to the geological problems posed by the intricate relationships observed at Alpha and other CRT uplands.

10.6 Phoebe and Tellus Regiones

Phoebe Regio is an elongate plateau and is the least distinct of the plateau-shaped highland massifs. It lies SSW of the southern end of Devana Chasma and is a region of generally radar-bright signatures that represent a massif of CRT upon which is superimposed a volcanic edifice (Fig.10.22). A modest gravity high is centred over this volcanic focus. A narrow, southeast-trending fracture system is the locus of at least one major volcanic edifice (Figure 10.23).

Tellus Regio is very similar to both Ovda and Thetis Regiones. Located about 3000 km to the north of Aphrodite, it has steep margins that border an upland of CRT. The tessera blocks are extremely complex, and have been termed *disrupted terrain* (Bindschadler & Head 1991). Near and lying parallel to the margins are compressional ridges; these comprise some of its highest terrain. In contrast to the other CRT massifs described, Tellus has negative gravity anomalies (–30 mgal) and also negative geoid (–30 m). It is predicted that apparent depths of compensation beneath this region may be as shallow as 25–50 km (Smrekar & Phillips 1991) and the crustal thickness a mere 20 km. Both plateaux show features that have been discussed in connection with Alpha Regio. The arguments concerning their formation apply here too.

10.7 Western Ishtar Terra

Ishtar Terra is unique among the elevated regions of Venus, for, although it has some of the characteristics of other plateau-shaped highlands, its perimeter stands several kilometres above the interior; this does not happen elsewhere. Furthermore, it exhibits far fewer extensional landforms than, say, Alpha, Ovda or Tellus Regiones, while the presence on its interior of two

Figure 10.23 Phoebe Regio summit region
Lobate lava flows with bright and intermediate radar signatures originate in a volcanic depression developed along a northeast–southwest fracture line. Magellan P-40844.

very large calderas is a further unusual feature. Ishtar, on average, stands at least 2 km above MPR and measures 3700 km from east to west and 1500 km from north to south. Western Ishtar comprises the huge plateau of Lakshmi Planum, tilted slightly towards the south, which is wrapped around by mountain belts and bounded by steep scarps; east of it rises the mountain massif of Maxwell Montes (Fig. 10.24). Plateau-like regions of tessera – complex ridged terrain – lie exterior to the bounding mountain belts of Lakshmi Planum, and lie at a lower elevation. East of 30°E lies eastern Ishtar, composed largely of the complex block of Fortuna Tessera across which there is a gradual fall in elevation towards the adjacent plains, over a distance of around 2000 km. This is one of the most extensive areas of complex ridged terrain on the planet.

Unfortunately, when Pioneer was conducting its gravity survey, the orbital height was over

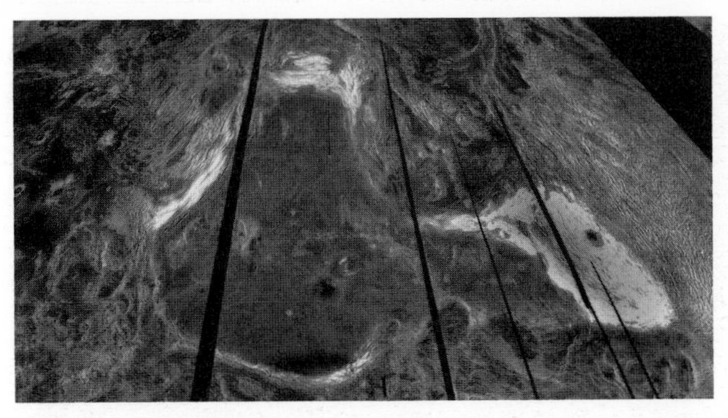

Figure 10.24 Western Ishtar Terra
A Magellan image showing the radar-dark plateau of Lakshmi Planum and surrounding radar-bright mountain belts. Maxwell Montes is very bright and lies towards the right of the image. Beyond Maxwell lies the complex CRT region of Fortuna Tessera. Magellan C2-MIDRP 60N333;2.

1000 km, with the result that available gravity data are of low resolution. Large positive anomalies in gravity (55 mgal) and geoid (60 m) are located to the south of Maxwell Montes, well away from the centre of Lakshmi Planum. It is difficult to assess how significant this offset may be.

10.7.1 Lakshmi Planum

Lakshmi Planum has been studied in considerable detail. Kaula et al. (1992) define several geological units covering the plateau surface: dominating the central regions are the two volcanic structures, Sacajawea and Colette, with their associated annular fractures; smaller calderae, such as Sissons, also occur. East of the prominent volcanic caldera of Sacajawea Patera, ridged terrain has an extremely complex structure, characterized by intersecting troughs and ridges such as those seen in Figure 10.25. Of widespread occurrence, this terrain type is heavily embayed by the volcanic flows that comprise another of the very extensive terrain types: smooth plains. Superposition/transection relationships indicate that the latter are consistently younger than the former. Smooth plains cover large areas, particularly in the lower, southern, sector of Lakshmi, and in places are incised by curvilinear troughs, many of which trend roughly parallel to either the plateau margins or the neighbouring mountain belts. Rangrid Fossae, located west of the western scarp of Maxwell Montes, constitutes one such example, but troughs are located along other sections of the plateau perimeter too.

10.7.2 Mountain belts

Surrounding Lakshmi are series of curvilinear mountain belts: Freyja, Akna, Danu and Maxwell Montes. All are terrains characterized by long ridges with intervening troughs, up to hundreds of kilometres long, that were interpreted to be compressional structures on the basis of pre-Magellan imagery (Campbell et al. 1983, Phillips & Malin 1984, Barsukov et al. 1986, Crumpler et al. 1986). In those parts that are highly deformed, the ridge spacing is

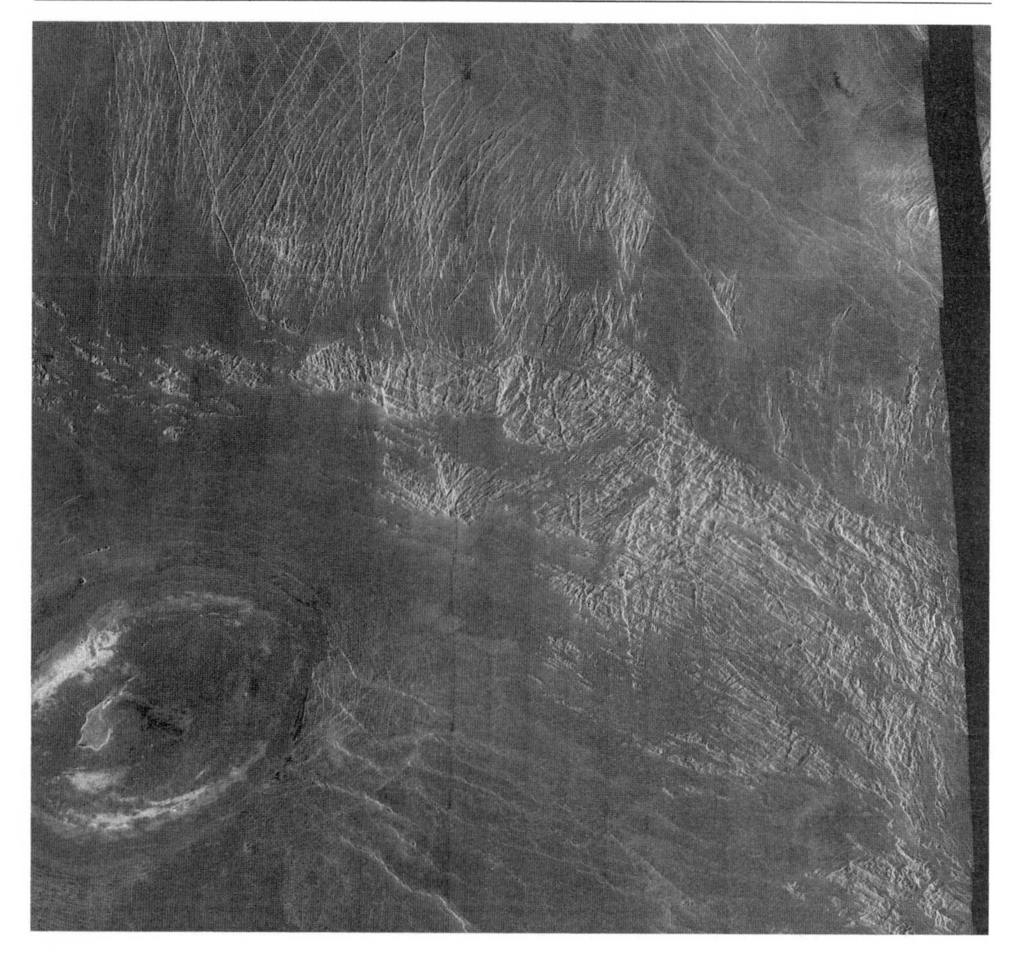

Figure 10.25 Central Lakshmi Planum

Ridged terrain is distinguished by extensional fractures which transect smooth, radar-dark plains units. Note also the volcanic caldera, Sacajawea Patera, located at the left of the image. Magellan image F-MIDRP 65N342.

variable, and, although typical spacings are 3–10 km, they range from 1 km to 20 km. Taking this into account and bearing in mind their typical length, Zuber (1987) suggests that their structure is consistent with a weak layer of accommodation just a few kilometres beneath the surface, whose intrinsic strength probably differs little from that of the underlying mantle.

Danu Montes

Danu Montes form the southern and southeastern perimeter of Lakshmi Planum, extending in all for a total of 1200 km. They rise at least 3 km above the adjacent plains. Lying on the south side of the western arm of the mountain belt is the narrow scarp of Vesta Rupes, which Magellan imagery has revealed to have an average slope of at least 20° (Fig. 10.26). The highest and radar-brightest section of the belt is that along the northeastern arm, and it is

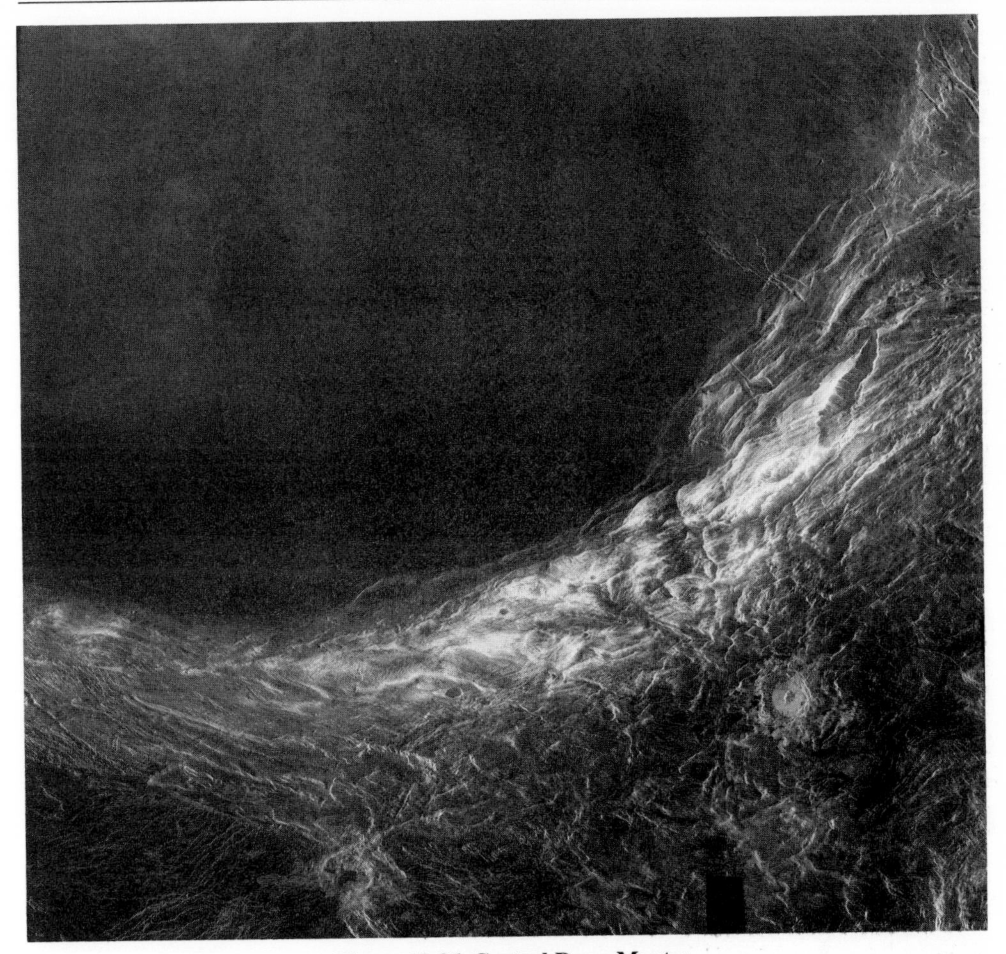

Figure 10.26 Central Danu Montes

The complex structure of Danu Montes is seen in this Magellan image. Vesta Rupes lies to the south. Magellan image F-MIDRP 60N334.

assumed that NW–SE directed compressional stresses were responsible for this, the most recent stage of uplift in this zone. Thrust faults and wrinkle ridges parallel the trend of the northeastern mountain arm on the plateau of Lakshmi, where its surface is tilted upwards towards the mountain barrier. These relationships imply that Danu Montes are younger than the Lakshmi plains units and that the deformation that generated the mountains involved also the Lakshmi volcanic plains, at least at their perimeter. This implies that crustal shortening processes must have continued into times more recent than those during which the youngest lava flows were extruded.

Freyja Montes

Freyja Montes, which run along the northwestern perimeter of Lakshmi Planum, have a general structure similar to Danu Montes, with many graben orientated normal to the regional slopes.

On the interior plains adjacent to them are several sinuous low scarps, which suggest low-angle thrusting at the edge of Lakshmi Planum. The smooth plains themselves play host to broad arches, sinuous ridges and troughs, all of which strike parallel to the mountain axis. The ridges, which may be up to 140 km in length, are spaced between 1 and 4 km apart, the spacing diminishing towards the mountains. The troughs, which appear to be graben, are most abundant in eastern Freyja, and have a somewhat wider spacing (3–10 km); they are a continuation of those which are observed on the slopes above, and extend over Lakshmi Planum for a distance of around 80 km. The generation of the ridges presumably accompanied the compressional stresses that were responsible for the uprise of the mountain belt. The fault troughs evidently had a rather different origin and may have opened as the unsupported plateau edge relaxed in the vicinity of what Kaula et al. (1992) call the "North Basin", an amphitheatre-shaped embayment in the plateau that lies due east of the eastern arm of Freyja Montes. Indeed, the rather different aspect of Freyja Montes in this region, where extensional structures dominate, suggests that, rather than active crustal shortening occurring hereabouts, the crust has been subject to gravitational collapse.

Of the various mountain belts, only Maxwell and Akna Montes have associated trenches, and these are significantly shallower than typical terrestrial trenches, being but 1 km deep. They are evidently not analogous with Earth-type oceanic trenches of the type associated with active island arc systems. Nevertheless, gradients on the front slopes of the belts are very steep, western Maxwell Montes and eastern Akna Montes supporting slopes that occasionally exceed 30°.

10.7.3 Volcanic features

The evidence for volcanism within the mountain belts of Ishtar is variable. Lava channels can be seen to have flowed down onto the interior plateau surface from Danu Montes, but not elsewhere. Volcanic landforms found near the northeast arm of Danu Montes include partly coalescent collapse pits and graben that trend either NW–SE or east–west, and transect those thrust faults that are present. This implies that volcanism post-dated much of the orogenesis. Volcanism is, however, much more widespread on the exterior slopes, where all of the mountain belts display a plethora of lava flows, sometimes with collapse depressions, spreading down their flanks. Interestingly, the only flows associated with Maxwell Montes appear to be those generated by impact melting during the formation of Cleopatra. This large impact crater, which lies east of the summit, has a marked effect upon the Maxwell uplands, the surface taking on a smoother appearance over a distance that varies between 130 and 210 km from its centre. The material is radar-bright, and may be ejecta. Cleopatra is the sole impact structure on the surface of Maxwell Montes.

In the case of both Danu and Freyja Montes, and Vesta Rupes, the presence of aligned collapse depressions and pits indicates that the rise of magma was facilitated by crustal extension. Indeed, extensional stresses clearly affected all of the belts, but the highest incidence of extensional features is to be found along Danu Montes and Vesta Rupes, while it is also evident that they affected the plateau margins. Smrekar & Solomon (1992) interpret the sets of narrow, closely spaced features that characterize these regions of high gradients, as graben and normal faults. Since many of the fault sets strike perpendicular to the downslope direction, the most likely explanation for their existence is gravitational spreading. The fact that, in several of the belts, extensional structures lie parallel to what appears to be the direction of shortening, implies also that this process was operating while mountain building was in progress. This kind of relationship can be observed in the Earth's Himalayan chain, and is a natural consequence of the tendency for thickened, elevated crust to spread under the ubiquitous influence of gravity.

10.7.4 Bounding scarps

Vesta Rupes, the 2–4 km high scarp that forms the southwestern boundary of the Lakshmi plateau, drops steeply (particularly in the east) southwestwards to the exterior plains, from a rim that borders the high plateau. Along its length, the strike of both graben and collapse features swings, suggesting that the tectonic regime evolved with time. Several normal faults run parallel to the trend of the scarp, indicating modest extensional tectonics. The relationships between the scarp and extensional features indicate that gravitational sliding may have occurred here also, and this is borne out by the development of compressional ridges on the plains some 200–600 km southwest of the Vesta Rupes escarpment (Bindschadler & Head 1991).

10.7.5 Maxwell Montes

The impressive mountain massif of Maxwell Montes, which lies at least 5 km above MPR, has a summit region over 11 km high. As we have noted, it borders Lakshmi Planum on its east and defines the eastern edge of western Ishtar. Most of the western flank of Maxwell is extremely bright on radar images, and is a complex of grooves and ridges (Kaula et al. 1992). Few, if any, graben can be identified here. In the higher regions of its northwest sector, the ridges have a saw-tooth pattern and are spaced 10–20 km apart. The pattern and nature of these features imply that there has been significant gravitational modification. In this respect this part of Maxwell resembles the other belts. The southern slope of Maxwell, on the other hand, although radar-bright, has a different small-scale structure from the ridge-and-groove pattern seen elsewhere (Fig. 10.27). Here, the structures present form both a radial and a concentric pattern, one set lying perpendicular to the maximum slope. The latter probably are a result of gravitational sliding. Sometimes the faults transect northeast-striking ridges;

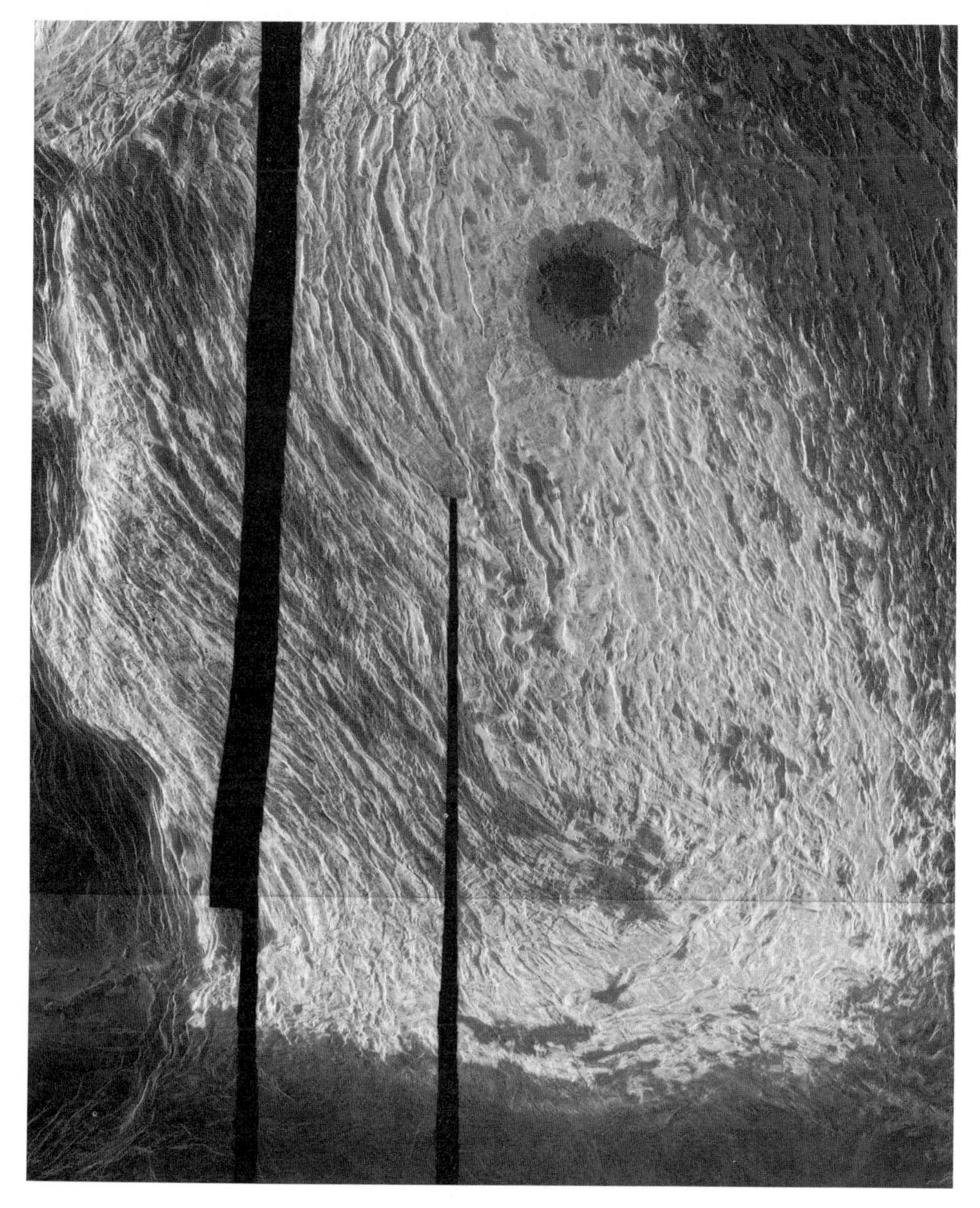

Figure 10.27 Southeastern Maxwell Montes

A set of NW-trending graben points in the downslope direction. Another NE-trending set is roughly perpendicular to the regional slope. Magellan images F-MIDRP 60N005;1 and 65N006;1.

on other occasions there is so much disruption to both that it becomes difficult to decipher exactly what has happened.

One of the problems in explaining Maxwell Montes is its tremendous height; how can such a large massif be compensated solely by a stable crustal root? The answer quite simply is that it cannot. Some kind of dynamic support is required (Vorder Bruegge & Head 1991). Kaula et al. (1992) feel that the obvious interpretation is that Maxwell is currently undergoing severe compression, directed in a WSW-ENE direction. Because of the geoid high known to exist here (although high-resolution gravity data are not currently available) they suggest that this compression, and the downflow associated with it, must penetrate to rather deep levels. Such mantle downflow may find an expression in the depressed zone that they identify to the west of a sinuous line that traverses western Maxwell; that is, this could be a modest trench-type feature. However, although the predicted high viscosity of Venus's upper mantle militates against this (Kaula 1990), finite-element modelling indicates that lithospheric downwelling can in fact occur on a short-term basis (Lenardic et al. 1992).

10.8 Eastern Ishtar Terra: the tesserae

External to all of the mountain belts – which represent some of the most marked crustal shortening to have taken place on Venus – are regions of tesserae that may extend outwards for anything between 100 and 1000 km. These highly deformed zones lie at a lower level than the mountain belts and eventually slope down towards the surrounding plains, or are terminated by steep scarps. Generally, the width of the tessera blocks corresponds to the height of the adjacent mountains.

The greatest extent of tessera lies to the east of Maxwell Montes, over the region of Fortuna Tessera. North of Freyja Montes lies the exterior CRT plateau of Itzpapalotl Tessera, while Clotho Tessera is located south of Lakshmi Planum. These have frequently been described as *outboard plateaux*. At around the 6 km contour, the eastern slope of Maxwell Montes gives way to tessera with an arcuate ridged structure that strikes roughly parallel with the ridges of Maxwell, as if wrapping around the massif (Fig. 10.28). Elsewhere the complex terrain has a more orthogonal pattern, particularly towards the east. However, between these two types the pattern of ridges is more disrupted and defines chevron patterns, while, in the south, a plumose pattern predominates.

The arcuate tessera belt is interpreted as having been generated by the same compressional forces that produced Maxwell; however, the more complex structures found farther east are seen as the result of crustal thickening, perhaps by the stacking of shear slices, perhaps by a more plastic crustal thickening process, accompanied by tectonic overprinting; in other words, some regions of CRT show evidence for much older tectonic activity.

(a) Fortuna Tessera

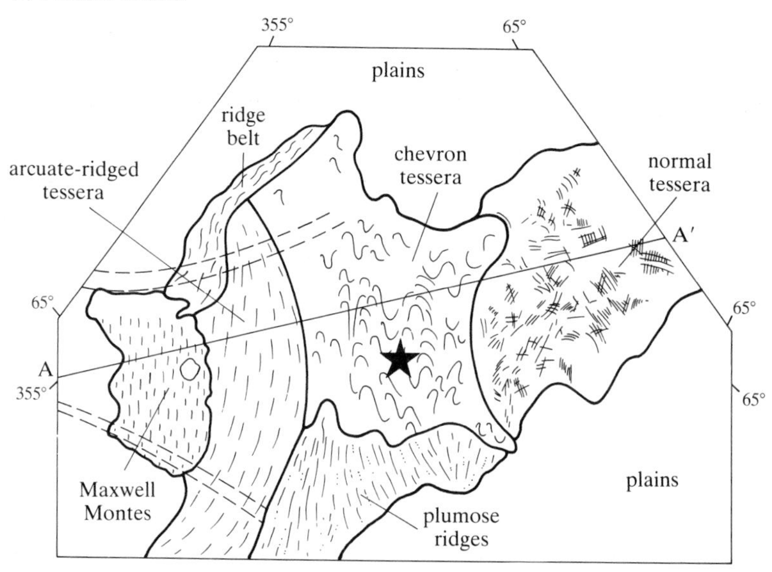

(b) Crustal thickness (Airy model)

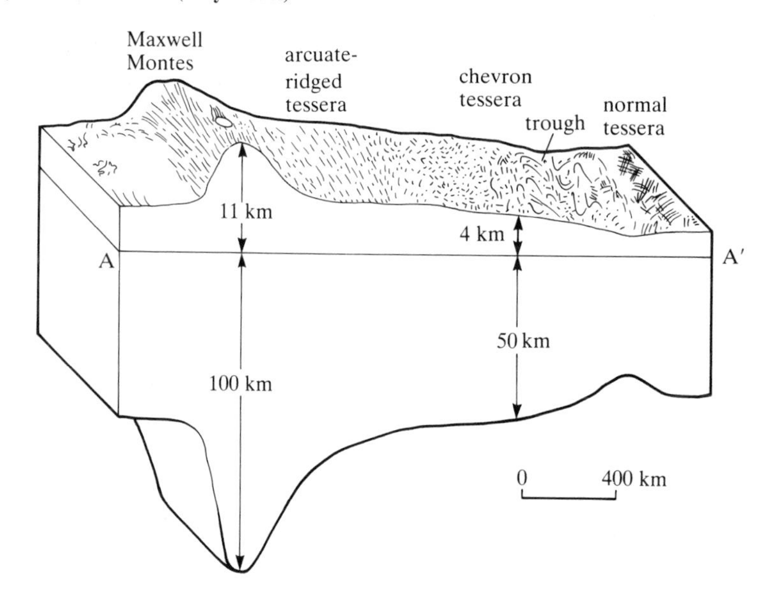

Figure 10.28 General geomorphological map of Fortuna Tessera
The dashed lines represent shear zones situated to the north and south of Maxwell Montes. The star within the area of chevron tessera indicates the supposed location of proto-Maxwell Montes. After Vorder Bruegge & Head (1989).

Viewed as a whole, Magellan images indicate that the majority of major tessera blocks exhibit a younger set of graben that are superimposed on more ancient ridged terrain, defining a clear-cut temporal sequence (Bindschadler & Head 1991). Thus, an earlier compressional phase, which effected variable degrees of crustal shortening, was followed by a more recent phase of lateral extension. As was noted by Solomon et al. (1991), during the earlier part of the Magellan mapping, regions of elevated topography and thickened crust show a tendency to strain in response to regional stress. This being so, tesserae simply may represent long-lived blocks of thickened crust that have been subjected to repeated episodes of deformation. The fact that they have remained elevated has also assured the survival of their deformational history, since volcanism has not overwhelmed them, as it has lower-lying areas. The extension of this logic is that, the older a tessera region, the more complex will be its deformational history.

10.9 Geological evolution of Ishtar Terra

There appears to be a general consensus that the marked geoid high, associated with the central part of Ishtar, is the result of vigorous, and possibly contemporary, convergence (Roberts & Head 1990, Kaula et al. 1992, Bindschadler et al. 1992a, Solomon et al. 1992). This manifests itself very clearly in the relatively simple structure of the western slope of Maxwell Montes. The greater degree of deformation characteristic of other regions within Maxwell, parts of the Fortuna Tessera and the exterior CRT plateaux encircling Lakshmi Planum, where considerable numbers of lineaments consistently cross-cut the dominant Maxwellian trend, implies more complex tectonic history, with different tectonic regimes having operated in the past, perhaps over a period of at least 500 Ma. The changing pattern of deformational structures observed across the region may reflect patterns within individual tessera blocks that have been brought together at different times, as crustal shortening has proceeded (Roberts & Head 1990).

Bindschadler et al. (1992a) assert that mantle downwelling beneath thickened crust can most effectively explain the geological features observed. In this respect they cite:

- the rugged, steep-sided topography that is characteristic;
- the observed relationships between gravity and topography, i.e. low apparent depths of compensation and topography:geoid ratio;
- the widespread compressional structures that are found at high elevations close to the plateau margins; and
- the fact that extensional structures are superimposed on compressional ones, a sequence that is consistent with deformation in response to downwelling mantle material (Bindschadler et al. 1990).

Vorder Bruegge & Head (1989) consider that Maxwell Montes formerly were located farther east than at present and were part of a longer, linear belt resulting from compressional deformation. They can interpret the present Maxwell structure to be the result of westward movement and deformation between two convergent shear zones. Tessera blocks, for instance Fortuna and Laima Tesserae, may have been brought closer together by these lateral movements, along shear faults. They suggest that eastern Ishtar grew by the accretion of tessera blocks driven by crustal convergence, a scenario depicted in Figure 10.29. Using a simple Airy isostatic model, they estimate that the crustal thickness beneath Maxwell is 95 km, but that it thins eastwards, becoming closer to 15 km beneath the adjacent plains.

They envisage the geological history of Lakshmi Planum as having begun with a major

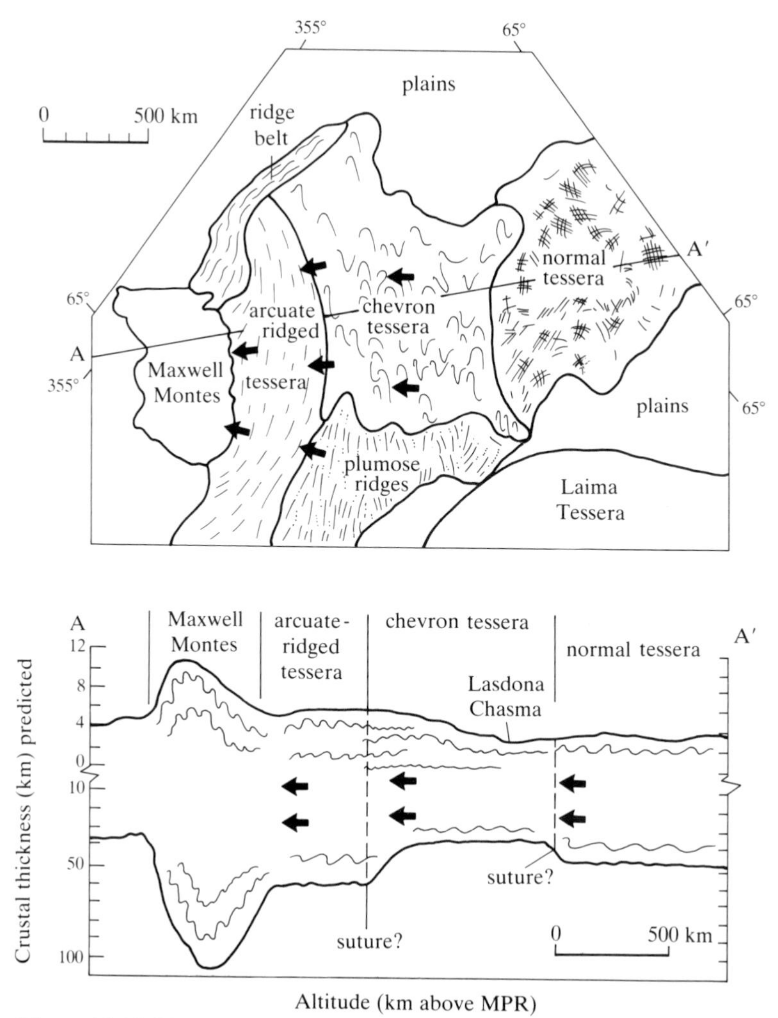

Figure 10.29 Model for accretion of tessera blocks in Eastern Ishtar Terra
Schematic diagram according to Vorder Bruegge & Head (1990).

tessera massif that was surrounded by low-lying plains. Subsequently, convergence and underthrusting of crustal segments generated peripheral mountain belts, with gradual thickening and uplift of the plateau core (Fig. 10.30). This thickened wedge of crust was then melted at its base, surface volcanism being coeval with the continued deformation of the plateau edge.

The arrangement of fracture families normal to the downslope directions within the elevated regions strongly suggests that gravitational spreading has occurred and probably ensued from crustal relaxation. Where graben families strike parallel to the shortening direction, the spreading evidently was coeval with orogeny. This is directly comparable to tectonic regimes known from terrestrial continental regions such as the Tibetan Plateau. Smrekar & Solomon (1992), using finite-element analysis to study viscoelastic relaxation under Venusian conditions, found that a plateau 1 km high with an underlying crust of thickness 20 km, a basal temperature of 770°C and a model width for the plateau–scarp– plains distance of 150 km, would relax by one-quarter over a period of 100 Ma. On this basis, rates of viscous relaxation clearly are substantially higher on Venus than on the Earth. Relaxation of exterior plateaux, where crustal compensation depths are predicted to be less, would be even quicker. Therefore, there is every

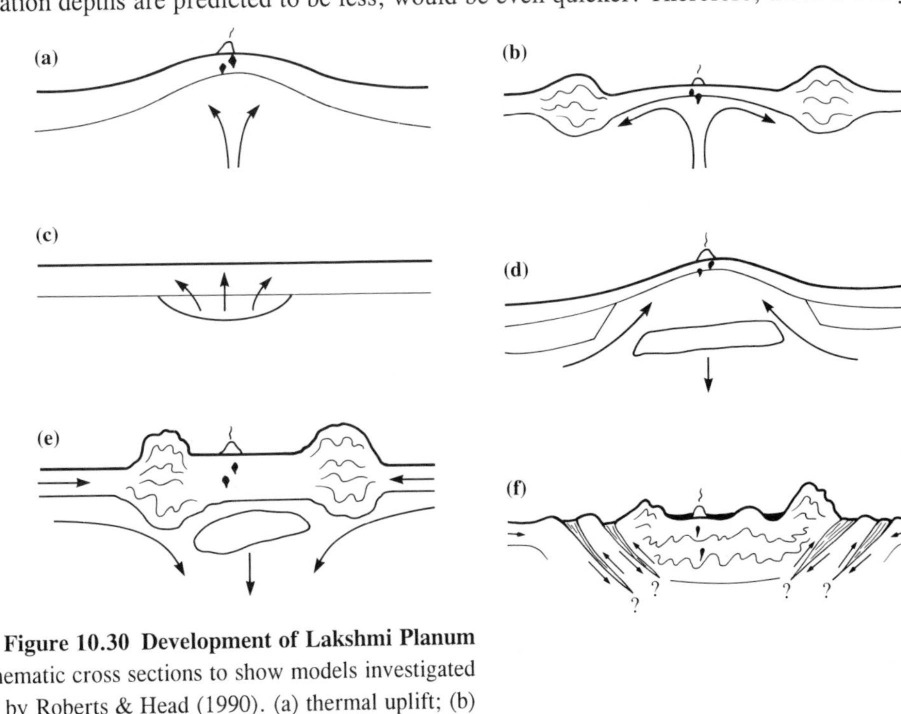

Figure 10.30 Development of Lakshmi Planum Schematic cross sections to show models investigated by Roberts & Head (1990). (a) thermal uplift; (b) thermal uplift with asthenospheric flow coupled to crust; (c) crustal underplating; (d) lithospheric delamination; (e) lithosperic delamination with mantle flow coupled to crust; (f) horizontal convergence and crustal thickening; (g) horizontal convergence against a pre-existing block of tessera (stippled pattern).

indication that some dynamic support is being supplied to the observed topography.

The models explored by Smrekar & Solomon also predict that regions of high slope (10° or so) will fail first, normal faulting taking place at the base of slopes, while scarps of lower gradient will fail first towards the tops. In reality, such a relationship is often obscured by subsequent deformation, but where only extensional structures are observed, for instance along the SE flank of Lakshmi Planum, the model relationships appear to hold true. The long anastomosing ridges that characterize the plains exterior to the scarps that border southern Lakshmi Planum are a manifestation of modest crustal shortening, also predicted by the modelling. The relaxation of the topography in this way gradually produces a profile reminiscent of a flexural bulge, a phenomenon that can be seen along the southeastern escarpment of Lakshmi Planum.

Because all of the superposition relationships observed in Magellan images show that the Lakshmi volcanic flows are younger than the fractured ground they embay, the phase of volcanism they represent presumably partially inundated an older cratonic block, the relative resilience of which forced tectonic strain to be diverted elsewhere. As noted by Kaula et al. (1992), it is very tempting to surmise that, beneath Ishtar Terra, both rising and sinking of the Venusian mantle are represented. Certainly the volcanism at Lakshmi Planum, and a widespread manifestation of extensional tectonics, suggests activity associated with rising mantle material; however, strangely, no major rifting is observed.

Among all this complexity, it is sometimes difficult to separate the wood from the trees. However, it does seem abundantly clear that the pattern of flow within the Venusian mantle must vary widely over relatively small areas. In this respect, Venus is quite different from the Earth, where huge lithospheric plates control what happens in the crust, giving a long-wavelength pattern to deformations at the surface. Furthermore, the high geoid at Ishtar, revealed by the gravity data, implies a mantle viscosity for Venus much higher than Earth's, and also implies that thermal outflow is much more regional than on Earth. This would readily account for the changing tectonic patterns that are observed over relatively small distances within Ishtar Terra and, indeed, elsewhere on the planet.

It remains to be established whether the impressive topography of Ishtar is supported by unusually strong (and probably cold) crust, or is maintained by mantle forces that have continued to operate until relatively recent times, i.e. < 10 Ma. The regional gravity solutions of Grimm & Phillips (1991) give a depth of compensation beneath Maxwell of 130–180 km, which is more easily accommodated by assuming active mantle upwelling than anything else. Furthermore, the widespread evidence for volcanism in Lakshmi Planum (and, indeed, along some of the bounding scarps) favours arguments for continued dynamic support, since the overlap of volcanism and extension implies a thermal gradient of more than 5 K km^{-1} (Smrekar & Solomon 1992). This being so, the timescale of relaxation implies that orogenic activity in at least parts of this region would probably have continued until the relatively recent past, say at most a few million years ago. Whether or not there is mantle upwelling beneath the plateau of Lakshmi Planum remains unclear at the present time. The decision to abort Magellan before the orbit was modified for gravity measurement probably has placed a major obstacle in the way of resolving this problem. However, the possibility remains that both mantle upflow and downflow are occurring beneath this highland region.

10.10 Venusian plains and ridge belts

As was discussed in an earlier section, a plethora of tectonic features characterizes the Venusian plains, including the global-scale ridge belts that traverse them. Most of the plains – which, it will be remembered, account for over 80% of the planet's surface area – show deformational structures that represent modest degrees of horizontal strain (i.e. 10^{-2} or less). These features include graben, linear and anastomosing ridges, and reticulate fracture families. Concentrated belts of deformation are termed *ridge belts* and *fracture belts,* and they exist either as broad ridges with an array of superposed fractures, as narrow to broad zones showing a concentration of long, parallel ridges, or as long, narrow belts consisting of a single sharply defined ridge with superposed smaller-scale structures, mainly ridges. Details of those belts developed in Lavinia and Vinmara Planitiae have been discussed in §6.4.

The circum-equatorial belt of volcanic rises, chasmata and tessera massifs that extends from western Eistla Regio to Atla Regio – believed by some workers to be a zone of "lithospheric weakness" – is but part of a global-scale network of linear deformation zones, which includes those that traverse the plains. The equatorial zone, generally elevated and cross cut by fractures, is some 25 000 km long. The "trans-polar" zone, which commences at one end in Atla Regio, runs along the 210° meridian. At the north pole it is about 200 km wide, but it gradually broadens, fanning out into a family of belts, and is at least 7000 km wide where it intersects the equatorial zone near Ulfrun Regio. In the other direction, it extends 9000 km from the north pole to Eistla Regio in three legs, each about 2900 km in length. This gives it a total length of 60 000 km (10 000 km longer than Earth's mid-oceanic ridge system).

Initial analyses suggest that, in those regions of concentrated deformation closely studied, the oldest structures associated with large-scale deformation are arcuate graben, implying early regionally distributed extensional stresses (Solomon et al. 1992). Although largely buried now, fragments of these faults may be preserved in some inliers of CRT. Ridge belts formed next, the ridges within these apparently having a compressional origin (Solomon et al. 1991, 1992). The very extensive plains units embay such ridge belts, indicating that the former's formation took place early in the sequence of plains generation and subsequent structural deformation. Much of the fracturing and ridge development within the plains took place part-way through plains evolution, while most, if not all, of the digitate lava flows that are observed on the plains were extruded more recently than this. Long, narrow lineations with regionally consistent trends apparently formed late in plains evolution, while wrinkle-type ridges seem to have formed throughout plains evolution.

In the case of Guinevere Planitia, the widespread broad graben of early formation imply an early phase of regional extension. Pre-tessera graben occur also in Lavinia Planitia, supporting a similar view here. However, in Lavinia there are no pervasive through-going faults as there are in Guinevere. Furthermore, in Lavinia there is extensive evidence for an early phase during which ridge belts were forming, implying not extension but compression, i.e. crustal shortening. At the present state of knowledge, it does not seem entirely clear whether such contrasting deformational regimes took place sequentially over large areas, or whether smaller parts of larger areas went through different early histories.

10.11 Summary

Deformational structures are more or less ubiquitous within the crust of Venus, and focused regional studies indicate quite clearly that deformation and volcanism took place both sequentially and contemporaneously. Thus, a sequence of alternating events may be observed in the rock units from any one region. The crust of Venus experienced stresses that operated in a coherent fashion over very large areas (hundreds of kilometres); these manifested themselves in modest degrees of strain, evidenced by landforms such as graben and wrinkle ridges with 1–20 km spacings. This implies that even local features are a reflection of the crust's response to mantle dynamic processes. More concentrated zones of deformation manifested themselves in major features such as mountain belts, which reveal evidence for not only lithospheric extension but also shortening and crustal thickening.

Coronas, the uniquely Venusian landforms, which range in diameter from 75 to 2600 km, are loci of high heat-flow, i.e. volcanism. They represent the first of two styles in which extensional tectonism expresses itself at the surface of Venus. The second is to be found in broad volcanic rises, often with much larger dimensions, with which are associated major volcanism and rifting, e.g. Alpha Regio. In only a few locations is there evidence for large-offset strike-slip faulting; in general, the limited horizontal shearing is taken up on a more local scale. In a few locations, extremely high gradients are observed, and it seems clear that active dynamic motions are occurring within the mantle beneath these regions, giving active support to the topography. This applies, particularly, to Maxwell Montes and probably also to the plateau of Ishtar Terra.

Finally, although there is some evidence for the existence of "mini-plates" and certainly of local movement of planitiae, and even limited obduction, the global pattern of deformation and volcanism does not resemble terrestrial-style oceanic plate tectonics. Indeed, the landforms and structures more closely resemble those developed in actively deforming continental regions.

11 What makes Venus tick?

11.1 Introduction

Several vital questions concerning the geological history of Venus remained unanswered prior to the Magellan Mission; some of these were discussed in §3.8. In particular, planetary geologists are anxious to understand how heat was transferred from the interior to the surface of Venus. This called for improved knowledge of the geophysics of Venus, particularly the internal density distribution and internal dynamics, and modelling specific landforms and structures with reference to known geophysical phenomena. A fuller understanding of how volcanism and tectonism has affected and modified the Venusian crust and lithosphere would complement this. Also of importance is the search for any scraps of evidence that point to there having been a different climatic regime in the distant past, which may have allowed volatiles to exist at or close to the surface of the planet, providing conditions suitable for terrestrial-type plate recycling processes.

The analysis of Magellan data is very much ongoing; the task is exciting, if daunting. Already it is clear that the data have provided new insights into the ways in which our sister planet has evolved. While many of the processes believed to have operated are familiar, the relative importance, rates and particular landscape features resulting from their activity have produced a planetary surface very different from the Earth's. In this final chapter will be found just a few ideas that currently are being explored, in an effort to explain how, when and why this hostile world, which lies but a rocket's journey away, attained its present configuration.

11.2 Characteristics of the surface of Venus

One of the most obvious features of the surface is that it is dominated by plains that are geomorphologically very complex. These relatively smooth areas have been modified by repeated episodes of tectonic deformation and volcanic activity, the most recent manifestation of which appears significantly to have modified the planet's impact record. Studies of the distribution and modification of impact craters imply that much of the surface of the planet is relatively young, of the order of 0.5×10^9 a. Deformational features within the plains, in particular the orthogonal pattern of ridges and faults, but also the widespread ridge and fracture belts, allow constraints to be put on the amount of horizontal strain experienced within the plains units.

The highland terrains are particularly diverse and complex, with individual highland regions being built from terrain elements that apparently originated from different processes, operating at either the same or different times. Some of the oldest crust appears to exist within these elevated regions and shows evidence of having occupied different positions in the geological past (Vorder Bruegge & Head 1990). This raises the question of what order of horizontal translational movements have taken place and whether or not these involve large-scale plate movements.

Tectonic deformation on Venus is widespread and often very intense. This is particularly true of the tesserae, where complex ridged terrains present some of the most complicated landscapes known in the Solar System. It is in these regions that superficial deposits appear to be most widespread. Evidently, the deformational upheavals have generated clastic debris that has been moved by mass wasting. This can be detected with emissivity and reflectance data (Tyler et al. 1991). In addition to the block-type tessera massifs, there are well defined linear belts of ridge-and-trough deformation in the highlands, which show features arising from compressional movements and low-angle thrusting. These are comparable in their broad-scale characteristics with terrestrial fold mountains, making Venus the only other planet to share this characteristic. Venus also shares with the Earth the characteristic of landforms generated by shearing movements. The recent discovery of widespread extensional fractures and faults within the mountain belts has inevitably led to the realization that passive relaxation of the Venusian crust has occurred and may be a widespread phenomenon within the highlands.

The patterns delineated by these deformed regions indicate there to be a close relationship between internal dynamics and surface deformation, as there is on the Earth and Mars. However, to date, no convincing evidence has been provided to support the idea that active plate tectonics is operational on Venus. Furthermore, unless water was more abundant previously – and thus far, little evidence exists that sheds any light on this issue – it is also unlikely to have occurred in the past. Doubtless there will be continued controversy regarding this issue for many years to come, but it has to be said that not only the tectonic features but also their distribution (and that of volcanic landforms), and the nature and distribution of regional slopes, militate against such a conclusion. However, it is clear that, like the Earth, Venus is geologically active at the present time.

Another characteristic that Venus shares with the other terrestrial planets is the development of basaltic volcanism. Extensive volcanic flooding is witnessed in the lowland areas, while volcanic shields on a variety of size scales, domes and lava channels are distributed widely. The development of features such as lava channels, collapse pits and flat-topped lobate flows suggests that plains-forming eruptions were of both high volume and low viscosity. Many regions of radar-dark plains within the highlands also exhibit tell-tale features of volcanism, showing that here, too, volcanic activity has occurred. Frequently this can be connected with extensional deformation.

On Venus, clusters of small shield volcanoes give rise to shield fields not typical of the Earth; furthermore, a considerable number of large flat-topped volcanic domes also represent a kind of un-Earth-like landform. Then again, the peculiarly Venusian landforms, coronas, novae and arachnoids, are also very numerous within the plains regions, and are believed to be surface manifestations of focused mantle-plume activity. Their mere abundance implies that the thermal outflow regime within Venus is very different from the Earth's.

11.3 The interior of Venus and its dynamic implications

The similarity in mass and density of Venus and the Earth has led to the assumption that the Venusian interior, in terms of both gross structure and density distribution, is similar to Earth's. In the absence of on-surface geophysical data, information regarding the interior, in particular the crust and lithosphere, is largely deduced from analysis of surface patterns of deformation and volcanism.

11.3.1 Of plumes and plate tectonics

The mean density of Venus is a mere 3% less than the Earth's. Owing to this close similarity, it is generally believed that, like the Earth, it has differentiated a core, mantle and crust, giving a density profile rather similar to the Earth's (Phillips & Malin 1983). However, because of the very elevated surface temperatures, the lithosphere of Venus ought to be significantly more buoyant than Earth's. This is an important fact, since it would militate against active subduction and terrestrial-style plate recycling. Such a conclusion is supported by the very distinctive unimodal hypsographic curve of Venus, which, compared with the Earth's (characterized by a distinct bimodality – reflecting the distribution of continents and ocean floors), shows a complete absence of typical ocean-floor and continental-interior profiles. The facts argue strongly against the notion that sea-floor spreading or processes akin to this can be taking place at the present time. Indeed, the topography of the Venusian surface in many ways is more closely similar to the terrestrial intra-plate continental or oceanic environment. Consequently, in the absence of a method of removing thermal energy from the interior by plate tectonics, mantle plume and hotspot activity ought to be more vigorous on Venus than on our own planet.

On the Earth, the physiography of the ocean floor illustrates how heat is removed along the cooling boundary of mantle convection. Thus, at divergent plate margins, new basaltic crust is being continually generated. Away from such margins, localized plume activity may give rise to volcanic intra-plate island groups, such as the Hawaiian–Emperor chain, but, by and large, intra-plate thermal activity is minimal over very large plate surfaces whose areas are measurable in millions of square kilometres. The continents, whose origin is less clear, also reflect the activity of lithospheric plates: suturing, thrusting and faulting occur along global-scale, linear, mobile zones.

Despite some similarities between the equatorial highlands of Venus and rifted sea-floor regions on Earth, explored by several workers as potential analogues (Brass & Harrison 1982, Kozak & Schaber 1989, Head & Crumpler 1990), the altimetric data provided by Magellan have not provided confirmatory evidence for rapidly spreading oceanic-type plate activity. In fact, the recent realization that the largest contiguous highland region, Aphrodite Terra, is actually built from a series of roughly circular volcanic rises and equidimensional tessera massifs, laced with a preponderance of quasi-circular structural patterns and associated with major plains- and flood-style volcanism, has weakened such an idea. Furthermore, gravity data indicate that the highlands are supported by low-density roots that penetrate to depths which, in some instances, must lie well below the lithosphere and show that the topography is being dynamically supported. Such a scenario means that the topography is maintained by thermal buoyancy in the mantle, that is, by upwelling mantle plumes. Such "hotspots" would either supply direct support or do so indirectly by the heating and consequent expansion of the overlying lithosphere. Nevertheless, there is some evidence that differential movements have caused transform-style offsets along the highland belt, but these cannot be along the lines of those that characterize mid-oceanic rises.

The fact that both isolated volcano–tectonic features (e.g. coronas and arachnoids) and larger-scale features (e.g. Beta Regio and the various components of the equatorial highlands) are characteristic of Venus makes it difficult to envisage any other process capable of generating the observed physiography than widespread plume activity. Such activity, well known from terrestrial experience, has the capacity to generate lithospheric swells, major rifting and associated volcanism (Fig. 11.1). This is exactly what is seen on Venus. Theoretical considerations point to the fact that, if coronas are produced above terrestrial-style hotspots, picritic magma could typify their cores, while tholeitic magma would form at the cooler margins. However, should plumes on Venus be 100–150°C hotter than their terrestrial counterparts, melts generated in the core regions of rising plumes could approach basaltic komatiite composition, with olivine–tholeiite towards the margins (Hess & Head 1990).

Ishtar Terra is the highland massif on Venus that many workers have considered most akin to a terrestrial continent. The western plateau, Lakshmi Planum, surrounded by arcuate zones of folded and thrusted rocks and bounded by steeply sloping scarps, bears some resemblance to terrestrial island arc regions. The comparable Venusian "mobile zones" and sutures are, to some peoples' minds, very clear to see. The massif is so high-standing that some measure of dynamical support has to be invoked to maintain its topography, a requisite implicit also in

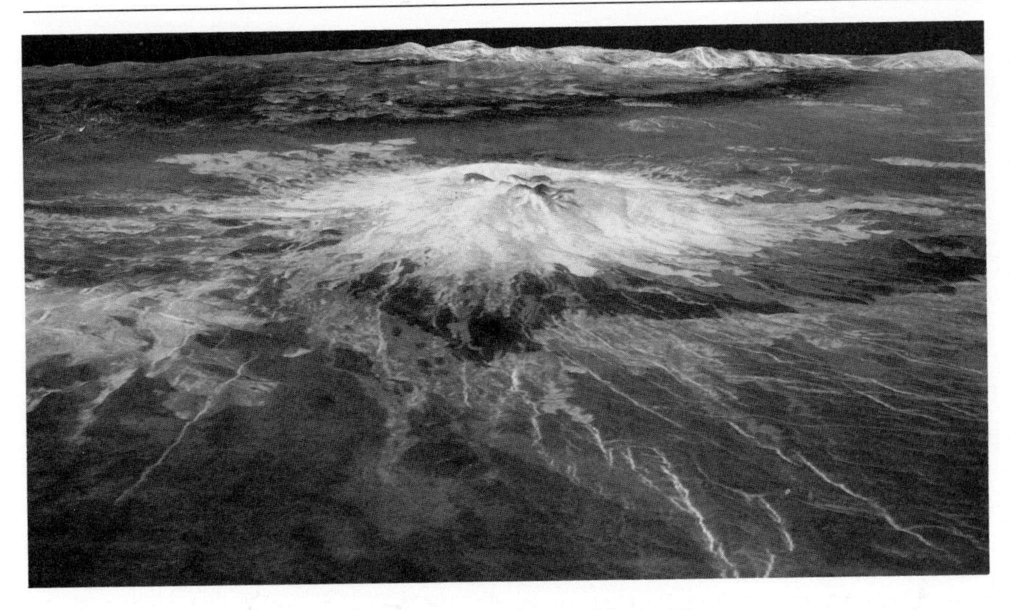

Figure 11.1 Sapas Mons and lava plains
Oblique view of large shield volcano from the south. The radar-bright lava flows and surrounding radar-dark lava plains are shown. Magellan image P-40701.

the gravity data. The problem that needs to be addressed is how can this be accommodated within the context of Venus tectonics and mantle processes.

On Earth, the secondary differentiation of primary crust is achieved by plate recycling, during which oceanic (basaltic) crust is hydrated, both by interaction with sea water and by incorporation of volatiles from sea-floor sediments, melted and differentiated (and possibly contaminated with more silicic materials). It then rises back to the surface to form island arcs, which develop behind subduction zones and associated deep trenches. Eventually these arcs will be accreted to adjacent arcs or existing continental margins in the way depicted in Figure 11.2. This is the modern view of how the Earth's continents grow over the aeons of time (Hess & Head 1990).

Such a process is practicable on the Earth since it has both an atmosphere and a hydrosphere and is not subject to the stifling, dehydrating, greenhouse effect that is a key feature of Venus. Nevertheless, if Ishtar is to be seen as a "continent" in the terrestrial sense, i.e. formed by plate processes, it becomes necessary to find some mechanism to differentiate the original mantle-derived materials. Thus far, it is not possible to envisage how this can have been achieved under present conditions; however, should there once have been lower temperatures on Venus, water could have been significantly more plentiful and plate recycling might have been able to occur.

Using very reasonable assumptions concerning the chemical make-up of the Venusian crust (i.e. anhydrous basaltic crust being subducted, with small content of released CO_2), Hess &

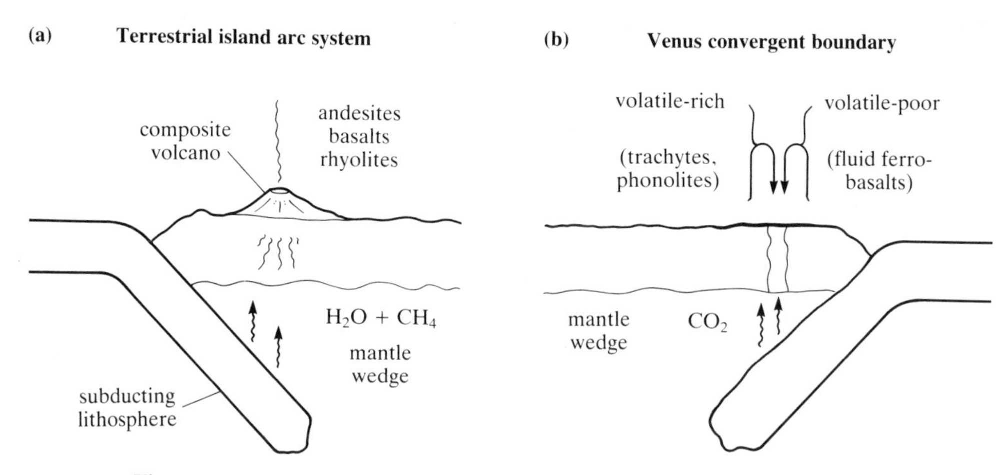

Figure 11.2 Generalized terrestrial and Venusian plate convergence zones
Schematic representation of island arc boundary on (a) the Earth and (b) Venus. After Hess & Head (1990).

Head (1990) predicted that subduction melts should be alkali-basaltic in type, rather than andesitic, as on Earth. This would probably give potential Venusian convergence zones rather different morphological signatures from their terrestrial counterparts, perhaps as envisaged in Figure 11.2.

Assuming such a scenario, Ishtar Terra would be seen as a manifestation of subduction and collision processes that occurred before the present runaway greenhouse was in place. The implications of such an idea for Ishtar are that it must be a relatively ancient feature. This seems unlikely, and, even if future research shows that it is a possibility, there remains the problem of the continued dynamical support needed to maintain its significant topography.

Long-term passive support can be ruled out – the high surface temperature of Venus means that the depth at which crustal rocks fail by ductile flow should be at lesser depths on Venus than on the Earth. Numerical models suggest that high-standing regions, or those with steep bounding slopes, should spread under the influence of gravity by ductile flow of the relatively weak Venusian lower crust on a timescale of the order of 10 million years (Smrekar & Solomon 1992). This implies that the geological processes that generate relief and steepen slopes must have been active during the past 10 million years.

11.3.2 The crust and lithosphere

Direct observational tests of mechanical and thermal models for the Venusian lithosphere are crucial if an understanding of the planet's dynamical history is to be reached. The simplest way to do this is to estimate the thickness of the elastic lithosphere from the flexural response to lithospheric loads (a technique that has been used with some success for the Tharsis Bulge on Mars). The pre-Magellan study of Head (1990b), who investigated the relationships between the northern plains and Ishtar Terra, identified possible lithospheric flexure where the northern plains appear to underthrust the Ishtar massif along Uorsar Rupes. Using the Venera 15/16 topographic profiles, Head estimated an elastic lithosphere of thickness 11–18 km, which is consistent with a thermal gradient of 15–25 K km^{-1} and a heat-flow of 50–70 mW km^{-1} (Solomon & Head 1990).

Using data collected before Magellan had completed its survey, Sandwell & Schubert (1992) arrived at an elastic lithosphere thickness of 10–25 km and a thermal gradient of 10–25 K km^{-1} for the same area of Uorsar Rupes and the circumferential fractures associated with the nearby corona, Eithinoha. For the very large Heng-O and Artemis coronas, and the smaller Latona structure, the same workers derived significantly greater thicknesses (i.e. 30–60 km) which give correspondingly lower thermal gradients of 3–9 K km^{-1}. This corresponds to values of conductive heat loss approximately one-half that of the expected average planetary value. They invoke a combination of differential thermal subsidence and lithospheric subduction to explain the observed features in this region.

The only other Magellan-based study in this respect is one of fracturing associated with Gula Mons in western Eistla Regio. Grimm & Phillips (1992), making the assumption that faulting in this region was generated in response simply to flexural loading, arrived at an estimated value of 50 km for the elastic lithosphere, at the time of graben formation. This equates with a thermal gradient of 33 K km^{-1}.

The larger values for thickness pose something of a problem to current ideas about the thermal structure of the planet, since the derived thermal gradients are surprisingly low, particularly for the coronas. On current thinking, these ought to be sites of high heat-flow; however, as Solomon et al. (1992) point out, the lower lithosphere and asthenosphere may be viscoelastic, while the Venusian crust and mantle are believed to be anhydrous. Both would have a fundamental effect upon the calculations, and, as a result, values for thermal gradient could be much higher.

The crust itself may be quite strong but, since the ambient temperature lies close to the point where crustal rocks would begin to lose their strength, the strong upper layer may be quite thin, say a few kilometres or less. Since the surface temperature on Venus falls with elevation at the rate of 8 K km^{-1} (Seiff et al. 1980), the highest regions may be 80 K cooler than the temperature at the MPR. Suppe & Connors (1992) have suggested that the relief of Venusian fold belts, which shows a remarkable linear dependence upon absolute elevation, may reflect an absolute elevation dependence of the depth of the brittle-plastic transition, which may be controlled by an isostatic coupling of elevation, lithosphere thickness and geothermal gradient.

Mantle material on Venus generally has been assumed to be olivine-rich and hence to possess much higher yield stress than the overlying crust. This will lead to a brittle–ductile failure envelope in the upper mantle (Zuber 1987, Banerdt & Golombek 1988). Another boundary

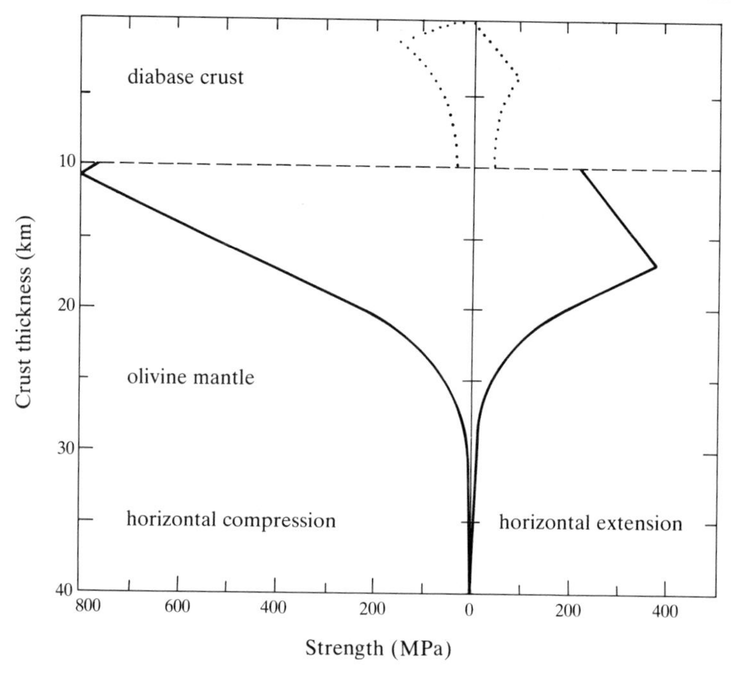

Figure 11.3 Hypothetical structure of the outer regions of Venus's interior

Important boundary layers will exist in the upper crust at a level where brittle failure gives way to ductile flow. A similar interface will exist in the upper mantle.

layer will exist in the upper crust, where brittle failure gives way to ductile deformation (Fig. 11.3).

On the basis of such a structure and from the known behaviour of assumed Venus materials, numerical modelling suggests that, once stress is removed, highland terrain characterized by thickened crust should have a lifetime of only 100 Ma. or so, because ductile flow would take over (Bindschadler & Parmentier 1990). The evidence for extensional deformation around the perimeter of Ishtar Terra bears witness to the tendency of the crust to adjust by ductile deformation.

11.3.3 The interior dynamics of Venus

In §10.5.2 it was shown that both the mantle upwelling (hotspot) and mantle downwelling (coldspot) models for the evolution of Venusian highland massifs face difficulties in their present, essentially formative, guises. This is in large part because the rheological properties of the Venusian upper crust, and the patterns of volcanism and tectonism, are both much more complicated than currently can be accommodated in existing models. Such models seek to reproduce the deformation and magmatic response of the crust to flow within the underlying mantle. The hotspot and coldspot models are distinguishable from each other since they predict different sequences of events and time-dependence between topography and gravity. It is important to stress that each operates on the basis of a number of premisses, some of which already have been mentioned above. If any or all of these are incorrect, then, naturally, the models may be less than adequate. The underlying premisses are as follows (Phillips et al. 1991):

(a) Venus's internal heat production is similar to Earth's.
(b) Owing to the high ambient temperatures prevailing at the surface, the Venusian elastic lithosphere is only a few tens of kilometres thick.
(c) The weak lower crust can become detached along a plane of ductile decollement.
(d) For crust 10–30 km thick there are two elastic lithospheres, one in the upper crust and one in the upper mantle (Zuber & Parmentier 1990).
(e) Mantle flow regimes will directly couple stress to the lithosphere.
(f) The crust will either thin or thicken in response to induced stresses on a geologically reasonable time-scale (500×10^6 Ma).
(g) Compressional deformation is related to downwelling lithosphere and extensional deformation to upwellings.
(h) The lithosphere cannot move large horizontal distances owing to the absence of a low-viscosity layer.
(i) Subduction zones driven by lithosphere with significant negative buoyancy do not exist on Venus, at least on a global scale.

The highlands, in particular Ishtar Terra, are difficult to interpret, but whatever processes are invoked to explain them must also account for the features of the plains and lowlands. Our modern understanding of the global structure shows that they form an integral geological system.

There is a general consensus that Beta-type highlands, that is volcanic rises, are sites of active mantle upwelling and that their associated volcanism occurs in response to pressure-release partial melting in the underlying mantle. Such volcanic rises, although much larger than most corona-type structures, do show a size overlap with them. Thus, Artemis Chasma and Heng-O coronas, 2600 km and 1060 km across respectively, compare closely with the 1500 km wide rise of Bell Regio; most coronas, however, seldom exceed 400 km diameter. It is also apparent that most volcanic rises show a greater diversity of volcanic features and greater amounts of uplift and extension (i.e. major rifting) than typical coronas.

If major rises and coronas originate in similar mantle processes, which is widely accepted, then the implication is that mantle upwelling activity within Venus occurs on different scales. The large apparent depths of compensation associated with volcanic rises and the very extensive volcanism imply that they are connected with much larger-scale mantle upwellings than

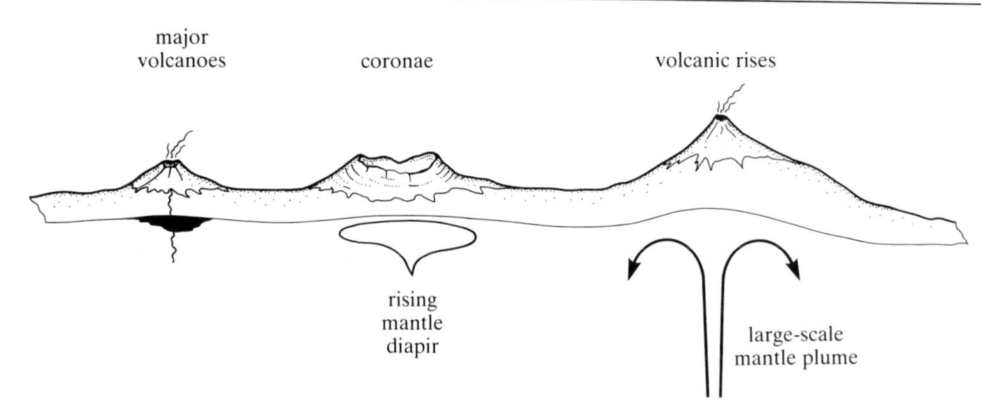

Figure 11.4 Cartoon illustrating different scales of mantle upwelling on Venus
Volcanic rises are thought to be due to major upwellings of mantle material, while large shield volcanoes are located above large bodies of relatively high-level magma. Coronae appear to lie between these two extremes. After Stofan et al. (1992).

are coronas. Since most coronas are smaller than rises, it may be reasonably assumed that coronas represent a smaller scale of upwelling, or weaker upflow of material, or flows that are shorter-lived along the lines depicted in Figure 11.4. The fact that there is a major concentration of volcanic rises, large shield volcanoes and coronas in the equatorial belt between Atla and Beta Regiones (shown in Plate 6) and along two principal tectonic trends, (i.e. Parga and Hecate Chasmata), which are characterized by extension, volcanism and uplift, suggests that both large- and small-scale plume activity originates in the mantle along hot regions encircling the core/mantle boundary. Unfortunately, at present, there is no way of knowing how long-lived are plumes beneath major rises.

Calculations by Stofan et al. (1992) suggest that most coronas can be generated by mantle **diapirs** 75–100 km across. They also can be produced with a temperature difference of around 300 K from the mantle temperature (Janes et al. 1992). On this basis, and bearing in mind that there are approximately 300 coronas on Venus, they calculate that the approximate heat loss rate through coronas is 4×10^{-3} of the estimated mantle heat production through global volcanism for Venus (Solomon & Head 1991). Thus, mantle plumes related to corona production probably account for but a small proportion of the heat transfer to the surface of Venus, a condition that is comparable with the estimates for terrestrial plumes, which generally are believed to account for around 10% of the Earth's internal heat loss (Sleep 1990).

What is not known is the relative contribution of intrusion and extrusion within the Venusian crust. On Earth, extrusive rocks account for between 10 and 20% of the total. Several factors argue that for Venus this proportion may be lower: first, high-density crust appears to be more extensive on Venus; secondly, the high surface pressure inhibits vesiculation within melts rising towards the surface. The latter would have the result of magma reaching neutral buoyancy zones at shallower levels than on Earth. This phenomenon may account for the apparent abundance of fractures and graben associated with volcanic units: such fractures may represent dykes.

The primordial crust of Venus is probably not to be found anywhere on the planet's surface and may have been rather like that of the lunar highlands. It is not known whether the Venusian crust that we do see is secondary, i.e. predominantly basaltic, or tertiary, i.e. reworked secondary material which might have a more evolved chemistry. The Venera 8 data suggest that some chemical differentiation has occurred, and this is supported by the now known occurrence of large steep-sided domes representing intrusions of more (evolved?) viscous magmas.

When applied to Ishtar, the hotspot model invokes lateral variations in the strength of the lithosphere to account for the development of mountain belts surrounding Lakshmi Planum. The coldspot scenario, on the other hand, states that Ishtar must overlie a vigorously downwelling mantle region, which accounts for the elevation of the high plateaux and the formation and continued existence of the encircling mountain belts. Any model, to be entirely appropriate, must also be capable of accounting for the extensive volcanism that has occurred over both Lakshmi Planum and the outboard plateaux. Furthermore, it must also accommodate the large amount of potential energy stored in the downwelling mantle plume that is required to satisfy the gravity field under the coldspot model. At this time, neither of the end-member models satisfactorily explains all the observations.

What is clear is that the Venusian crust behaves more like the Earth's continents – where blocks of relatively thick stable crust are separated by actively deforming zones – than ocean floors – where relatively stronger crust is being created and spreading from divergence zones. This is in large part because the elevated ambient temperatures prevailing at the surface and the predicted lithospheric thermal gradients, and the strong dependence of ductile flow on temperature, all push the Venusian lithosphere towards responding in a weak manner for crustal thicknesses comparable with those of the Earth's continents.

11.3.4 Plains deformation and its implications

Now that global high-resolution imagery is available, it can be shown that horizontal crustal shortening and extension have both occurred widely on scales that range from 1 km to over 1000 km. Shearing movements also took place, particularly in ridge-and-groove belts and in mountain belts. However, the shearing appears to have been distributed rather broadly and also to have been accompanied by horizontal stretching or shortening, rather than by large offset strike-slip faulting. In this characteristic its development is un-Earth-like.

Both large- and small-scale linear deformation belts extend through the Venusian plains, the different scales of deformation reflecting the rather complicated mechanical and dynamical structure of the planet. From what has been said about the detailed structure of the outer layers of Venus, deformation within the range 10–30 km is seen as a response of the strong upper crustal layer. Deformation on smaller scales can be perceived as due to either internal deformation of the strong upper crust, or to some form of disruption of a thin superficial stratum that is decoupled either thermally or mechanically from that layer. Deformation on the scale of a few hundred kilometres represents deformation of the strong upper mantle. Scaling up – from a few hundred to a few thousand kilometres – mantle convection comes into play. This is where long-wavelength gravity correlates with major topographic features.

The global distribution of such deformation belts has implications for global stress configuration. In particular, the prominent ridge-belt fan that characterizes the northern hemisphere must be accounted for. Sukhanov & Pronin (1989) proposed that the fan represents a zone of crustal spreading, an idea supported subsequently by Kozak & Schaber (1989). The latter also suggested that the fan formed a part of a trans-polar tectonic zone connecting with an equatorial extension zone running from Eistla to Beta Regio. The spreading interpretation was based largely on the observation of a broad bilateral symmetry of features around an axial ridge-belt structure (Pandrosa Dorsa). However, recent work indicates that there are other longitudinal features that do not share this symmetry; furthermore, certain of the structures formerly interpreted as extensional now appear to have a compressional origin (Frank & Head 1990). One way to test this hypothesis is to search for evidence that units gradually increase in age with distance from the axis. Unfortunately, the cratering record and its derived statistics are insufficiently good either to prove or disprove this idea.

While the plains themselves exhibit a variety of tectonic features, a distinct orthogonality is characteristic (Solomon et al. 1991, Squyres et al. 1992). This has been observed particularly in the plains of Lavinia and Atalanta Planitiae – the first plains regions to be imaged by Magellan. The wide extent and uniformity of this tectonic pattern implies that the near-surface rocks have been subject to a stress field of significant spatial and temporal regularity (Squyres et al. 1992). The simplest interpretation is to think of the characteristic tectonic features – wrinkle ridges and grooves – as manifestations of thrusting and normal faulting respectively. Indeed, such an interpretation may be correct, in which case there are some interesting implications.

On the basis of well tried methods of stress/faulting prediction (Anderson 1951), it can be shown that orthogonal thrust faults and graben may both be formed, but only if the two stress components in the plane of the planet's surface maintain their relative values but vary appropriately with respect to the vertical stress. Such a condition could be attained either if the stresses within the plane varied with time relative to the vertical stress or if initiation of the two kinds of features occurred at differing depths, i.e. near-surface for thrust faults and deeper down for grabens. On this basis, in the former case the vertical stress is σ_3, while in the latter it is σ_1.

The 100–150 km wide ridge belts that extend from hundreds to thousands of kilometres across the plains, and with elevations of the order of 1 km or less, have radar-bright signatures owing to high back-scatter induced by rough deformed surfaces. Associated fracture belts are also associated with raised relief and comprise belts of fractures and fault troughs spaced 5–20 km apart. Both Zuber (1990) and Squyres et al. (1992) suggest that the large-scale spacing of ridge belts 300–400 km) within Lavinia Planitia are best explained by compressive stresses, while the short-wavelength ridges and grooves within them (10–20 km) are likely to be both compressional and extensional.

Squyres and his colleagues also explore the formation of both ridge and fracture belts in Lavinia Planitia. The first problem encountered is that of whether or not ridge belts and fracture belts originate from the same process. In favour of a common origin are the observations that both types of deformation belt have similar widths, topography and broad geometry. Furthermore, instances can be cited where a ridge belt may change into a fracture belt along its length. However, if some underlying common process is responsible for both, the surface tectonic expressions of the two types of belt are quite dissimilar in detail.

In view of the gross similarities described above, one reasonable hypothesis is that they have been generated by crustal shortening and thickening, the stresses being applied normal to the strike of the belts. Such an argument is straightforward in the case of the ridge belts, since the ridges can simply be seen as folds formed by buckling of the strong upper crustal layer in a direction perpendicular to the maximum compressive stress. However, in the case of the fracture belts, all of which stand higher than the surrounding terrain, it is only the altimetry that suggests that they are produced by crustal thickening of the relatively thin buoyant crust. In fact, their nature and alignment suggest quite the opposite: they ought to be explained by crustal extension. One way out of this dilemma is to view fracture belts as having grown in response to stretching of brittle rocks across arch-like uplifts.

Although we have noted that there must have been great uniformity in the stress regime responsible for the widespread orthogonal pattern in the plains, this cannot have been true of that which generated the deformation belts, which characteristically show considerable irregularity, in both form and orientation. Should there be one underlying belt-formation process, the variable orientations of the belts with respect to the tectonic pattern of the plains indicate that it cannot share a simple relationship with the near-surface stresses, which find their expression in the plains deformation. Nevertheless, the morphological expression of individual belts can be shown frequently to depend on the trend of the belt with respect to the tectonic pattern of the adjacent plains. Thus, ridge belts are found where deformation is perpendicular to the regional compressive stress, while fracture belts are found where the belt trend is perpendicular to the regional extensional stress. The most obvious source of the stress is some form of mantle convection.

Because it is widely held that the Venusian mantle is olivine-rich in composition but that an asthenosphere is lacking, convection has the inherent ability to couple mantle flow strongly to the overlying lithosphere (Phillips 1990). Consequently it could engender deformation of the Venusian lithosphere at wavelengths comparable to the spacing of the tectonized belts in Lavinia Planitia (Zuber 1987, 1990). However, the precise way in which the convective stresses are transferred will be dependent upon the rheology of the crustal layer.

This line of argument again points to the widely held belief that Venus has a basalt-like crust overlying a peridotitic mantle. This being so, again we arrive at the following layered configuration: (a) strong upper crust, (b) ductile lower crust, (c) strong upper mantle, and (d) ductile mantle at greater depth (Banerdt & Golombek 1988).

In an effort to try to understand those processes responsible for the two scales of deformation in Lavinia Planitia, Squyres et al. (1992) used finite-element models to investigate quantitatively the relationship between mantle convection and crustal deformation. Their particular need was to account for the spacing of the belts themselves, and of the spacing of the linear features within the belts. Their modelling indicates that convective motions within the Venusian mantle can couple to the crust and generate horizontal stresses of sufficient magnitude to induce the formation of Lavinia-type deformation. Sufficient compression or extension can be generated if a weak lower crust is sandwiched between a relatively strong upper mantle and an upper crust of considerable strength. The high degree of correlation between long-wavelength topography and gravity already noted suggests that the planet lacks a lower-mantle low-viscosity zone like the Earth's, in which case mantle convective motions should readily couple with the overlying lithosphere, producing recognizable tectonic and topographic patterns at the surface.

Analysis of the kind of crustal-flow activity induced by convection (Bindschadler & Parmentier 1990) indicates that the Venusian crust can be regarded as responding in two different ways to downwelling: (a) flexural response, during which the surface and crust-mantle interface effectively deform together, and (b) a subsequent phase during which the crust responds by thickening, whereupon the surface and crust/mantle boundary gets out of phase. It is on this basis that Simons et al. (1991) proposed that deformation belts may form during the transition between initial flexuring and the later phase of crustal thickening.

The modelling exercises of Squyres et al. (1992) indicate that, after an initial stage of downwelling and creation of topographic lows, the topography gradually grows, owing to an isostatic contribution from the thickened crust. Consequently, if the deformation belts are indeed due to compressional stress induced by mantle flow, then the belts will be produced at or before the time of maximum horizontal compressive stress.

Squyres and his colleagues also deduce that the small-scale wavelengths of deformation observed within individual ridge belts can be utilized to place a lower limit on the Venusian thermal gradient, at least in the area of Lavinia Planitia that they studied. Using appropriate values for surface temperature (720 K), thermal gradient (15 K km^{-1}) and strain rate (10^{-15} s^{-1}), they show how strength increases linearly with depth near the surface (brittle regime) because of increasing lithospheric stress; it then decreases exponentially (ductile regime) owing to the thermal gradient (Fig. 11.5). For this reason, for a given compressive stress, only those depths with a strength exceeding the stress applied (between Z_t and Z_b) will support the entire stress load elastically. Above Z_t will support only a part of the load elastically and will undergo brittle failure. At depths less than Z_b only a part of the load can be supported and it will undergo ductile creep. This rheological layering could account for the short-wavelength deformation observed in Lavinia Planitia.

A further rheological transition will occur at the crust/mantle boundary; here again, another brittle–ductile failure envelope will exist, this one in the upper mantle. In the same way as happens with the short-wavelength deformation, the long wavelengths of deformation probably are related to the presence of this deeper strong zone.

Because of the rapid fall in lithospheric strength, the folding wavelength is not particularly sensitive to crustal thickness, particularly for thicknesses of > 10 km. This is due to ductile creep. Now the bottom of a thick crust can contribute little to the overall strength of the whole crust, and therefore has little influence on the folding that occurs. However, as crustal thickness diminishes, the wavelength of folding becomes ever more dependent, the thinnest crust giving rise to the shortest wavelengths. This confirms the line of argument followed earlier.

The wavelength of folding also will vary according to changes in thermal gradient and strain rate. Thus, increasing the thermal gradient decreases the wavelength, and vice versa. On the basis of the observed geometry of deformation belts in Lavinia Planitia, Squyres et al. (1992) derive a value of around 30 K km^{-1}, a somewhat larger figure than that obtained in earlier

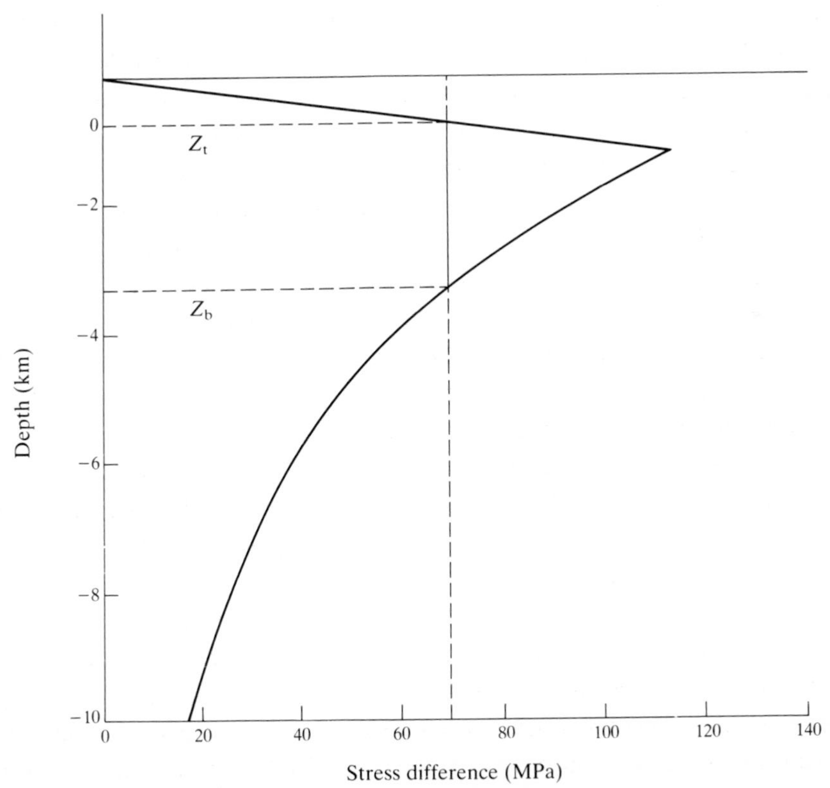

Figure 11.5 Plot of depth versus strength for the Venusian crust
After Squyres et al. (1992).

work (e.g. Zuber 1987). The difference is a result of the greater resolution power of Magellan, which can resolve deformational features of shorter wavelengths than those resolvable by earlier imaging systems. It is similar to the values for thermal gradient derived for Guor Linea by Grimm & Phillips (1992).

11.3.5 General observations on deformation

The intensity of deformation and the state of preservation of tectonic features on Venus can now convincingly be shown to have a strong dependence on local topographic relief. Many elevated regions thus can be perceived as areas of thickened crust, in which case they also represent a layer of material that is relatively weak and therefore capable of ductile flow (i.e. the lower crustal layer is thicker). The upshot is that these regions will behave as concentrators of regional strain within the lithosphere and, as such, are the regions most likely to retain evidence for more than one phase of regional deformation. It is for just this reason that the tesserae, with their highly complex structure, are now viewed as having been formed by contributions from different processes, e.g. crustal convergence at linear rises or mantle upwelling, where crustal thickening has occurred owing to convergent movements in the lithosphere, or because of gravitational relaxation and gravity sliding. Their characteristic diversity of structure implies that more than one of these activities may have contributed to their present form, which has the further implication that they are relatively ancient massifs.

11.4 Volcanism and crustal growth on Venus

The distribution of Venusian volcanic features is not confined to linear zones, as on the Earth; rather there is a concentration along the equatorial latitudes, particularly in the region of Beta–Atla–Themis Regiones, which covers 20% of the area of the planet, and to the south of Alpha Regio. Furthermore, there is a deficiency of features such as coronas in the lowlands, which most likely is attributable to a combination of elevation-dependent eruption conditions and partial or complete burial by younger volcanic units. Magellan data, now available for the whole planet, show that, in the time period represented by the present Venus surface, volcanic activity has occurred, at one time or another, virtually everywhere. It is one of the dominant geological phenomena on Venus.

One of the major problems that still needs to be addressed – and this has an important implication for the volcanic history of the planet – will be the relative dating of individual regions. The complexity of the geology and the relative paucity of impact craters imposes severe limitations on stratigraphic studies. It is difficult to see how this can be improved in the foreseeable future. With regard to the resurfacing rates due to volcanic activity and tectonism, crater distribution studies suggest that it has been episodic and spatially inhomogeneous. Crater statistics appear to indicate that the mean age of the surface of the planet is between a few hundred million and one billion years.

The style of impact cratering is rather different, however, from that seen on other terrestrial planets. This is largely a function of the very dense atmosphere, which plays various rôles, not only in filtering the incoming meteoroids but also in their break-up during atmospheric passage, and in the nature and distribution of their associated ejecta. Because of this, it is often difficult to make accurate comparisons between Venusian cratering statistics and those derived from airless worlds.

While surficial processes do operate on Venus, it appears that the production of sedimentary material is limited, much more so than on either the Earth or Mars. Some of the largest concentrations appear to be within the tesserae. The planet-wide lack of sediments is in part due to the shielding effect of the dense atmosphere, which prevents constant bombardment of the surface rocks by micrometeorites. Furthermore, the absence of water and the apparent lack of a global recycling process akin to the Earth's severely limit physical and chemical weathering. Locally, however, large impacts and explosive volcanism may provide sources of fragmental material that may be moved by the wind. Generally speaking, however, it is unlikely that the percentage of the surface veneered in fragmental deposits is very large (Pettengill et al. 1988).

The observed crater population on Venus leads to three possible end-member models for the resurfacing of the planet: (a) catastrophic resurfacing, (b) "leaky planet" global resurfacing, and (c) regional resurfacing. These have been discussed in detail by Phillips et al. (1992). Figure 11.6 shows in cartoon form the characteristics of each.

The *catastrophic resurfacing model* considers the observed crater tally as a production population and invokes an episode of rapid resurfacing about 500×10^6 a. ago, of sufficient depth to obliterate the pre-existing crater record and provide a pristine surface on which the current production population accumulated (Schaber et al. 1992). Such a model views subsequent volcanism as minimal in both volume and extent. In support of this idea, Schaber and his colleagues cite the relatively small number of impact craters highly modified by volcanic processes. However, what is not clear is both the volume and method of emplacement of many of the plains units found on Venus that cannot yet be directly attributed to volcanicity. Where they are not obviously related to specific volcanic foci, they may have formed in response to catastrophic events, or equally well may have been related to extrusion of variable volumes of lavas from different loci and at different times, i.e. they were formed sequentially, over the past few hundred million years.

The *leaky planet* scenario sees volcanism as having occurred relatively uniformly in time, and having taken place almost everywhere at the same time. By this method the Venusian crust would be thickening gradually with time, and impact craters constantly being either modified or obliterated. However, the lack of impact craters in intermediate to advanced stages

(a) Catastrophic resurfacing

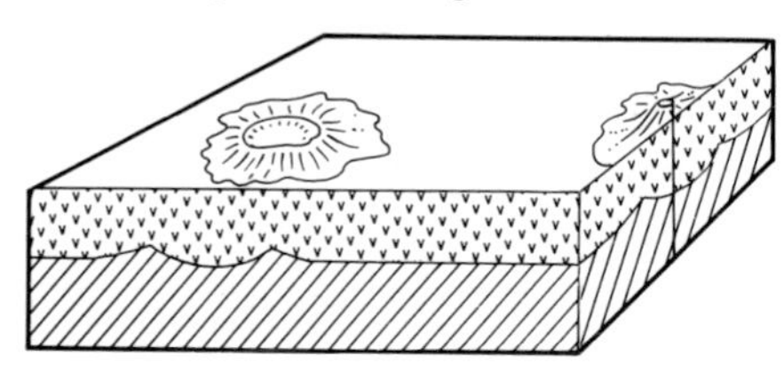

- resurfacing takes place infrequently at extremely high rates
- resurfacing rate very low between events
- production crater population
- almost all craters pristine

(b) Vertical equilibrium: "leaky planet"

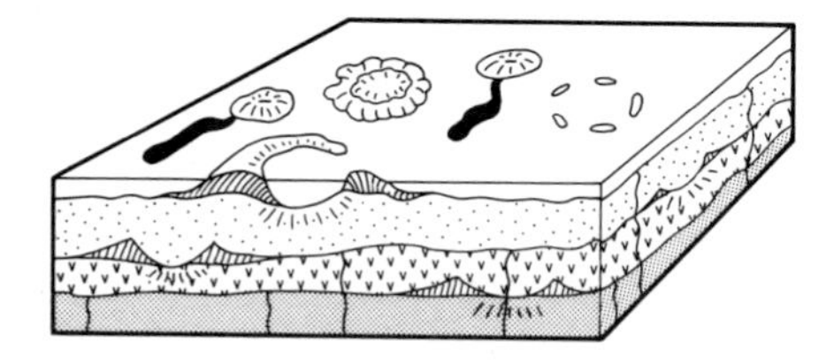

- resurfacing events very widespread globally
- moderate rates of resurfacing
- widespread volcanism modifies and obliterates craters
- produces wide range of crater degradation states

(c) Regional resurfacing: "collage"

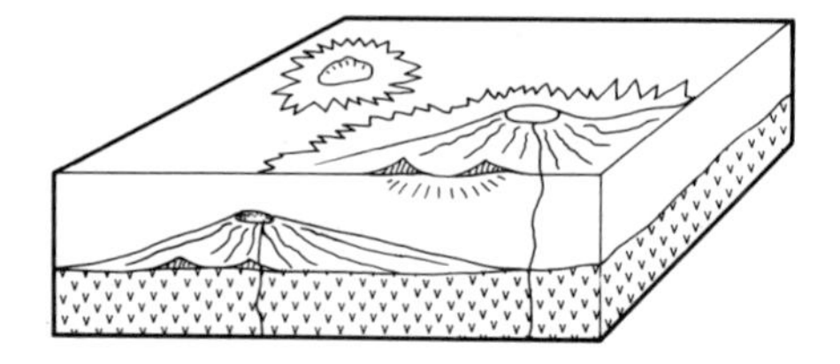

- resurfacing events local and sufficiently large to obliterate craters completely
- resurfacing patch size, frequency and integrated resurfacing rate specified by crater population
- vast majority of impact craters are unmodified

Figure 11.6 Cartoons showing different resurfacing models
(a) Catastrophic resurfacing (takes place infrequently but at extremely high rates); (b) "Leaky planet" model (resurfacing events very widespread globally); (c) Regional resurfacing (local resurfacing events but sufficiently high volume to obliterate craters). After Head et al. (1992).

of volcanic burial tends to militate against this theory.

The extent of volcanic deposits that are easily mappable over specific regions individually cover areas in excess of 125 000 km^2; furthermore, emplacement of volcanic units took place serially, probably on both local and regional scales. Such a process means that, although individual volcanic episodes occurred in sequence, over long periods of time these could have contributed to general resurfacing of the whole planet, or, at least, very large parts of it. This *regional resurfacing model* has sometimes been described as the "cookie-cutter" model, an analogy whose meaning will not be lost, even on non-cooks! It envisages resurfacing as having proceeded by the infilling of topographic lows at different times, with the occasional obliteration of impact craters as an integral part of the same process. The preponderance many different types of volcanic feature related to mantle instabilities on the scale of a few hundred kilometres (e.g. coronas, shield fields, large shields, arachnoids, etc.) suggests that this is true, and that magmatism on Venus is linked with pressure-release partial melting associated with plumes and hotspots. The fact that the total volume of volcanic deposits predicted by the model is greater than that observed does not invalidate it; as has been observed above, large areas of plains units undoubtedly have been deposited by processes as yet not understood.

It is the stratigraphy and method of formation of the extensive Venusian plains that pose one of the greatest remaining problems in Venus geology. An understanding of how individual units were laid down (most, presumably, by volcanic processes), on what scale and over what timescale, can only be achieved by detailed mapping using the recently acquired high-resolution Magellan imagery. At present, this work is in its infancy. To me, the regional or serial resurfacing model has the most appeal, but it should be remembered that Venus is a world very different from the Earth. Geologists imbued with the notion of uniformitarianism will be excused for finding any kind of catastrophic resurfacing model unappealing. However, the thermal evolution of Venus may have been very different from that of the Earth, and current work seeks to refine resurfacing models, so that the stratigraphic history of the planet can be written.

The actual heat budget of Venus is obviously an important factor in discussing volcanic resurfacing: measurements made by Venera landers have revealed bulk values for K, Th and U which suggest that the planet's radiogenic heat budget may be similar to that of Earth. However, values for ^{40}Ar are only one-quarter to one-third that of the Earth, a difference that can be attributable easily only to degassing differences (Donahue & Pollack 1983). Assuming that Venus loses heat at much the same rate as does the Earth, the planet's mean heat flux would be 70 $mW\ m^{-2}$ (Solomon & Head 1991). On the further assumption that most of this emanated from below the crust, this would supply more than enough heat to fuel vigorous convection. It would be equivalent to an average conductive thermal gradient of 15–30 $K\ km^{-1}$. On the Earth, fractional heat loss due to cooling of the planet's interior is around 50%; this fraction may be smaller on Venus if its mantle cooled more due to more efficient mantle convection, or because it is just slightly less massive. If this were so, then both the average heat-flow and thermal gradient could be smaller.

11.5 Epilogue

Even before Magellan had reached the Venus environment, it had become clear that Venus shared many of the geological characteristics of the Earth. Such differences as were observed – for instance the widespread development of coronas (features not found elsewhere in the Solar System) and ridge and fracture belts (unlike any terrestrial regional features) – if not comparable with terrestrial landforms, could be envisaged as having their origins in internal processes familiar from experience of the Earth, Moon and Mars. At present it seems clear that the dominant mechanism of thermal transfer from the interior is conduction, a feature Venus shares with the smaller rocky planets. In the past, however, this may not always have

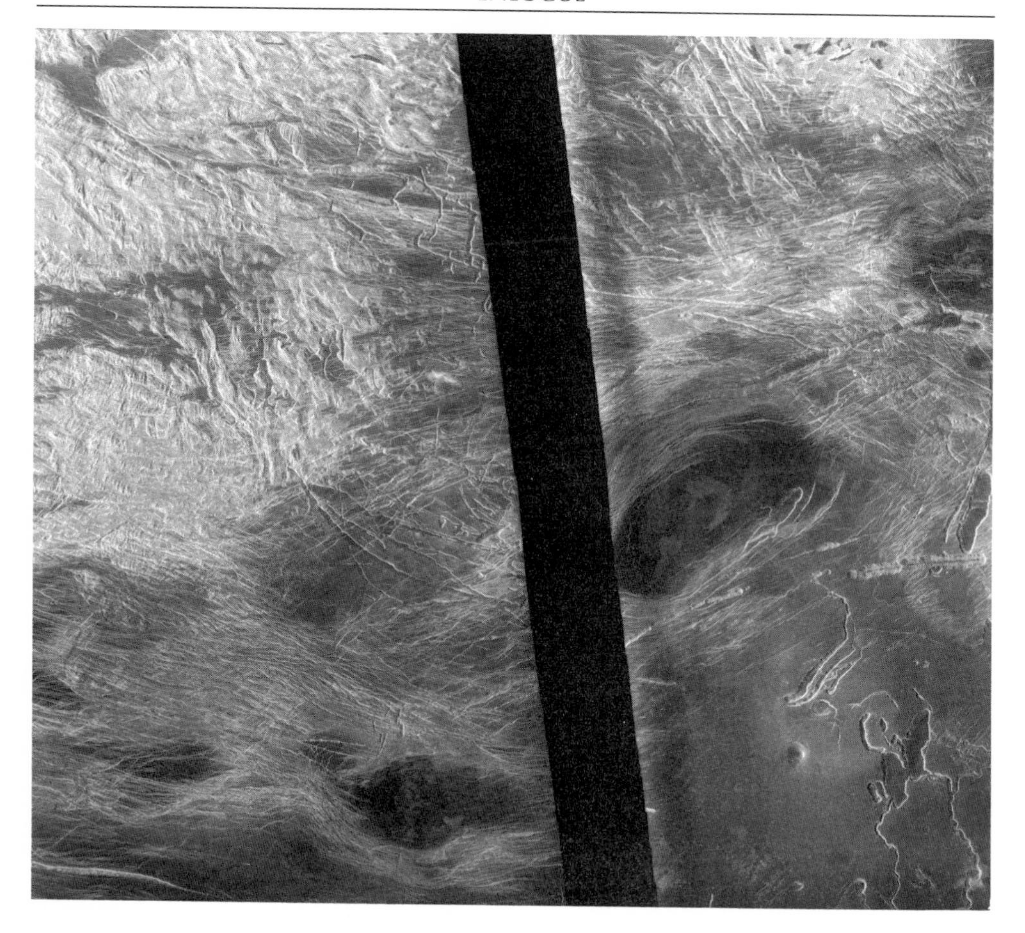

Figure 11.7 Epilogue

A Magellan image of a part of Ovda Regio that epitomizes the complex nature of the surface of Venus. To the north, complex ridged terrain is partly embayed by radar-dark volcanic plains traversed by lava channels. Farther south an elongate caldera structure, 150 km across, has associated annular fractures and a floor that has been flooded by lavas of different radar signature. On the volcanic plains to its south, rows of coalescent pits and prominent anastomosing lava channels cross the smooth surface, upon which sits a prominent volcanic cone with summit pit. Magellan image F-MIDRP 10S087;1.

been so, and evidence is being sought of clues that point to earlier crustal spreading and recycling activities. As more detailed studies are made of Magellan data, a more complete picture of Venusian geology will undoubtedly emerge. In particular it should be possible to refine our ideas concerning such important topics as volcanic flux, lithospheric strain rates and thermal budget. Detailed mapping should also improve knowledge of the stratigraphy of those units present at the surface. Together, the new and ever-more detailed geological maps that will emerge, geophysical analysis and the theoretical modelling that will accompany it, will contribute to an ever-improving understanding of how Venus has evolved.

Appendix 1

Venus Data

Mean distance from the Sun	108 200 000 km	Mean surface pressure	90 bar
Sidereal period	224.701 days	Mean density (water = 1)	5.2 g cm^{-3}
Orbital inclination	3°.394	Surface gravity (Earth = 1)	0.88
Orbital eccentricity	0.00678	Escape velocity	10.36 km s^{-1}
Mean synodic period	583.92 days	Mean surface temperature	470°C
Axial rotation period	243.01 days	Mean visual opposition magnitude	–4.4
Radius	6051 ± 2 km	Albedo	0.76
Mass (Earth = 1)	0.815	Atmospheric main constituents	CO_2, N_2
Volume (Earth = 1)	0.88		

Chronology of Venus missions

Launch date	Mission	Type of mission
27 August 1962	Mariner 2	Encounter at 34.745 km distance
12 June 1967	Venera 4	Atmospheric entry
14 June 1967	Mariner 5	Flyby at 4 023 km
5 January 1969	Venera 5	Parachute descent
10 January 1969	Venera 6	Parachute descent
17 August 1970	Venera 7	Soft landing
27 March 1972	Venera 8	Soft landing/orbiter
3 November 1973	Mariner 10	Encounter only
8 June 1975	Venera 9	Soft landing/orbiter
14 June 1975	Venera 10	Soft landing/orbiter
20 May 1978	Pioneer 12	Radar-mapper
8 August 1978	Pioneer 13	Multiprobe descents
9 September 1978	Venera 11	Flyby/lander
14 September 1978	Venera 12	Flyby/lander
31 October 1981	Venera 13	Lander/orbiter
4 November 1981	Venera 14	Lander/orbiter
2 June 1983	Venera 15	Radar-mapper
7 June 1983	Venera 16	Radar-mapper
15 December 1984	Vega 1	Descent probe and balloon probe
21 December 1984	Vega 2	Descent probe and balloon probe
4 May 1989	Magellan	Radar-mapper

Appendix 2

Units

a	year
C	Celsius or centigrade
cal	calorie
cm	centimetre
G	giga
J	joule
K	kelvin
kb	kilobar
km	kilometre
m	metre
Ma	million years
mb	millibar
mgal	milligal
mm	millimetre
N	newton
Pa	pascal
s	second
W	watt
μm	micron

Glossary

accretion the process of planet growth from smaller bodies by collisions

adiabatic occurring without heat entering or leaving the system

adiabatic gradient natural increase in temperature experienced in moving towards the centre of a planet

ambient surrounding

apapsis the point at which a satellite, orbiting in an elliptical orbit, is at the greatest distance from the primary body

apparent depth of compensation the depth to the base of a freely floating crustal block above a denser mantle layer that will compensate for a recorded gravity anomaly

asthenosphere the region of a planet immediately beneath its lithosphere where it is fluid enough to allow convection

atmophile an element that occurs in an uncombined state in the atmosphere, as a gas (e.g. nitrogen)

basalt dark-coloured lava composed of the minerals pyroxene and plagioclase feldspar, of relatively low silication and high density

carapace chilled outer layer of extrusive volcanic dome or flow

carbonaceous chondrite a type of primitive meteorite containing carbon and peculiar inclusions termed chondrules

chalcophile a chemical element normally occurring as a sulphide

condensation chemical reaction where a solid crystallizes or a liquid precipitates from a gas

core the dense central regions of a planet or star

crust the outermost shell of a chemically differentiated planet

diagenesis the physical and chemical changes (excluding metamorphism) that affect sedimentary rocks after burial

diapir a vertical intrusion of rock melt, usually bulbous in shape

Doppler effect the difference in wavelength of radiation received from an approaching (shorter and "bluer") or receding (longer and "redder") object, compared with a stationary one

ejecta material thrown out of the cavity excavated by a cosmic body such as a meteorite or comet

emissivity the ratio of the radiant flux from a body to that of a perfect black body at the same kinetic temperature

felsic a light-coloured rock or mineral, e.g. feldspar or quartz

granite coarse-grained igneous rock consisting of alkali- and plagioclase feldspars and quartz; it is the characteristic material of the Earth's continental crust

gravity anomaly the corrected difference between the gravity measured above a point on a planet's surface and the theoretical value for an idealized ellipsoid

greenhouse effect the effect whereby a planet's atmosphere acts to maintain the surface temperature well above that expected for a black body

Hadley cell response of a planetary atmosphere to differential latitudinal heating
hotspot the surface expression of a mantle plume

isostasy a state of equilibrium in which segments of a planet's crust stand at levels determined by their density and thickness
isotope one of the forms of a chemical element that share the same number of protons in the nucleus, but different numbers of neutrons

komatiite dense ancient terrestrial lava types that contain skeletal crystals of olivine

lithophile an element with an affinity for silicate
lithosphere the relatively rigid outer layer of a planetary body; in the Earth's case, above the asthenosphere

mafic rock containing more than 50% ferromagnesian minerals
mantle planetary layer between the crust and the core
matrix material, usually fine-grained, interstitial between the larger grains of a rock

olivine dense dark silicate of iron and magnesium
outgassing process whereby a planet loses its more volatile elements to the atmosphere or space

pahoehoe basaltic lava with ropy surface crust
partial melt a melt generated by the liquefaction of minerals with low melting points within a rock body, owing to changes in either temperature or pressure
periapsis the point in an elliptical orbit when a satellite is closest to the primary body
plate segment of lithosphere that effectively floats on the underlying mantle layer
plate tectonics unifying theory of planetary dynamics in which the lithosphere is created and destroyed by recycling back into the mantle
primary atmosphere the original atmosphere of a planet, prior to subsequent modification
primordial primitive, existing from the beginning of time
pyroxene dense dark aluminium silicate of iron, calcium and magnesium

radiogenic associated with radioactivity
refractory element chemical element with high melting point or condensation temperature
regolith fragmented layer formed at surface of planetary body

secondary atmosphere planetary atmosphere produced by slow outgassing of volatiles and other modifying processes
seismic related to earthquakes or waves generated by natural explosions or those caused by humans
serpentine a greenish hydrated magnesium silicate mineral
sialic term denoting crustal material rich in silica and alumina
siderophile an element soluble in metallic iron
silicate mineral consisting of silicon and oxygen in combination either with each other or with other elements
simatic crustal material rich in silica and magnesia
subduction process whereby a lithospheric plate plunges beneath another and is eventually recycled
syenite an igneous rock composed primarily of feldspars and hornblende
tectonic pertaining to the deformation of rocks

triple point the unique condition of pressure and temperature at which a chemical compound can exist as gas, liquid and solid; for water this occurs at 273.16 K
volatile element element vaporized at relatively low temperature

References

Anderson, E. M. 1951. *The dynamics of faulting and dyke formation with applications to Britain*, 2nd edn. Edinburgh: Oliver & Boyd.

Apt, J. & R. M. Goody 1979. Infrared image of Venus at the time of the Pioneer Venus encounter. *Science* **203**, 785–7.

Arvidson, R. E., V. R. Baker, C. Elachi, R. S. Saunders, J. A. Wood 1991. Magellan: initial analysis of Venus surface modification. *Science* **252**, 270–75.

Arvidson, R. E., R. Greeley, M. C. Malin, R. S. Saunders, N. Izenberg, J. J. Plaut, E. R. Stofan, M. K. Shepard 1992. Surface modification of Venus as inferred from Magellan observations of plains. *Journal of Geophysical Research* **97**, 13303–17.

Aubele, J. C. & E. N. Slyuta 1990. Small domes on Venus: characteristics and origin. *Earth, Moon and Planets* **50/51**, 493–532.

Avduevskii, V. S., S. L. Vishnevetskii, I. A. Golov, Yu. Ya. Karpeiskii, A. D. Lavrov, V. Ya. Likhushin, M. Ya. Marov, D. A. Mel'Nikov, N. I. Pomogin, N. N. Pronina, K. A. Razin, V. G. Fokin 1977. Measurement of wind velocity on the surface of Venus during operation of stations Venera 9 and 10. *Cosmic Research* **14**, 622–5.

Bagnold, R. A. 1941. *The physics of blown sand and desert dunes*. London: Methuen.

Banerdt, W. B. & M. P. Golombek 1988. Deformational models of rifting and folding on Venus. *Journal of Geophysical Research* **93**, 4759–72.

Barsukov, V. L. 1982. New Venera results. *13th Lunar and Planetary Science Conf.* Houston, Texas, March 1982.

Barsukov, V. L. and 29 other authors 1986. The geology and geomorphology of the Venus surface as revealed by the radar images obtained by Veneras 15 and 16. *Journal of Geophysical Research* **91**, 378–98.

Basilevsky, A. T. & J. W. Head 1988. The geology of Venus. *Annual Review of Earth and Planetary Sciences* **16**, 295–317.

Basilevsky, A. T., A. A. Pronin, L. B. Ronca, V. P. Kryuchkov, A. L. Sukhanov, M. S. Markov 1986. Styles of tectonic deformation on Venus: analysis of Venera 15 and 16 data. *Journal of Geophysical Research* **91**, D399–411.

Basilevsky, A. T., B. A. Ivanov, G. A. Burba, I. M. Chernaya, V. P. Kryuchkov, O. V. Nopkolaeva, D. B. Campbell, L. B. Ronca 1987. Impact craters of Venus: a continuation of the analysis of data from the Venera 15 and 16 spacecraft. *Journal of Geophysical Research* **92**, 12869–901.

Bindschadler, D. L. & J. W. Head 1986a. Characterization of Venera 15/16 units using Pioneer reflectivity and RMS slope. *17th Lunar and Planetary Science Conf.*, Abstr., 50–51.

Bindschadler, D. L. & J. W. Head 1986b. Pioneer Venus radar altimetry of the parquet terrain. *17th Lunar and Planetary Science Conf.*, Suppl., 1025–26.

Bindschadler, D. L. & J. W. Head 1988. Distribution of tesserae on Venus using Pioneer Venus and Venera data. *Lunar and Planetary Science* **19**, 824–5.

Bindschadler, D. L. & J. W. Head 1991. Tessera terrain, Venus: characterization and models for origin and evolution. *Journal of Geophysical Research* **96**, 5889–907

Bindschadler, D. L. & E. M. Parmentier 1990. Mantle flow tectonics: the influence of a ductile lower crust and implications for the formation of topographic upland on Venus. *Journal of Geophysical Research* **95**, 21 329–44.

Bindschadler, D. L., J. W. Head, J. B. Garvin 1986. Vega landing sites: Venera 15/16 analogs from Pioneer Venus reflectivity and RMS slope data. *Geophysical Research Letters* **13**, 1415–18.

Bindschadler, D. L., G. Schubert, W. M. Kaula 1990. Mantle flow tectonics and the origin of Ishtar Terra, Venus. *Geophysical Research Letters* **17**, 1345–8.

Bindschadler, D. L., A. deCharon, K. K. Beratan, S. E. Smrekar, J. W. Head 1992a. Magellan observations of Alpha Regio: implications for formation of complex ridged terrains on Venus. *Journal of Geophysical Research* **97**, 13563–77.

Bindschadler, D. L., G. Schubert, W. M. Kaula 1992b. Coldspots and hotspots: global tectonics and mantle dynamics of Venus. *Journal of Geophysical Research* **97**, 13495–532.

Brass, G. W. & C. G. A. Harrison 1982. On the possibility of plate tectonics on Venus. *Icarus* **49**, 86–96.

Cameron, A. G. W. 1973. Accumulation processes in the primitive solar nebula. *Icarus* **18**, 407–50.

Campbell, B. A. & D. B. Campbell 1992. Analysis of volcanic surface morphology on Venus from comparison of Arecibo, Magellan and terrestrial airborne radar data. *Journal of Geophysical Research* **97**, 16 293–314.

Campbell, D. B. & B. A. Burns 1980. Earth-based radar imagery of Venus. *Journal of Geophysical Research* **85**, 8271–81.

Campbell, D. B., L. B. Dyce, R. P. Ingall, R. P. et al. 1972. Venus: topography revealed by radar data. *Science* **175**, 514–16.

Campbell, D. B., J. W. Head, J. K. Harmon, A. A. Hine 1983. Identification of banded terrain in the mountains of Ishtar Terra, Venus. *Science* **221**, 644–7.

Campbell, D. B., J. W. Head, J. K. Harmon, A. A. Hine 1984. Venus volcanism and rift formation in Beta Regio. *Science* **226**, 167–70.

Campbell, D. B., J. W. Head, A. A. Hine, J. K. Harmon, D. A. Senske, P. C. Fisher 1989. Styles of volcanism on Venus: new Arecibo high resolution radar data. *Science* **246**, 373–7.

Campbell, D. B., D. A. Senske, J. W. Head, A. A. Hine, P. C. Fisher 1991. Geologic character and age of terrains in the Themis–Alpha–Lada region. *Science* **251**, 180–3.

Cattermole, P. J. 1987. Sequence, rheological properties and effusion rates of volcanic flows at Alba Patera, Mars. *Journal of Geophysical Research* **92**, 553–60.

Cattermole, P. J. 1989. *Planetary volcanism*. Chichester: Ellis Horwood; and New York: John Wiley (Halsted Press).

Craib, K. B. 1972. Synthetic aperture SLAR systems and their application for regional resources analysis. In *Remote sensing of Earth resources*, F. Sharokhi (ed.), 1152–78. Tullahoma: University of Tennessee Space Institute.

Crumpler, L. S. 1990. Eastern Aphrodite Terra on Venus: characteristics, structure and mode of origin. *Earth, Moon and Planets* **50/51**, 343–88.

Crumpler, L. S., J. W. Head, D. B. Campbell 1986. Orogenic belts on Venus. *Geology* **14**, 1031–4.

Dewey, J. F. & K. Burke 1974. Hot spots and continental break-up: implications for collisional orogeny. *Geology* **2**, 57–60.

Donahue, T. 1993. Pioneer data bolsters case for a once-wet Venus. *Lunar and Planetary Science Bulletin* **67**, 6–7

Donahue, T. M. & J. B. Pollack 1983. Origin and evolution of the atmosphere of Venus. See Hunten et al. (1983), 1003–36.

Elsasser, W. M. 1963. Early history of the Earth. In *Earth science and meteoritics*, J. Geiss & E. Goldberg (eds), 1–30, Amsterdam: North Holland.

Esposito, P. B., W. L. Sjogren, N. A. Mottinger, B. G. Bills, E. Abbott 1982. Venus gravity: analysis of Beta Regio. *Icarus* **51**, 448–59.

Fegley Jr, B., R. G. Prinn 1989. Estimation of the rate of volcanism on Venus from reaction rate measurements. *Nature* **337**, 55–8.

Florensky, C. P., L. B. Ronca, A. T. Burba, O. V. Nikolaeva et al. 1977. The surface of Venus as revealed by Venera 9 and 10. *Geological Society of America, Bulletin* **88**, 1537–45.

Florensky, K. P., A. T. Basilevsky, G. A. Burba, O. V. Nokolayeva, A. A. Pronin, A. S. Selivanov, M. K. Narayeva, A. S. Panfilov, V. P. Chemodanov 1983. Panorama of Venera 9 and 10 landing sites. See Hunten et al. (1983), 137–53.

Ford, P. G. & G. H. Pettengill, G. H. 1992. Venus topography and kilometre-scale slopes. *Journal of Geophysical Research* **97**, 13103–14.

Frank, S. L. & J. W. Head 1990. Ridge belts on Venus: morphology and origin. *Earth, Moon and Planets* **50/51**, 421–70.

Goldstein, R. M., R. R. Green, A. C. Rumsey 1976. Venus radar images. *Journal of Geophysical Research* **81**, 4807–17.

Goldstein, R. M., R. R. Green, A. C. Rumsey 1978. Venus radar brightness and altitude images. *Icarus* **36**, 334–52.

Greeley, R. 1986. Aeolian landforms: laboratory simulations and field studies. In *Aeolian geomorphology*, W. G. Nickling (ed.), 195–211. Boston: Allen & Unwin.

Greeley, R., J. Iverson, R. Leach, J. Marshall, B. White, S. Williams 1984a. Windblown sand on Venus: preliminary results of laboratory simulations. *Icarus* **57**, 112–24.

Greeley, R., J. R. Marshall, R. N. Leach 1984b. Microdunes and other bedforms on Venus: wind tunnel simulations. *Icarus* **60**, 152–60.

Greeley, R., J. R. Marshall, J. B. Pollack 1987. Physical and chemical modification of the surface of Venus by windblown particles. *Nature* **327**, 313–15.

Greeley, R., R. E. Arvidson, C. Elachi, M. A. Geringer, J. J. Plaut, R. S. Saunders, G. Schubert, E. R. Stofan, E. J. P. Thouvenot, S. D. Wall, C. M. Weitz 1992a. Aeolian features of Venus: preliminary Magellan results. *Journal of Geophysical Research* **97**, 13319–45.

Greeley, R., N. Lancaster, S. Lee, P. Thomas 1992b. Martian aeolian processes, sediments and features. In *Mars*, Tucson: University of Arizona Press.

Grimm, R. E. & R. J. Phillips 1991. Gravity anomalies, compensation mechanisms, and the dynamics of western Ishtar Terra, Venus. *Journal of Geophysical Research* **96**, 8305–24.

Grimm, R. E. & R. J. Phillips 1992. Anatomy of a Venusian hot spot: geology, gravity and mantle dynamics of Eistla Regio. *Journal of Geophysical Research* **97**, 16 035–54.

Grossman, L. 1972. Condensation of the primitive solar nebula. *Geochimica Cosmochimica Acta* **36**, 597–619.

Grossman, L. & J. Larimer 1974. Early chemical history of the solar system. *Reviews of Geophysics and Space Physics* **12**, 71–101.

Guest, J. E., M. H. Bulmer, J. Aubele, K. Beratan, R. Greeley, J. W. Head, G. Michaels, K. Weitz, C. Wiles 1992. Small volcanic edifices and volcanism in the plains of Venus. *Journal of Geophysical Research* **97**, 15949–66.

Hartmann, W. K. 1977. Relative crater production rates on planets. *Icarus* **31**, 260–76.

Head, J. W. 1990a. Assemblages of geologic/morphologic units in the northern hemisphere of Venus. *Earth, Moon and Planets* **50/51**, 391–408.

Head, J. W. 1990b. Processes of crustal formation and evolution on Venus: an analysis of topography, hypsometry and crustal thickness variations. *Earth, Moon and Planets* **50/51**, 25–55.

Head, J. W. & D. B. Campbell 1982. *Identification of banded terrain in the mountains near Ishtar Terra, Venus*. NASA TM-85127, 80–2.

Head, J. W. & L. S. Crumpler 1990. Venus geology and tectonics: hot spot and crustal spreading models and questions for the Magellan mission. *Nature* **346** (6284), 525–33.

Head, J. W. and Burns, D. B. 1982. Identification of banded terrain in the mountains near Ishtar Terra. Venus, NASA TM-85127, 80–82.

Head, J. W. & L. Wilson 1986. Volcanic processes and landforms on Venus: theory, predictions and observations. *Journal of Geophysical Research* **91**, 9407–46.

Head, J. W. & L. Wilson 1992. Magma reservoirs and neutral buoyancy zones on Venus: implications for the formation and evolution of volcanic landforms. *Journal of Geophysical Research* **97**, 3877–903.

Head, J. W., B. B. Campbell, C. Elachi, J. E. Guest, D. McKenzie, R. S. Saunders, G. G. Schaber, G. Schubert 1991. Venus volcanism: initial analysis from Magellan data. *Science* **252**, 276–88.

Head, J. W., L. S. Crumpler, J. C. Aubele 1992. Venus volcanism: classification of volcanic features and structures, associations, and global distribution from Magellan data. *Journal of Geophysical Research* **97**, 13153–97.

Herrick, R. R. & R. J. Phillips 1990. Blob tectonics: a prediction for western Aphrodite Terra, Venus. *Geophysical Research Letters* **17**, 2129–32.

Hess, P. C. & J. W. Head 1990. Derivation of primary magmas and melting of crustal materials on Venus: some preliminary petrogenetic considerations. *Earth, Moon and Planets* **50/51**, 57–80.

Hess, S. C. 1975. Dust on Venus. *Journal of Atmospheric Science* **32**, 1076–8.

Hunt, G. E. & P. Moore 1982. *The planet Venus*. London: Faber and Faber.

Hunten, D. M., L. Colin, T. M. Donahue, V. I. Moroz (eds) 1983. *Venus*. Tucson: University of Arizona Press.

Ivanov, B. A., A. T. Basilevsky, V. P. Kryuchkov, I. M. Chernaya, 1986. Impact craters of Venus: analysis of Venera 15 and 16 data. *Journal of Geophysical Research* **91**, D412–30.

Ivanov, B. A., I. V. Nemchikov, V. A. Svetsov, A. A. Provalov, V. M. Khazins, R. J. Phillips 1992. Impact cratering on Venus: physical and mechanical models. *Journal of Geophysical Research* **97**, 16 167–82.

Iverson, J. D. & B. R. White 1982. Saltation threshold on Earth, Mars and Venus. *Sedimentology* **29**, 111–19.

Iverson, J. D., R. Greeley, J. B. Pollack 1976. Windblown dust on Earth, Mars and Venus. *Journal of Atmospheric Science* **33**, 2425–9.

Janes, D. M., S. W. Squyres, D. L. Bindschadler, G. Baer, G. Schubert, V. L. Sharpton, E. R. Stofan 1992. Geophysical models for the formation and evolution of coronae on Venus. *Journal of Geophysical Research* **97**, 16 055–67.

Jannle, P., D. Jannsen, A. T. Basilevsky 1987. Morphologic and gravimetric investigations of Bell and Eistla Regiones on Venus. *Earth, Moon and Planets* **39**, 251–73.

Jannle, P., D. Janssen, A. T. Basilevsky 1988. Tepev Mons on Venus: morphology and elastic bending models. *Earth, Moon and Planets* **41**, 127–39.

Jurgens, R. F., R. M. Goldstein, H. R. Rumsey, R. R. Green 1980. Images of Venus by three-station radar interferometry – 1977 results. *Journal of Geophysical Research* **85**, 8282–94.

Kaula, W. M. 1990. Venus – a contrast in evolution to Earth. *Science* **247**, 1191–6.

Kaula, W. M., D. L. Bindschadler, R. E. Grimm, V. L. Hansen, K. M. Roberts, S. E. Smrekar 1992. Styles of deformation in Ishtar Terra and their implications. *Journal of Geophysical Research* **97**, 16 085–120.

Keldysh, M. V. 1977. Venus exploration with Venera 9 and Venera 10 spacecraft. *Icarus* **30**, 605–26.

Kiefer, W. S. & B. H. Hager 1992. Geoid anomalies and dynamic topography from convection in cylindrical geometry: application to mantle plumes on Earth and Venus. *Geophysical Journal International* **108**, 198–214.

Kozak, R. C. & G. G. Schaber 1989. New evidence for global tectonic zones on Venus. *Geophysical Research Letters* **16**, 175–8.

Kryuchkov, V. P. 1988. Ridge belts on the plains of Venus, 1. *19th Lunar and Planetary Science Conf.* Abstr., 649–50.

Kuiper, G. P (ed.) 1952. *The atmospheres of the Earth and planets*, 2nd edn, 306-405. Chicago: University of Chicago Press.

Lenardic, A., W. M. Kaula, D. L. Bindschadler 1992. Maxwell and the Andes: analogous structures? *23rd Lunar and Planetary Science Conf.*, 773–4.

Marchenkov, K. I., V. N. Zharkov, A. M. Nikishin 1990. The stress state of Venusian crust and variations of its thickness: implication for tectonics and geodynamics. *Earth, Moon and Planets* **50/51**, 81–98.

Marshall, J. R., R. Greeley, D. W. Tucker 1988. Aeolian weathering of Venusian materials: preliminary results from laboratory simulations. *Icarus* **74**, 495–515.

Masursky, H., E. Eliason, P. G. Ford, G. E. Pettengill, G. G. Schaber, G. Schubert 1980. Pioneer Venus radar results: geology from images and altimetry. *Journal of Geophysical Research* **85**, 8232–60.

McGill, G., S. Steenstrup, C. Barton, P. Ford 1981. Continental rifting and the origin of Beta Regio, Venus. *Geophysical Research Letters* **8,** 737–40.

McKenzie, D., P. G. Ford, C. Johnson, B. Parsons, D. Sandwell, S. Saunders, S. C. Solomon 1992a. Features on Venus generated by plate boundary processes. *Journal of Geophysical Research* **97**, 13533–44.

McKenzie, D., P. G. Ford, Fang Liu, G. H. Pettengill 1992b. Pancake-like domes on Venus. *Journal of Geophysical Research* **97**, 15967–76.

McNamee, J. B., N. J. Borderies, W. L. Sjogren 1993. Venus: global gravity and topography. *Journal of Geophysical Research* **98**, 9113–28

Nikishin, A. M. 1990. Tectonics of Venus: a review. *Earth, Moon and Planets* **50/51**, 101–25.

Nozette, S. & J. S. Lewis 1982. Venus: chemical weathering of igneous rocks and buffering of atmospheric composition. *Science* **216**, 181–3.

O'Keefe, J. D. & T. J. Ahrens 1982. Cometary impacts of planetary surfaces. *Journal of Geophysical Research* **87**, 6668–80.

Petrov, G. I. & V. P. Stulov 1975. The motion of large bodies in atmospheres of planets [in Russian]. *Kosm. Issled.*, **13**, 587–94.

Pettengill, G. H., E. Eliason, P. G. Ford, G. B. Loriot, H. Masursky, G. E. McGill 1979. Venus: preliminary topographic and surface imaging results from the Pioneer Orbiter. *Science* **205**, 90–3.

Pettengill, G. H., E. Eliason, G. G. Ford, G. B. Loriot, H. Masursky, G. E. McGill 1980. Pioneer-Venus radar results: altimetry and surface properties. *Journal of Geophysical Research* **85**, 8261–70.

Pettengill, G. H., P. G. Ford, R. J. Wilt 1992. Venus radiothermal emission as observed by Magellan. *Journal of Geophysical Research* **97**, 13103–14.

Pettengill, G. H., P. G. Ford, B. D. Chapman 1988. Venus: surface electromagnetic properties. *Journal of Geophysical Research* **93**, 14881–92.

Phillips, R. J. 1990. Convection-driven tectonics on Venus. *Journal of Geophysical Research* **95**, 1301–16.

Phillips, R. J. & M. C. Malin 1983. The interior of Venus and tectonic implications. See Hunten et al. (1983), 159–214.

Phillips, R. J. & M. C. Malin 1984. Tectonics on Venus. *Annual Review of Earth and Planetary Sciences* **12**, 411–43.

Phillips, R. J., W. Kaula, G. McGill, M. C. Malin 1981. Tectonics and evolution of Venus. *Science* **212**, 879–87.

Phillips, R. J., R. E. Grimm, M. C. Malin 1991. Hot-spot evolution and the global tectonics of Venus. *Science* **252**, 651–8.

Phillips, R. J., R. F. Raubertas, R. E. Arvidson, I. C. Sarkar, R. R. Herrick, N. Izenberg, R. E. Grimm 1992. Impact craters and Venus resurfacing history. *Journal of Geophysical Research* **97**, 15923–48.

Pieters, C. M., J. W. Head, W. Patterson, S. Pratt, J. Garvin, V. L. Barsukov, A. T. Basilevsky, I. L. Khodakovsky, A. S. Selivanov, A. S. Panfilov, Yu. M. Gektin, Y. M. Narayeva 1986. The colour of the surface of Venus. *Science* **234**, 1379–83.

Pronin, A. A. & E. R. Stofan 1990. Coronae on Venus: morphology, classification and distribution. *Icarus* **87**, 452–74.

Raitala, J. & K. Kauhanen 1991. Ridge belt tectonics of Ganiki Planitia on Venus. *Earth, Moon and Planets* **53**, 127–48.

Raitala, J. & T. Törmanen 1989. Small planitiae on Venus: tensional or compressional tectonics? *Earth, Moon and Planets* **45**, 237–63.

Raitala, J. & T. Törmanen 1990. Cytherean ridge belts connected with tessera areas: tensional or compressional structures? *Earth, Moon and Planets* **49**, 57–83.

Reasenberg, R. D. & Z. M. Goldberg 1992. High-resolution gravity model of Venus. *Journal of Geophysical Research* **97**, 14681–90.

Roberts, K. M. & J. W. Head 1990. Lakshmi Planum, Venus: characteristics and models of origin. *Earth, Moon and Planets* **50/51**, 193–249.

Roberts, K. M., J. E. Guest, J. W. Head, M. J. Lancaster, 1992. Mylitta Fluctus, Venus: rift-related, centralized volcanism and the emplacement of large-volume flow units. *Journal of Geophysical Research* **97**, 15991–16 015.

Roddy, D. J., R. O. Pepin, R. B. Merrill (eds) 1977. *Impact and explosion cratering*. Oxford: Pergamon.

Rogers, A. E. & P. R. Ingalls 1969. Venus: mapping the surface reflectivity by radar interferometry. *Science* **165**, 797–9.

Rogers, A. E. & P. R. Ingalls 1970. Radar mapping of Venus with interferometric resolution of the Range–Doppler ambiguity. *Radio Science* **5**, 425–33.

Rumsey, H. C., G. A. Morris, R. R. Green, R. M. Goldstein 1974. A radar brightness and altitude image of a portion of Venus. *Icarus* **23**, 1.

Sagan, C. 1975. Windblown dust on Venus. *Journal of Atmospheric Science* **32**, 1079–83.

Sagan, C. A., J. Veverka, P. Fox, R. Dubisch, J. Lederberg, E. Levinthal, L. Quam, R. Tucker, J. B. Pollack, B. A. Smith 1972. Variable features on Mars, 2: Mariner 9 global results. *Journal of Geophysical Research* **78**, 4163–96.

Sandwell, D. T. & G. Schubert 1992. Flexural ridges, trenches, and outer arc rises around coronae on Venus. *Journal of Geophysical Research* **97**, 16 069–84.

Saunders, R. S. & M. C. Malin 1977. Geologic interpretation of new observations of the surface of Venus. *Geophysical Research Letters* **4**, 547–50.

Saunders, R. S., R. E. Arvidson, J. W. Head, G. G. Schaber, E. R. Stofan, S. C. Solomon 1991. An overview of Venus geology. *Science* **252**, 249–52.

Saunders, R. S. & Magellan Mission Team 1992. *Journal of Geophysical Research* **97**, E8, Special Magellan Issue,

Schaber, G. G. 1982. Venus: limited extension and volcanism along lines of lithospheric weakness. *Geophysical Research Letters* **9**, 499–502.

Schaber, G. G., R. G. Strom, H. J. Moore, L. A. Soderblom, R. L. Kirk, D. J. Chadwick, D. D. Dawson, L. R. Gaddis, J. M. Boyce, J. Russell 1992. Geology and distribution of impact craters on Venus: what are they telling us? *Journal of Geophysical Research* **97**, 13257–301.

Schubert, G. 1983. General circulation and the dynamical state of the Venus atmosphere. See Hunten et al. (1983), 681–765.

Schultz, P. H. 1992. Atmospheric effects of ejecta emplacement and crater formation on Venus from Magellan. *Journal of Geophysical Research* **97**, 16 183–248.

Seiff, A. 1983. Thermal structure of the atmosphere of Venus. See Hunten et al. (1983), 215–79.

Seiff, A., D. B. Kirk, R. E. Young, R. C. Blanchard, J. T. Findlay, G. M. Kelly, S. C. Sommer 1980. Measurements of thermal structure and thermal constraints in the atmosphere of Venus and related dynamical observations: results from the four Pioneer Venus probes. *Journal of Geophysical Research* **85**, 7903–33.

Selivanov, A. S., Yu. M. Getkin, M. K. Naraeva, A. S. Panfilov, A. B. Fokin 1983. Dynamic phenomena detected in panoramas of the surface of Venus transmitted by the Venera 13–14 spacecraft [in Russian]. *Kosm. Issled.* **21**, 200–204.

Senske, D. A. 1990. Geology of the Venus equatorial region from Pioneer-Venus radar imaging. *Earth, Moon and Planets* **50/51**, 305–27.

Senske, D. A., J. W. Head, E. W. Stofan, D. B. Campbell 1991. Geology and structure of Beta Regio, Venus: results from Arecibo radar imaging. *Geophysical Research Letters* **18**, 1159–62.

Senske, D. A., G. G. Schaber, E. R. Stofan 1992. Regional topographic rises on Venus: geology of western Eistla Regio and comparison with Beta Regio and Atla Regio. *Journal of Geophysical Research* **97**, 13395–420.

Settle, M. 1978. Volcanic eruption clouds and thermal power output of explosive eruptions. *Journal of Volcanology and Geothermal Research* **3**, 309–24.

Sharpton, V. L. & J. W. Head 1986. A comparison of the regional slope characteristics of Venus and Earth: implications for geologic processes on Venus. *Journal of Geophysical Research* **91**, 7545–54.

Shaw, H. R. & D. A. Swanson 1970. Eruption and flow rates of flood basalts. *Proc. 2nd Columbia River Basalts Symp.* E. H. Gilmous & D. Stradling (eds), 271–99. Cheney: Eastern Washington College Press.

Simons, M., S. C. Solomon, B. H. Hager 1991. Dynamic models for ridge belt formation on Venus. *21st Lunar and Planetary Science Conf.* Abstr., 1263–4.

Sjogren, W. L., R. J. Phillips, P. W. Birkeland, R. N. Wimberly 1980. Gravity anomalies on Venus. *Journal of Geophysical Research* **85**, 8295–302.

Sleep, N. H. 1990. Hotspots and mantle plumes: some phenomenology. *Journal of Geophysical Research* **95**, 6715–36.

Slyuta, E. N., O. V. Nikolaeva, M. A. Kreslavsky 1988. Distribution of small domes on Venus: Venera 15/16 radar data [in Russian]. *Astron. Vestnik.* **22**, 287.

Smrekar, S. E. & R. J. Phillips 1991. Venusian highlands: geoid to topography ratios and their implications. *Earth and Planetary Science Letters* **107**, 582–97.

Smrekar, S. E. & S. C. Solomon 1992. Gravitational spreading of high terrain in Ishtar Terra, Venus. *Journal of Geophysical Research* **97**, 16 121–48.

Solomon, S. C. & J. W. Head 1982. Mechanisms for lithospheric heat transport on Venus: implications for tectonic style and volcanism. *Journal of Geophysical Research* **87**, 9239–46.

Solomon, S. C. & J. W. Head 1990. Lithospheric flexure beneath the Freyja Montes foredeep, Venus: constraints on lithospheric thermal gradient and heat-flow. *Geophysical Research Letters* **17**, 1393–6.

Solomon, S. C. & J. W. Head 1991. Fundamental issues in the geology and geophysics of Venus. *Science* **252**, 252–60.

Solomon, S. C., J. W. Head, W. M. Kaula, D. McKenzie, B. Parsons, R. J. Phillips, G. Schubert, M. Talwani 1991. Venus tectonics: initial analysis from Magellan. *Science* **252**, 297–312.

Solomon, S. C., S. E. Smrekar, D. L. Bindschadler, R. E. Grimm, W. M. Kaula, G. E. McGill, R. J. Phillips, R. S. Saunders, G. Schubert, S. W. Squyres, E. R. Stofan 1992. Venus Tectonics: an overview of Magellan observations. *Journal of Geophysical Research* **97**, 13199–255.

Squyres, S. W., D. G. Jankowski, M. Simons, S. C. Solomon, B. H. Hager, G. E. McGill 1992. Plains tectonism on Venus: the deformation belts of Lavinia Planitia. *Journal of Geophysical Research* **97**, 13579–99.

Stofan, E. R. & J. W. Head 1990. Coronae of Mnemosyne Regio: morphology and origin. *Icarus* **83**, 216–43.

Stofan, E. R., J. W. Head, D. B. Campbell, S. H. Zisk, A. F. Bogomolov, O. M. Rzhiga, A. T. Basilevsky, N. Armand 1989. Geology of a rift zone on Venus: Beta Regio and Devana Chasma. *Geological Society of America, Bulletin* **101**, 143–56.

Stofan, E. R., D. L. Bindschadler, J. W. Head, E. M. Parmentier 1991. Corona structures on Venus: models of origin. *Journal of Geophysical Research* **96**, 20 933–46.

Stofan, E. R., V. L. Sharpton, G. Schubert, G. Baer, D. L. Bindschadler, D. M. Janes, S. W. Squyres 1992. Global distribution and characteristics of coronae and related features on Venus: implications for origin and relation to mantle processes. *Journal of Geophysical Research* **97**, 13347–78.

Sukhanov, A. L. 1986. Parquet: regions of areal plastic dislocation. *Geotectonics* **20**, 294–305.

Sukhanov, A. L. & A. Pronin 1989. Ridge belts on Venus as extensional features. *Proc. 19th Lunar and Planetary Science Conf.*, 335–48.

Sukhanov, A. L. & 11 other authors 1989. Geomorphic/geologic map of part of the northern hemisphere of Venus. Map I-2059, USGS.

Suppe, J. & C. Connors 1992. Critical taper wedge mechanics of fold-and-thrust belts on Venus: initial results from Magellan. *Journal of Geophysical Research* **97**, 13545–61.

Surkov, Yu. A., F. F. Kirnozov, V. K. Khristianov, B. N. Glazov, F. Ivanov, B. N. Korchuganov 1977. Investigations of the density of the Venusian surface rocks by Venera 10. *COSPAR Space Research* **17**, 651–7.

Surkov, Yu. A., F. F. Kirnozov, V. N. Glazov, A. G. Dunchenko, L. P. Tatsil, O. P. Sobornov 1986a. Uranium, thorium and potassium in the Venusian rock at landing sites of Vegas 1 and 2. *17th Lunar and Planetary Science Conf.* Abstr., 847–8.

Surkov, Yu. A., L. P. Moskalyova, V. P. Kharyukova, A. D. Dudin, G. G. Smirnoff, S. Ye. Zaitseva 1986b. Venus rock composition at the Vega 2 landing site. *17th Lunar and Planetary Science Conf.*, Abstr., 849–50.

Taylor, F. W., D. M. Hunten, L. V. Ksanfomaliti 1983. The thermal balance of the middle and upper atmosphere of Venus. See Hunten et al. (1983), 650–80.

Thomas, P. & J. Veverka 1979. Seasonal and secular variation of wind streaks on Mars: analysis of Mariner 9 and Viking data. *Journal of Geophysical Research* **84**, 8131–46.

Thomas, P. J., J. Veverka, S. Lee, A. Bloom 1981. Classification of wind streaks on Mars. *Icarus* **45**, 124–53.

Thornhill, G. D. 1993. Theoretical modelling of eruption plumes on Venus. *Journal of Geophysical Research* **98**, 9107–11.

Tomasko, M. G. 1963. The thermal balance of the lower atmosphere of Venus. See Hunten et al. (1983), 604–31.

Turekian, K. & S. P. Clark 1969. Inhomogeneous accumulation of the Earth from the primitive solar nebula. *Earth and Planetary Science Letters* **6**, 346–8.

Tyler, G. L., P. G. Ford, D. B. Campbell, C. Elachi, G. H. Pettengill, R. A. Simpson 1991. Magellan: electrical and physical properties of Venus's surface. *Science* **252**, 265–70.

Tyler, G. L., R. A. Simpson, M. J. Maurer, E. Holmann 1992. Scattering properties of the Venusian surface: preliminary results from Magellan. *Journal of Geophysical Research* **97**, 13115–39.

Urey, H. C. 1962. Evidence regarding the origin of the Earth. *Geochimica Cosmochimica Acta* **26**, 1–13.

USGS 1989. *Maps of the northern hemisphere of Venus*. USGS Misc. Inv. Series, Maps I-2041 and I-2059.

Veverka, J., P. Geirasch, P. Thomas 1981. Wind streaks on Mars: meteorological control of occurrence and mode of formation. *Icarus* **45**, 154–66.

Vinogradov, A. P, Yu. A. Surkov, F. F. Kirnozov 1973. The content of uranium, thorium and potassium in the rocks of Venus as measured by Venera 8. *Icarus* **20**, 253–9.

Von Zahn, U., S. Kumar, H. Niemann, R. Prinn 1983. Composition of the Venus atmosphere. See Hunten et al. (1983), 299–430.

Vorder Bruegge, R. W. & J. W. Head 1989. Fortuna Tessera, Venus: evidence of horizontal convergence and crustal thickening. *Geophysical Research Letters* **16**, 699–702.

Vorder Bruegge, R. W. & J. W. Head 1990. Orogeny and strike-slip faulting on Venus: tectonic evolution of Maxwell Montes. *Journal of Geophysical Research* **95**, 8357–81.

Vorder Bruegge, R. W. & J. W. Head 1991. Processes of formation and evolution of mountain belts on Venus. *Geology* **19**, 885–8.

Walker, G. P. L. 1973. Lengths of lava flows. *Royal Society of London, Philosophical Transactions* **274**A, 107–18.

Watson, A. J., T. M. Donahue, J. C. G. Walker 1982. The dynamics of a rapidly escaping atmosphere: applications to the evolution of Earth and Venus. *Icarus* **48**, 150–66.

Zuber, M. T. 1987. Constraints on the lithospheric structure of Venus from mechanical models and tectonic surface features. *Proc. 17th Lunar and Planetary Science Conf.*; *Journal of Geophysical Research* **92**, E541–51.

Zuber, M. 1990. Ridge belts: evidence for regional and local-scale deformation on the surface of Venus. *Geophysical Research Letters* **17**, 1369–72.

Zuber, M. T. & E. M. Parmentier 1990. On the relationship between isostatic elevation and the wavelengths of tectonic surface features on Venus. *Icarus* **85**, 290–308.

Index